# Spectroscopic Membrane Probes

## Volume I

Editor

**Leslie M. Loew, Ph.D.**
Professor of Physiology
University of Connecticut Health Center
Farmington, Connecticut

CRC Press
Taylor & Francis Group
Boca Raton  London  New York

CRC Press is an imprint of the
Taylor & Francis Group, an **informa** business

# DEDICATION

To my parents.

# ACKNOWLEDGMENTS

I am pleased to acknowledge my wife, Helen, for putting up with late-night word processing in our bedroom and for her selfless support throughout my career.

I thank those contributors to this work who were able to provide their manuscripts within 6 months after the deadline. My thanks are offered not only for their punctuality but also for their tolerance of their more tardy colleagues. I also wish to thank all of the contributors for their willingness to work with me on this project and for the high quality of each of their chapters.

# PREFACE

The optical spectra of molecular membrane probes can be interpreted in terms of the structure, dynamics, and physiological state of the membrane. The general picture we have of membranes and of the properties of the proteins imbedded in them, has, arguably, emerged directly from probe studies over the last 20 years. This work is designed to make these techniques accessible to a broad audience of cell biologists.

The techniques discussed revolve primarily around the fluorescence of membrane probes, but applications of light absorption and Raman scattering are included. In addition to reviews of the major applications, most chapters include information on the required apparatus, experimental design, data analysis, and potential pitfalls. Thus, significant attention has been given to practical considerations involved in the use of these methods. The aim is to provide the novice with some appreciation of the real requirements, with respect to both expertise and equipment, for the implementation of these methods.

Specific methods that are represented include fluorescence anisotropy, resonance energy transfer, fluorescence recovery after photobleaching, digital video microscopy, total internal reflection fluorescence, near-field imaging, multisite optical recordings of electrical activity, and resonance Raman spectroscopy. Chapter 1 gives an overview of molecular photophysics. It serves to introduce Chapters 2 to 13 which cover dynamic membrane structure, membrane fusion assays, and studies of the properties of specific membrane proteins. Chapter 14 is a practical and concise guide to the characteristics of potentiometric membrane dyes and introduces Chapters 15 to 22 which describe specific applications. These include measurements of membrane potential in preparations ranging from isolated energy-transducing organelles, through cell suspensions and individual cells, to complex neuronal systems. The result is a most comprehensive examination of probe methods for the study of membrane potential, which can serve as a convenient source for the design of new physiological applications of these dyes.

# THE EDITOR

**Leslie M. Loew, Ph.D.,** is Professor of Physiology at the University of Connecticut Health Center in Farmington.

Dr. Loew received his B.S. degree from the City College of the City University of New York in 1969. Under the tutelage of Professor C. F. Wilcox, he obtained his M.S. and Ph.D. degrees in 1972 and 1974, respectively, from the Department of Chemistry, Cornell University, Ithaca, N.Y. After a year's postdoctoral training with F. H. Westheimer at Harvard University, he joined the Department of Chemistry at the State University of New York at Binghamton in 1974 as Assistant Professor. He became an Associate Professor of Chemistry in 1979 at SUNY Binghamton. As a result of a steady shift in research interest from physical organic chemistry to membrane biophysics, Dr. Loew moved to the University of Connecticut, where he was appointed Associate Professor of Physiology in 1984 and Professor in 1987. He has retained his association with SUNY Binghamton by holding the position of Adjunct Associate Professor of Chemistry.

Dr. Loew is a member of the American Association for the Advancement of Science, the American Chemical Society, the Biophysical Society, and Phi Beta Kappa. He was a Research Career Development Awardee of the National Institutes of Health (1980 to 1985) and has been a visiting faculty member at Cornell University (1984), Bar Ilan University (1983), and the Weizmann Institute of Science (1981, 1983, 1985).

He has been the recipient of research grants from the National Institutes of Health, the American Chemical Society, the Research Corporation, and private industry. He has published more than 50 papers. His current research interests are in the design of potentiometric membrane dyes and the study of the electrical properties of membranes.

# CONTRIBUTORS

## Volume I

**Barbara Baird, Ph.D.**
Associate Professor
Department of Chemistry
Baker Laboratory
Cornell University
Ithaca, New York

**Joe Bentz, Ph.D.**
Assistant Professor
Departments of Pharmacy and
  Pharmaceutical Chemistry
University of California
San Francisco, California

**Robert Blumenthal, Ph.D.**
Section on Membrane Structure and
  Function, LTB
National Cancer Institute
National Institutes of Health
Bethesda, Maryland

**Diane Bradley**
Laboratory for Experimental
  Neuropathology
National Institutes of Health
Bethesda, Maryland

**Sarina A. Cavalier**
Department of Chemistry
University of Oregon
Eugene, Oregon

**Nejat Düzgüneş, Ph.D.**
Associate Professor
Cancer Research Institute and Department
  of Pharmaceutical Chemistry
University of California
San Francisco, California

**Carter C. Gibson**
Biomedical Engineering and
  Instrumentation Branch
National Institutes of Health
Bethesda, Maryland

**David Holowka, Ph.D.**
Senior Research Associate
Department of Chemistry
Baker Laboratory
Cornell University
Ithaca, New York

**Bruce Hudson, Ph.D.**
Professor and Associate Member,
  Institute of Molecular Biology
Department of Chemistry
University of Oregon
Eugene, Oregon

**Alan Kleinfeld, Ph.D.**
Associate Professor
Department of Physiology and Biophysics
Harvard Medical School
Boston, Massachusetts
Division of Membrane Biology
Medical Biology Institute
La Jolla, California

**Barry R. Lentz, Ph.D.**
Associate Professor
Department of Biochemistry and Nutrition
University of North Carolina
Chapel Hill, North Carolina

**Stephen J. Morris, Ph.D.**
Professor
Division of Molecular and Cellular
  Neurobiology
School of Basic Life Sciences
University of Missouri at Kansas City
Kansas City, Missouri

**Paul D. Smith**
Biomedical Engineering and
  Instrumentation Branch
National Institutes of Health
Bethesda, Maryland

**David E. Wolf, Ph.D.**
Senior Scientist
Endocrine Reproductive Biology Group
  for Experimental Biology
Worcester Foundation
Shrewsbury, Massachusetts

# TABLE OF CONTENTS

## Volume I

## Volume II

### Volume III

Chapter 1

# INTRODUCTION TO FLUORESCENCE AND ITS APPLICATION TO MEMBRANE STRUCTURE AND DYNAMICS

**Leslie M. Loew**

## TABLE OF CONTENTS

# I. INTRODUCTION

This chapter presents an overview of the molecular basis of fluorescence spectroscopy and a physical description of the fluorescence-based processes that can be used to probe the cell surface. It is not intended as a comprehensive or mathematically rigorous treatment of fluorescence. On the contrary, the intention is to provide a qualitative basis for the more detailed discussions in the chapters that follow. For the sake of brevity, several quantum chemical concepts will be summarized in the form of equations, but these will be introduced via simple physical arguments, rather than full derivations. Also, several key equations relating experimental fluorescence parameters to the physical state of the fluorophore or its environment are described in order to serve as a convenient central reference. A thorough treatment of fluorescence spectroscopy, with special emphasis on applications to biological macromolecules and molecular assemblies, is available in the comprehensive text by Lakowicz.[1]

Interspersed with the general discussions in this chapter are brief mentions of the topics covered in the rest of this volume. This will serve to direct the reader to specific problems in membrane biology that can be tackled by fluorescence. It will also indicate where additional detail can be obtained on the implementation of the various techniques and the theory behind them.

# II. FLUORESCENCE AND OTHER PATHWAYS FOR EXCITED STATE DECAY

## A. Light Absorption

The absorption and emission of visible or ultraviolet light involves transitions between the ground and electronic excited states of a chromophore, usually consisting of a delocalized $\pi$-system. The purpose of this discussion is to provide a molecular rationale for the probability of an electronic transition. This will allow the reader to understand the structural features which control both the extinction coefficient and quantum yield (i.e., fluorescence intensity) of a chromophore.

The state of a molecule can be divided into electronic, vibrational, and rotational components. To a good approximation these are independent, allowing one to think of the state as a simple superposition of the components. The electronic component is described by a wave function which defines the spatial and spin distributions of all the electrons around a fixed array of nuclei. The square of this wave function represents the probability of finding an electron from the molecule at any point in space and identifies it with one of the two possible spin quantum numbers. The related facts that a molecule cannot be in two states simultaneously and that the electrons within a given state must be found somewhere in the cartesian space, are summarized, respectively, by the following equations:

$$\int \psi_m \chi_m \psi_n \chi_n d\tau dS = 0 \tag{1}$$

$$\int (\psi_m \chi_m)^2 d\tau dS = 1 \tag{2}$$

In these equations, the integration is over the cartesian ($\tau$) and spin (S) coordinates of all the electrons, $\psi_j$ is the spatial part of the wave function, and $\chi_j$ is the spin part of the wave function for the state "j". The implication is that the spatial and spin components of the wave function are completely independent and that the equations hold for these individual components separately; this approximation is not valid for molecules composed of "heavy" atoms, i.e., atoms beyond the first two rows of the periodic table.

A transition from the ground state to an excited electronic state of a molecule can be achieved by the appropriate interaction with light, effectively circumventing the exclusivity implied by Equations 1 and 2. There are two requirements for this interaction. First, the energy of the incident photons, h$\nu$, must match the difference between the energy levels of the ground, $E_g$, and excited $E_e$, states:

$$E_{es} - E_{gs} = h\nu = \frac{hc}{\lambda} \tag{3}$$

where h is Planck's constant, c is the speed of light, and $\nu$ and $\lambda$ are the frequency and wavelength of the light, respectively. Second, the oscillating dipole of the electric component of the electromagnetic light wave must couple the electron waves in the two states. This second requirement can be summarized by the following equation:

$$\vec{T} = \int\psi_g\chi_g\vec{R}\psi_e\chi_e d\tau dS = \int\psi_g\vec{R}\psi_e d\tau \int\chi_g\chi_e dS \tag{4}$$

where $\vec{R}$ represents the displacement vector of the oscillating electric field. This quantity has the units of a dipole moment and is generally referred to as a transition moment; its square is directly proportional to the extinction coefficient of the absorption band.

Of course, typical absorption spectra reflect transitions which extend over a range of wavelengths rather than the sharp lines which might be expected from Equation 3. This is because rotational and vibrational sublevels within the excited state serve to effectively multiply the target energies for the incoming photons. The spectra of rigid aromatic hydrocarbons will still display relatively sharp vibrational structures superimposed on the electronic transition, but more complex molecules will have featureless absorption bands. In addition, for chromophores in solution, one can expect a distribution of molecular solvation interactions with both the ground and excited states to further contribute to absorption band broadening.

Some implications of Equation 4 also need to be considered in order to develop an appreciation for the factors which control the probability of a transition. Because the integral over the spin coordinates can be factored out and is therefore not influenced by the oscillating electric field of the light wave, the possibility of a transition which involves a change in the electron spin (i.e., a singlet-triplet transition) is excluded; this forbiddenness is based on the approximate independence of the spin and spatial components of the wavefunctions in Equation 1, and is relaxed for molecules which contain heavy atoms. Focusing now on the spatial part of Equation 4, it is clear that $\psi_g$ and $\psi_e$ must have significant amplitudes which overlap in space in order for the extinction coefficient to be high. Since at ordinary UV or visible light energies, transitions involve just a single electron, we need only consider the overlap between the molecular orbital from which the electron is removed in the ground state and that in which it has been placed in the excited state. For a large delocalized $\pi$ system and a transition sending an electron from a $\pi$-bonding orbital to a $\pi^*$-antibonding orbital, the integral can have very large values, corresponding to extinction coefficients of $10^4$ to $10^5$ AM$^{-1}$ cm$^{-1}$. The lowest energy transition of the most common organic dyes and of simple aromatics, such as naphthalene or pyrene, are of this type. On the other hand, a transition from a nonbonding orbital (i.e., an unshared electron), to a $\pi^*$ orbital involves very poor overlap, as the former is localized with its amplitude in the molecular plane, while the latter is usually quite delocalized and has a node in the molecular plane; thus, n $\rightarrow$ $\pi^*$ transitions usually have extinctions of only 100 AM$^{-1}$ cm$^{-1}$. Carbonyl compounds all have these weak transitions — as do azo dyes (the high extinction of azo dyes originates from additional $\pi \rightarrow \pi^*$ states). The last thing to notice about Equation 4 is that the transition moment is a vector with a fixed direction with respect to the atomic coordinates of the molecule. The practical consequence of this is that the molecule absorbs that component of the incident light which is polarized parallel to the transition moment.

## B. Light Emission

Now that we have discussed the production of the excited state by the absorption of light, we may turn to a consideration of emission. A molecule can be exicted, in general, to any one of a number of singlet states, with a distribution of vibrational levels superimposed on each. These excited states are, as a rule, closely spaced in energy — much closer than the spacing between the ground state and the lowest of the excited states. This permits a cascade of relaxation via nonradiative decay to quickly (<picoseconds) populate the lowest vibrational level within the lowest energy excited singlet. Thus, fluorescence can only occur from this state. (The only well-documented exception to this generality is the nonbenzenoid aromatic molecule azulene which displays emission from both the lowest and second lowest excited states. This exceptional molecule has not found any application to biological problems, however.)

The probability of fluorescence emission, just like that of absorption, is given by the square of the transition moment. The relevant physical parameter for emission is a rate constant, rather than an extinction coefficient, however. Because they are both linked to the same theoretical origins, the rate of radiative decay, $k_f$, of a population of excited chromophores to the ground state can be directly calculated from the molar extinction coefficient, $\epsilon$, of the lowest energy absorption band of the molecule:

$$k_f = 1.07 \times 10^{-40} n^2 \nu_m^2 \int \epsilon d\nu \tag{5}$$

where n is the refractive index of the solvent and $\nu_m$ is the mean frequency of the band. The inverse of $k_f$ is the natural radiative lifetime of the excited state, $t_f$. Equation 5, better known as the Strickler-Berg equation, predicts that emission from a $\pi \rightarrow \pi^*$ state in the visible region of the spectrum should have a natural lifetime in the nanosecond range. If the lowest excited state of the molecule is $n \rightarrow \pi^*$, emission will be several orders of magnitude slower, irrespective of whether that state was populated directly or by initial excitation into a higher energy $\pi \rightarrow \pi^*$ state; as noted above, this is because relaxation to the lowest excited state is always faster than any radiative process.

Also, even if the incident light is of the proper wavelength to populate the lowest energy-excited state directly, the emitted light will be of longer wavelength (lower energy). This is fortunate since otherwise it would be difficult to distinguish fluorescence from scattered light. There are several origins of this shift in the emission spectrum with respect to absorption, the Stokes shift. The most universal of these is the fact that while absorption orignates with the lowest vibrational level within the ground state and can terminate in an upper vibrational level of the excited state, emission originates from the lowest vibrational level of the excited state and can terminate in an upper level of the ground state (Figure 1A). Thus, the long wavelength edge of the absorption band can overlap the short wavelength edge of the emission band, but the centers of the bands are well separated. In the case of rigid nonpolar chromophores, this is usually the only source of the Stokes shift and the absorption and emission spectra often appear as mirror images of each other. When the chromophore is not confined within a rigid ring system, the excited state may relax to a twisted geometry which is more stable than the arrangement of atoms during the light absorption event — i.e., the ground state geometry. Emission will then occur to a ground state geometry which is distorted from that of the most stable form; this state of affairs is schematized in Figure 1B. The third origin of the Stokes shift is most important for polar chromophores in polar solvents. If a significant reorganization of electronic charge accompanies excitation, a reorientation of solvent molecules around the chromophore to stabilize the excited state will follow; this situation is analogous to the geometry relaxation and may be understood by reference to Figure 1C.

Nonradiative decay to the ground state can compete with emission, resulting in a dimin-

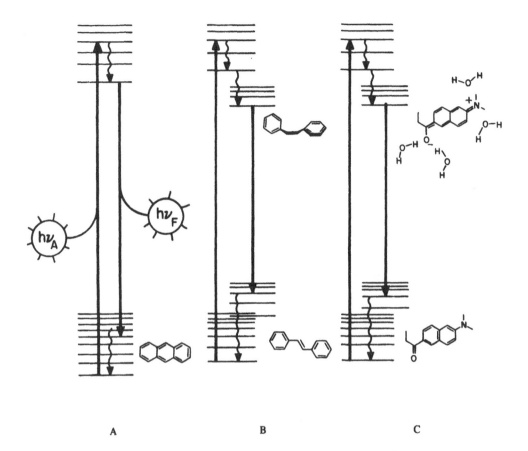

A                                    B                                    C

FIGURE 1. The origins of the Stokes shift. The vertical axis is energy and the horizontal direction represents variations in molecular geometry. Electronic states are represented as groups of closely spaced vibrational levels. Absorption and emission of light are represented by heavy up and down arrows, while radiationless transitions are wavy arrows. (A) Rigid chromophores like anthracene have small Stokes shifts. (B) Stilbene is planar in the ground state and is excited directly into a metastable planar excited state. This relaxes to a twisted conformation before emission can occur. (C) Prodan[2] has a dipolar excited state which experiences strong solvation in polar solvents like water. The vertically excited state produced immediately upon absorption, however, has the same weak solvation as the ground state.

ished fluorescence intensity. The quantum yield for fluorescence, Q, is defined as the fraction of the excited molecules which decay via light emission and is equal to the rate of emission divided by the sum of all decay rates:

$$Q = \text{photons emitted/photons absorbed} = k_f/(k_f + k_{nr}) \qquad (6)$$

The experimentally observed lifetime of the excited state, $t_e$, usually measured by monitoring the decay of fluorescence after a short exciting laser pulse, is simply $(k_f + k_{nr})^{-1}$. Using Equation 5 to find $k_f$, one can then derive Q. An experimentally simpler approach is to compare the integrated emision intensity of a compound to that of a standard with a known quantum yield; unless this comparison is made with samples of the same optical density at a given excitation wavelength, corrections for variations in exciting light intensity and/or optical density must be included.

## C. Nonradiative Decay Mechanisms

The simplest mechanism for nonradiative decay is simply loss of the excitation energy to heat. This can only happen, however, if the electronic energy can somehow be converted into internal vibrational or rotational motions of the ground state or of the surrounding solvent molecules. Therefore, rigid chromophores or chromophores in viscous solvents generally display high quantum yields. A pair of well-known dye molecules, phenolphthalein and fluorescein, nicely illustrate this point:

P                    F

Both molecules have similar chromophores with high extinction $\pi \rightarrow \pi^*$ absorptions. Phenolphthalein, Structure P, has several internal rotational degrees of freedom and has undetectable fluorescence (Q <0.0001). Fluorescein, arguably the world's most popular fluorophore (Q ~0.5), has, as its main structural distinction from phenolphthalein, a bridging oxygen atom to prevent rotation of two of the three aromatic rings. The idea that $t_e$ can be used to probe domains of varying microviscosity within a membrane is explored in Chapters 3 and 8.

A more complex pathway for depopulating the lowest excited singlet state involves a spin flip to the corresponding triplet. As noted in the discussion of absorption, such spin flips are generally forbidden radiative transitions, so that once a triplet is formed it will be likely to decay to the ground state by nonradiative pathways much faster than it can emit a photon. (Emission from a triplet is "phosphorescence" and is only observed for organic molecules in glasses at low temperatures.) The rate, $k_{st}$, of the transformation of the excited singlet to the triplet (i.e., "intersystem crossing") is inversely related to the energy difference between these two states and is enhanced if heavy atoms are part of the chromophore. Thus, 1-fluoronaphthalene has a Q of 0.85 while 1-bromonaphthalene has a Q of 0.002;[3] these molecules have identical $k_f$s so that the difference in Q must be attributable to a difference in the rate of nonradiative decay.

Intermolecular processes in solution can promote nonradiative decay of the singlet, thereby quenching the fluorescence. The simplest of these involves the induction of intersystem crossing by a solvent or cosolute which bears a heavy atom or an unpaired electronic spin (e.g., oxygen quenching). A more theoretically involved process is collisional quenching with excitation energy transfer by an exchange mechanism which requires orbital overlap. Quenching also results if a stable complex can form between the excited state of one molecule and the ground state of another; the resultant complex is called an excimer if it is composed of a dimer and an exciplex if its components are two different chromophores. All of these processes are strongly dependent on the concentration of the quencher and on the viscosity of the solvent.

Several assays employing fluorescence quenching can be applied to the study of membrane fusion and are detailed in Chapters 4 and 5. The fluorophore and quencher can be confined to either the aqueous interiors (Chapter 4) or the lipid bilayers (Chapter 5) of separate liposome populations and the fluorescence decrease after mixing reflects the rate of fusion.

There is, of course, the trivial case where the emitted photon is reabsorbed by the quencher before it can emerge from the sample volume to be detected; this is only important if the quencher has an optical density in the sample of >0.01 in the wavelength range of the emission. On the other hand, a decidedly nontrivial mechanism exists for the quenching of fluorescence by a quencher whose absorption spectrum overlaps the emission spectrum of a fluorophore. This process involves transfer of the excitation energy from the fluorophore to the quencher over distances of tens of angstroms by a direct resonance of the transition moments. Resonance energy transfer is a particularly useful tool for membrane studies because of its characteristic distance dependence and will be discussed in greater detail later.

## III. FLUORESCENCE ANISOTROPY

If plane polarized light is used to excite a sample, the scalar product of the electric vector of the light wave and the transition moment of the chromophore determine the interaction. Since the probability of excitation depends on the square of the transition moment, absorption of polarized light by a rigid and well-oriented sample (e.g., a single crystal), is characterized by a cosine-squared angular dependence. Likewise, if emission is detected through an analyzing polarizer, the intensity will depend on the cosine squared of the angle between the polarizer and the transition moment of the emitting state. In the case where both excitation and emission are passed through parallel polarizers, a rigid oriented sample will display fluorescence with an intensity proportional to the fourth power of cosine of the angle between the polarizers and the direction of the collection of transition moments. The fluorescence is then a direct measure of the anisotropic arrangement of molecules in the sample.

The next level of complexity in fluorescence experiments with polarized light, is the case where the sample is rigid, but the distribution of chromophores is isotropic. Here we may wish to consider cases where the excitation and emission polarizers pass light with either parallel or perpendicular electric field vectors. These paired orientations of the polarizers are considered in relation to each other, since there is no intrinsic directionality associated with the sample. It suffices to present the well-known result of such an experiment: the intensity of fluorescence from the case where the polarizers are parallel, $F_{\parallel}$ is a factor of 3 greater than the fluorescence measured when they are perpendicular, $f_{\perp}$. The fluorescence anisotropy, r, defined formally according to Equation 7, is, then, equal to 0.4

$$r = \frac{F_{\parallel} - F_{\perp}}{F_{\parallel} + 2F_{\perp}} \tag{7}$$

in the present case of a rigid isotropic sample. In general, intramolecular relaxations leading to small geometry changes within the excited state can change the direction of the transition moment a bit, so that this limiting anisotropy is usually somewhat lower than 0.4.

This fluorescence anisotropy physically derives from the fact that the fluorophores are anisotropically excited by the polarized exciting light, despite their random orientations. Since the molecules are in a rigid, albeit isotropic, matrix, the cosine-squared distribution of excited states produced by the polarized exciting light, is frozen until emission can occur. Emission from one of these excited states will, in turn, display a cosine-squared intensity distribution, but about the direction of its own transition moment, rather than the direction of the polarized excitation. The analyzing polarizer, then, measures the resultant anisotropy of emission from the entire population of excited states. It is not the intention here to provide a mathematical derivation for this situation, but the integrations that are involved to average the angular dependence of the absorption and fluorescence over all possible molecular orientations are quite straightforward.

The term rigid, as used in the context of fluorescence anisotropy, really requires only that molecular rotations be slow on the timescale of the excited state lifetime, $t_e$. This condition is not met, however, for most fluorophores in common solvents at ambient temperatures. In fact, the most important application of fluorescence anisotropy measurements is to probe molecular rotations. The extent that the fluorescence is depolarized from the limiting value of the anisotropy, $r_0$, is directly interpretable in terms of motions of the fluorophore during its excited state lifetime. For a spherical fluorophore, the rate at which the anisotropy decays after an excitation pulse is given by a simple exponential:

$$r(t) = r_o e^{-t/t_c} \qquad (8)$$

The parameter $t_c$ is the rotational correlation time and is related to the molecular volume of the solute, V, and the viscosity of the solvent, $\eta$, according to Equation 9:

$$t_c = V\eta/kT \qquad (9)$$

where k is the Boltzmann constant and T is the absolute temperature. A steady state determination of r is equivalent to the integral to infinite time of the exponential in Equation 8 weighted with the exponential decay of the excited state. The resultant steady state anisotropy, a much more experimentally accessible quantity (Equation 7), still contains the information on the rotational motion of the fluorophore:

$$\frac{r_o}{r} = 1 + \frac{t_e}{t_c} = 1 + \frac{t_e kT}{\eta V} \qquad (10)$$

Clearly these beautifully simple relationships can be of tremendous power in biophysical studies of macromolecules. In membrane studies, fluorescence anisotropy measurements have had a major impact on our understanding of the properties of membranes as two-dimensional solvents.[4] These applications are detailed in Chapters 2 and 3 of this volume. It should be emphasized that Equations 8, 9, and 10 are all based on the assumption of a spherical fluorophore. For many fluorescent probes. especially in the anisotropic microenvironment of a membrane, this assumption is clearly invalid. Chapter 2 explores some of the implications of this assumption and presents more general theoretical treatments of fluorescence anisotropy which are particularly suitable for membrane studies; the time-resolved anisotropy for several models of probe motion is explicitly treated. Chapter 3 reviews the application of a particular fluorescent fatty acid to the study of membrane structure and dynamics.

## IV. RESONANCE ENERGY TRANSFER

As mentioned at the end of Section II, resonance energy transfer (RET) is one of several mechanisms for intermolecular quenching of fluorescence. The excitation energy is transferred from the originally excited fluorophore, termed the "donor" in this context, to the quencher or "acceptor". In fact, it is common to observe emission from the acceptor concomitant to donor quenching, even though the acceptor is excited by this indirect route. The reason that RET is such an attractive phenomenon from a biophysical standpoint is that it occurs over distances which can be precisely determined from theory and within a range, 10 to 70 Å, that is not easily accessible by other "wet" techniques.

The theory was originally proposed by Förster[5] and is based on the idea that the coupling of the donor and acceptor transition moments can be treated as the interaction of two dipoles (recall the discussion of Equation 4). Classically, the interaction of two dipoles depends on the inverse third power of the distance between them, $R_{DA}$; the probability of this transfer is proportional to the square of this coupling so that the rate constant for RET has an inverse sixth power dependence on the separation of the donor and acceptor. Since the transfer is based on the coupling of transition moment vectors, a factor based on their relative orientations, $\kappa$, must appear in the theory along with a measure of the spectral overlap of the two transitions. Finally, since the squared transition moments for the donor and acceptor are proportional to $k_f$ and $\epsilon$, respectively, it would be convenient to use these experimental quantities directly in the equation for the RET rate constant. Based on these considerations, the rather imposing equation derived by Förster does indeed appear to make qualitative sense:

$$k_{RET} = \frac{8.71 \times 10^{-5}\kappa^2 k_f}{n^4 R_{DA}^6} \frac{\int F(\lambda)\epsilon(\lambda)\lambda^4 d\lambda}{\int F(\lambda) d\lambda} \tag{11}$$

$$\kappa = \cos\theta_{DA} - 3\cos\theta_{DR}\cos\theta_{AR} \tag{12}$$

For convenience, Equation 11 is written with $R_{DA}$ in angstroms and $\lambda$ in nanometers. The angles in Equation 12 are between the donor and acceptor transition moments, $\theta_{DA}$, and between each of the moments and the line joining their centers, $\theta_{DR}$ and $\theta_{AR}$, respectively. In cases where both donor and acceptor rotate freely on the timescale of fluorescence, $\kappa^2$ averages to a value of two thirds. This approximation is frequently employed and seems to work well, even where the assumption of free rotation is not fully justifiable.

A transfer efficiency, TE, may now be defined as the fraction of the donor-excited state which undergoes RET. It is related to the relative donor quantum yields in the presence, $Q_{DA}$, and absence, $Q_D$, of acceptor:

$$TE = \frac{k_{RET}}{k_f + k_{nr} + k_{RET}} = 1 - \frac{Q_{DA}}{Q_D} = \frac{R_{DA}^{-6}}{R_{DA}^{-6} + R_o^{-6}} \tag{13}$$

The ratio of quantum yields in Equation 13 is equal to the corresponding ratio of donor emission intensities, so that TE is a readily determined quantity. $R_o$ is the "critical transfer distance" at which RET and other donor decay processes are equally probable. From Equations 11 and 6 it is easily shown that:

$$R_o^{-6} = \frac{8.71 \times 10^{-5}\kappa^2 Q_D}{n^4} \frac{\int F(\lambda)\epsilon(\lambda)\lambda^4 d\lambda}{\int F(\lambda) d\lambda} \tag{14}$$

The most common application of RET is as a "spectroscopic ruler".[6] Different donor-acceptor pairs can have $R_0$ values ranging, typically, from 20 to 55 Å and can be chosen to optimize the sensitivity for a particular measurement. These chromophores can be intrinsic to a macromolecule or may be linked to it via labeling reactions.

For a homogeneous solution of donors and acceptors, TE will simply be a function of the acceptor concentration; in fact, the value of $R_0$ may be determined experimentally by measuring the concentration of acceptor at which TE = 0.5, $[A]_o$:

$$R_o = 7.35/([A]_o)^{1/3}\text{Å} \tag{15}$$

Equation 15 just expresses the idea that when TE = 0.5, each molecule of donor will be, on average, $R_0$ away from an acceptor; this is equivalent to saying that a sphere of radius $R_0$ will contain an average of one acceptor.

RET has been of particular value in membrane studies because the membrane serves to concentrate lipophilic donor-acceptor pairs so that they come within transfer distance, even when their bulk concentrations are quite low. Chapters 6 and 7 of this volume include descriptions of several RET-based assays for membrane fusion which use the dilution or concentration of donor-acceptor probe pairs in the lipid bilayer. In addition, questions relating to the structure of membrane proteins and their association with the lipid bilayer can be approached by RET if the theory is expanded to account for special geometrical constraints that may be imposed on the donor or acceptor probes. In Chapter 5, the geometry of a membrane receptor-ligand complex is probed with fluorescently labeled monoclonal antibodies by RET; the proximity of various sites on the complex to the membrane surface are also probed by employing membrane-associated acceptors and an elaboration of the theory for the case of a two-dimensional acceptor distribution. The tertiary structure of membrane proteins can be obtained by RET from tryptophan residues to acceptor probes dissolved in the membrane at specific depths within the lipid bilayer (Chapter 4); for multiple tryptophan donors, a Monte Carlo approach can be used to obtain the best structure to fit the data.

## V. FLUORESCENCE MICROSCOPY

Coupling fluorescence excitation and detection to a microscope allows the application of the aforementioned techniques to the membranes of individual cells. There are several additional implications and consequences, however, which need to be considered in order to fully appreciate the power of microfluorometry.

Microfluorometry can be incredibly sensitive because of the high collection efficiencies inherent in the optics and because of the small sample volume. Fluorescence intensity is proportional to the extinction coefficient, quantum yield, concentration, and pathlength of the sample, the incident light intensity, and the light collection efficiency of the instrument. A conventional fluorometer is capable of detecting a highly fluorescent substance down to ca. $10^{-12}\,M$; for a 1 m$\ell$ sample volume, this corresponds to a detection limit of $10^9$ molecules. A microscopic sample may have a very short pathlength, say 1 $\mu$m, so that the concentration of a fluorophore may have to be several orders of magnitude higher in order to be detectable; however, because the sample volume decreases as the cube of the pathlength, this translates to a much-improved detection limit, calculated on the basis of number of molecules. A microfluorometer also concentrates all the light from a conventional light source or a laser onto the microscopic specimen and, for high numerical aperture objectives, can gather the emission from a large solid angle; thus, both collection efficiency and incident light intensity are higher than in a conventional fluorometer. Chapter 10 shows how a single molecule labeled with about 40 fluorophores can be detected and tracked along the surface of a cell.

Indeed, the heterogeneity on a microscopic level of cell surface labels is the basis for several fluorometric methods for following the lateral motions and kinetics of cell surface molecules. The most direct of these is fluorescence-correlation spectroscopy (Chapter 9) which analyzes the fluctuations in the fluorescence signal on a small patch of membrane to determine the rate of some equilibrium process, such as the diffusion of fluorophores in and out of the patch. Another approach to measuring lateral diffusion on the membrane surface is to create a well-defined spatial heterogeneity in the distribution of a labeled cell surface component and follow the subsequent relaxation to the equilibrium distribution. This is most often achieved by photochemically destroying a fluorophore with a focused high-intensity laser pulse and monitoring the return of fluorescence in the bleached spot. Chapters 8 and 9 detail the experiment for fluorescent lipid analogs and labeled membrane proteins, respectively. Chapter 11 shows how one can confine the excitation of fluorophores to a region within a wavelength of an interface between two media via total internal reflection; thus, the diffusion or reaction kinetics of molecules at the interface can be studied independently of their behavior in the bulk solution. This approach can be used independently or in conjunction with photobleaching or correlation spectroscopy to probe dynamic processes on cell surfaces.

The resolution of light microscopy is limited by the wavelength of light so that molecular features cannot ordinarily be distinguished or monitored. Since fluorescence can, in principle, be obtained from single molecules (cf. Chapter 10), fluorescence microscopy does indirectly provide molecular resolution. As mentioned above, Chapter 11 employs total internal reflection to limit fluorescence excitation to a few hundred nanometers along the optical axis. A radically different approach toward increasing the resolution of the light microscope is described in Chapter 12. The usual resolution limits are defeated by employing submicron apertures brought to within a few hundred angstroms of the specimen. Fluorescence can be excited and collected through these apertures with normal microscope optics. This technique promises the ability to probe the organization of molecular components on membranes under physiological conditions and with molecular resolution.

## REFERENCES

1. **Lakowicz, J. R., Ed.,** *Principles of Fluorescence Spectroscopy,* Plenum Press, New York, 1983.
2. **Weber, G. and Farris, F. J.,** Synthesis and spectral properties of a hydrophobic fluorescent probe: propionyl-2-(dimethylamino)naphthalene, *Biochemistry,* 18, 3075, 1979.
3. **Turro, N. J.,** *Molecular Photochemistry,* W. A., Benjamin, Reading, Mass., 1965.
4. **Shinitzky, M., Ed.,** *Physiology of Membrane Fluidity,* CRC Press, Boca Raton, Fla., 1984.
5. **Förster, T.,** Intermolecular energy transfer and fluorescence, *Ann. Phys.,* 2, 55, 1948.
6. **Stryer, L.,** Fluorescence energy transfer as a spectroscopic ruler, *Annu. Rev. Biochem.,* 47, 819, 1978.

Chapter 2

# MEMBRANE "FLUIDITY" FROM FLUORESCENCE ANISOTROPY MEASUREMENTS

**Barry R. Lentz**

## TABLE OF CONTENTS

## I. MEMBRANE "FLUIDITY"

The term "membrane fluidity" is one that has been greatly misused and consequently greatly maligned. The principal source of malcontent with this term has been the tendency for it to be used in an imprecise way. Thus, our concepts of a fluid membrane may vary greatly, depending on the particular aspect of membrane dynamics that draws our focus. From the point of view of bilayer thermodynamics, a "fluid" membrane exists at temperatures above the "lamellar-extended" to "lamellar-disordered" phase transition. In the eyes of a cell biologist, a fluid membrane might contain protein and lipid molecules capable of rapid and extensive diffusive motions in the plane of the membrane bilayer. As used by a spectroscopist, the term "fluidity" often assumes a more narrow and specific meaning. For instance, deuterium magnetic resonance spectroscopy represents a fluid membrane as one in which a given C–D bond experiences, on average, substantial excursions from its orientation in a fully extended chain, with the averaging process occurring over 10 to 100 $\mu$sec. By contrast, infrared spectroscopy presents a view of membrane "fluidity" in terms of the ensemble- or long-time average number of *gauche* C–C rotational conformations per phospholipid acyl chain.

This article also stresses a narrow definition of "fluidity", namely as it pertains to the rate and extent of phospholipid acyl chain excursions away from some initial chain orientation during the lifetime of the fluorescence-excited state, usually several nanoseconds. It is this aspect of overall membrane dynamics that one hopes to quantify with the techniques and fluorescent probes reviewed here. In addition, this review rejects a view of membrane "fluidity" that assumes isotropic probe motion in an isotropic liquid hydrocarbon bilayer. While this view has revealed many useful empirical relationships,[1-4] it makes assumptions that are of questionable validity (see Sections IV and V). It is this author's view that a more detailed molecular approach is much more likely to extract information about fluorescent probe motion that will advance our knowledge of membrane dynamics and structure.

The purpose of this article, then, is to summarize and critically review what is known about the detailed molecular behavior in a membrane of several popular fluorescent probes of membrane fluidity. Rather than provide a compilation of all the instances or ways in which these probes have been used since the last major reviews,[3,4] my aim is to evaluate the usefulness of different probes through an evaluation of probe motion and location within the bilayer, probe photophysical properties, probe partitioning between bilayers or bilayer domains, and probe perturbation of bilayer structure. Before beginning this evaluation in Sections IV and V, I summarize in Section II some basic principles of fluorescence polarization and then, in Section III, propose a classification of fluidity probes about which the remainder of the review article is organized.

## II. FLUORESCENCE ANISOTROPY

A common feature of all the probes reviewed here is that the depolarization of their fluorescence is monitored as a means of estimating membrane fluidity. Fluorescence depolarization is a phenomenon described in detail in Cantor and Schimmel's *Biophysical Chemistry* text[7] and in a recent monograph on fluorescence.[8] For the sake of a complete description of membrane fluidity probes, a brief summary of the phenomenon is given here.

The ground and excited state electronic distributions of any fluorophore define directions within the molecule in which the probability of absorption or emission of a photon is greatest. These are the directions of the excitation and emission dipole moments, the magnitudes of which determine the maximal probabilities of absorption or emission. When polarized exciting light (see Figure 1) is directed toward a chromophore molecule whose excitation dipole moment is aligned with the electric vector of the exciting light, the chromophore will

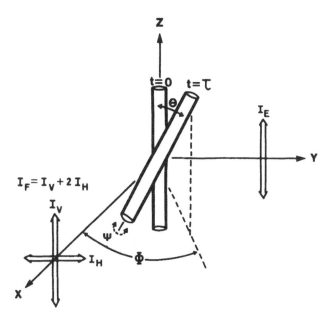

FIGURE 1.   Fluorescence anisotropy of a cylindrically symmetric fluorophore whose excitation and emission dipoles are parallel to the symmetry
axis. Vertically polarized light ($I_E$) from the Y-axis vertical polarizer (not
shown) selectively excites (photoselection) a population of fluorophores
whose excitation dipoles are assumed to be symmetrically distributed about
the Z-axis. After a time, photons ($I_F$) are emitted (electric vector parallel
to the emission dipole) with probability $e^{-t/\tau}$ ($\tau$ = the excited-state lifetime)
and detected after analysis by vertical and horizontal X-axis polarizers (not
shown). Rotation of the probe during the time $\tau$ diminishes the verticle
component ($I_V$) and enhances the horizontal component ($I_H$) of analyzed
light in a way described by the time-dependence of fluorescence anisotropy
[r(t)]. The assumptions of cylindrical probe symmetry and of symmetrical
Z-axis distribution simplify the description of r(t) to one involving an
expansion of the angular distribution of probe in terms of the Legendre
polynomials ($P_n$) in the cosine of the single angle $\theta$. The condition of
parallel excitation and emission dipoles allows for a single ($P_2$ approximation to describe the probe angular distribution) or triple ($P_2 + P_4$ approximation to the angular distribution) exponential behavior of r(t). Multiple
exponentials are obtained if this condition is violated, even if a simple $P_2$
approximation is made for the probe angular distribution.

preferentially absorb this light. Since the absorption process is so much faster than molecular
rotations, the use of oriented exciting light creates a population of preferentially oriented
excited fluorophores. This is referred to as photo selection. Since the emission of a photon
by the excited fluorophore requires a much longer time (the lifetime, $\tau$, of the excited state)
than does absorption, the fluorophore can often reorient before emission occurs. This is
illustrated in Figure 1. If such a situation occurs (i.e., the rotational correlation time of the
excited molecule is less than or on the order of the excited state lifetime), the emitted photon
will no longer be polarized parallel to the exciting photon, even if the molecular excitation
and emission dipoles are parallel within the fluorophore. The resulting depolarization of
fluorescence is often defined in terms of the steady-state fluorescence anisotropy:

$$r = \frac{I_V - I_H}{I_V + 2\,I_H} \tag{1}$$

As illustrated in Figure 1, $I_V$ and $I_H$ are the intensities measured in directions parallel and perpendicular to the electric vector of the exciting light. From Equation 1 and Figure 1, it is clear that the greater the extent of reorientation of the fluorophore during the lifetime of its excited electronic state, the smaller will be the observed fluorescence anisotropy, to the extent that $r = 0$ for complete fluorophore reorientation ($I_V = I_H$). It is in this sense that fluorescence anisotropy provides a measure of membrane fluidity. It can also be shown[8] that, for a rigidly fixed fluorophore,

$$r = r_0 = 1/5(3\cos^2\alpha - 1) \tag{2}$$

where the intrinsic anisotropy ($r_0$) is determined only by the electronic distributions of the excited and ground states, since these determine the angle $\alpha$ between the excitation and emission dipoles of the fluorophore. From Equation 2, we see that the maximum value for the fluorescence anisotropy is 0.4 ($r_0 = 0.4$, $\alpha = 0°$).

While on the subject of fluorescence depolarization, it is necessary to point out a potential pitfall in measuring fluorescence anisotropy in membrane suspensions. Light scattering by highly turbid solutions depolarizes both the exciting and emitted light and can, thereby, dramatically lower the apparent anisotropy of a fluorescence signal.[8a] This can be empirically corrected through successive dilution of the turbid fluorescent sample.[8b,8c]

In addition to measurements of steady-state fluorescence anisotropy, researchers in the field of membrane fluidity often take advantage of the additional information contained in measurements of the real-time decay of fluorescence anisotropy, r(t). This quantity is defined exactly as in Equation 1, except that the intensities $I_V(t)$ and $I_H(t)$ are both measured as functions of time following excitation by a short burst of polarized light. For a simple, isotropically and freely rotating fluorophore with a single (or degenerate) moment(s) of rotation, r(t) decays in exponential fashion with time, with a rotational correlation time $\phi$ ($\phi = 1/6\,D_{rot}$, $D_{rot}$ = rotational diffusion coefficient):

$$r(t) = r_0 e^{-t/\phi} \tag{3}$$

Integrating r(t) weighted by the total fluorescence intensity (also an exponential with characteristic time, $\tau$), we obtain the well-known Perrin equation for the steady-state anisotropy:

$$r = r_0/(1 + \tau/\phi) \tag{4}$$

Thus, measurements of fluorescence anisotropy and excited-state lifetime can provide, in the simplest of cases, the fluorophore rotational correlation time, which we assume reflects the reorientational correlation time of phospholipid acyl chains in a membrane. It is precisely in this sense, and with the assumptions of free, isotropic rotation, that the early treatments of fluorescence anisotropy of membrane-associated probes provided a measure of membrane fluidity.[1-4] For fluorophores experiencing anisotropic or hindered rotation, or displaying more than one rotational correlation time or fluorescence lifetime, it becomes much more difficult and often impossible to obtain a description of fluorophore rotational motion in terms of the steady-state fluorescence anisotropy or even the measured real-time decay of fluorescence anisotropy [r(t)]. These complications and their implications will be considered in more detail in Sections IV and V.

## III. TYPES OF PROBES

Several hydrophobic fluorescent molecules have been used, with varying degrees of success, as probes of membrane fluidity. In general, the approach with most probes has

## 1. Linear Probes

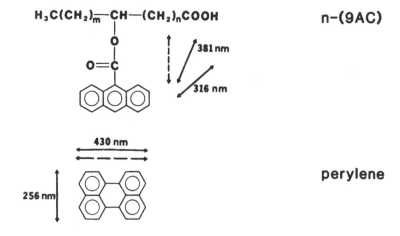

FIGURE 2. Fluorescent probes of membrane fluidity. The relative orientations of the excitation (solid lines) and emission (dashed lines) dipoles are indicated, as derived from measured $r_0$ values (see Equation 2). The orientation of the emission dipole relative to the molecular axes is fairly well-established for DPH[30] and is estimated for n-(9AC)[78] and perylene[79] from the normal photophysical properties of aromatics.

been the same, namely, to detect the extent or rate of acyl chain orientational excursions by monitoring the anisotropy of probe fluorescence. The approach makes the fundamental assumption that the probe rotational motions that result in depolarization of fluorescence are tightly coupled to acyl chain orientational fluctuations. This assumption will be examined for each of the probes reviewed.

Broadly, the fluorescent molecules used to monitor membrane fluidity fall into two general categories. The probes in these two categories differ essentially in their shapes and these differences in shape result in differences in their positions and movements within the lipid bilayer. The first category includes what I call the linear probes (see Figure 2). These are long rigid molecules of roughly the shape and size of fully extended phospholipid acyl chains. These fluorophores have excitation and emission dipoles that are roughly colinear with their long molecular axes. For this reason, rotations about the long molecular axis make essentially no contribution to the depolarization of polarized fluorescence, while rotations about the two remaining rotational axes (mutually orthogonal and orthogonal to the long axis) are equally capable of depolarizing fluorescence. Probe molecules falling into this category include *cis-* and *trans-*parinaric acid[9] and phospholipids containing the parinaric acids; and 1,6-diphenyl-1,3,5-hexatriene (DPH)[1] and molecules containing the DPH moiety, such as 1-[4-(trimethylamino)phenyl]-6-phenylhexa-1,3,5-triene (TMA-DPH),[10] (2-carboxyethyl)-1,6-diphenyl-1,3,5-hexatriene (CE-DPH),[11] and 1-palmitoyl-2-[[2-[4-(6-phenyl-*trans*-1,3,5-hexatrienyl)phenyl]ethyl] carboxyl]-3-*sn*-phosphatidylcholine (DPHpPC).[12] The parinaric acids and their derivatives are reviewed elsewhere in this volume[14] and, therefore, will not be considered further here. The properties of DPH and its derivatives will be reviewed in Section IV.B.

The second category of fluorescent fluidity probes is comprised of nonlinear disk-like probes (see Figure 2). In general, the excitation and emission dipoles of these molecules are not colinear, nor do they coincide with the principal molecular rotational axis. These considerations make interpretation of observed fluorescence depolarization in terms of probe rotational motion quite difficult. In addition, because these probes do not resemble phospholipids or their acyl chains, translation of a description of probe motion into information about acyl chain order and dynamics requires broad leaps of interpretation. Despite these difficulties, the probes in this category have enjoyed wide popularity, especially the 9-anthroyloxy-labeled fatty acids or phospholipids,[15-17] anthracene-labeled fatty acids,[18] and perylene.[19,20] The fluorescence properties of these probes as well as their behavior in membranes will be considered in Part V of this review.

## IV. LINEAR PROBES

### A. Overview and Summary

In this section, the motion and physical properties of membrane-associated DPH and DPH-related probes are reviewed. In the first part of this section, a rather complex literature is distilled in order to obtain a detailed physical picture of what we know or do not know about the motion of these probes in a membrane. For the benefit of those readers who do not wish to wade through the details of this somewhat involved subject, a brief summary is given here.

The simplest treatment that will account for the anisotropic motion of DPH in a membrane views the probe as undergoing free rotation in a cone-shaped volume symmetrically distributed about an axis normal to the plane of the membrane bilayer. As applied to the observed time dependence of DPH fluorescence anisotropy, this treatment yields both the rate of DPH rotational diffusion and the orientational order parameter of the probe in the membrane. This simple treatment accounts for the nonzero long-time limit of the fluorescence anisotropy decay curve but cannot account for its nonexponential approach to this limit. Two possible explanations for the nonexponential character of the decay curve have been proposed and are reviewed in this section. First, it has been proposed that DPH is distributed not only about the bilayer normal but also in the plane of the bilayer. A mathematical treatment of this model, when used to fit the observed anisotropy decay curve, yielded an additional orientational order parameter, the fourth rank order parameter, to account for the distribution of probe in the plane of the bilayer. A second possible explanation for the nonexponential decay is also considered. In this view, the contributions of different rotating species of DPH can account for the multiexponential behavior of the decay of fluorescence anisotropy. At present, neither explanation has been shown to be clearly superior.

In the second part of this section, the properties of DPH and DPH-related probes are reviewed. First, the anomalous photophysical properties of DPH are discussed in terms of the possible existence of two distinct excited states of the DPH molecule. This is consistent with reports of two fluorescence lifetime components for DPH as well as with the proposal that multiple fluorescent forms can account for the multiexponential decay of DPH fluorescence anisotropy. Second, the problem of locating the DPH probe in the bilayer is considered. Third, the ability of DPH and its derivatives to partition between membranes or between different domains within a membrane is summarized. In this regard, DPH and its derivatives are reported to behave quite differently. Fourth, the different perturbing effects of DPH, DPHpPC, or TMA-DPH on membrane structure are discussed. Finally, this section concludes with a description of the different types of information about membrane order that are provided by DPH and its two common derivatives, DPHpPC and TMA-DPH

## B. Probe Motion
### 1. Free Rotation Approximation; "Microviscosity"

Information about membrane fluidity can be extracted from the fluorescence properties of a membrane-associated probe only in the context of an adequate theoretical description of probe motion. In the early work of Shinitzky and co-workers[1,21] with DPH and perylene, probe motion was assumed to be that of a free isotropic rotator. With this assumption, Einstein's equation relating the rotational diffusion rate ($\rho$) to sample viscosity ($\eta$) and effective rotational molar volume ($\overline{V}$),

$$1/\phi = \rho = kT/\eta\overline{V} \tag{5}$$

could be combined with the Perrin equation (Equation 4 above) in order to obtain[22] a simple expression for the "microviscosity" within a membrane bilayer. The actual estimation of "microviscosity" values required calibration of the effective rotational molar volume of the probe in a viscous, but isotropic, hydrocarbon oil.[21]

These early studies opened the field of membrane "fluidity" and provided many useful qualitative insights.[3,4] However, especially as applied to the linear probe, DPH, they have several quantitative shortcomings. Briefly, these shortcomings are

1.  A long linear molecule such as DPH is not likely to undergo isotropic rotation, i.e., rotation whose angular rate is independent of rotational azimuth.[22] Indeed, Brand and co-workers[5,23] have demonstrated that the real-time decay of membrane-associated DPH fluorescence anisotropy could not be described by a single exponential but required, even for a presumably isotropic paraffin oil solution, a second exponential and, in membranes, a constant term to reflect the fact that the anisotropy did not return to zero at long times after excitation.[23]

2.  Aside from the complication of anisotropic motion in a membrane, Brand's detection of two rotational correlation times for the decay of DPH fluorescence anisotropy, even in paraffin oil, questions another simplifying assumption made in the definition of "microviscosity". While the origin of the second correlation time is unclear, Tao's treatment of isotropically rotating fluorophores predicts a single exponential for r(t) only for chromophores that can be considered cylindrically symmetrical molecules with their absorption and emission dipoles parallel to the symmetry axis.[24] While such a view may be a reasonable approximation for DPH, it *is* an approximation. DPH is rigorously a planar, slightly nonlinear molecule (see Figure 2) which has symmetry properties described by the point group $C_{2h}$, while a cylinder has higher-order symmetry described by the point group $D_{\infty h}$. Even free isotropic rotation of a molecule such as DPH could, in theory,[24] result in a small contribution to r(t) of up to four additional exponentials. While Brand and co-workers chose not to attempt such model-dependent interpretations of their data, their results demonstrated clearly that a single exponential treatment is not adequate to describe the rotational motion of DPH in a membrane even when dual excited-state lifetimes are taken into account.

3.    In addition to the issue of anisotropic rotation, the molar volume calibration procedure using hydrocarbon oils has been questioned[6,24a,25] on the basis that such oils have molecular order distinctly different from that within a membrane. Indeed, different calibration media have been shown to give different values of $\overline{V}$.[25] Nonetheless, a comprehensive comparison of different calibration solvents as well as probes (DPH vs. an electron spin resonance probe) has demonstrated that consistent values of "microviscosity" can be estimated by these procedures for bilayers well above their order-to-disorder phase transition,[25] while estimates by different procedures varied by as much as 30% in membranes only slightly above their phase transition. Thus, as would be expected, the assumption of isotropic probe rotation breaks down most severely in more highly ordered membranes.

It is now well-accepted that "fluidity", i.e., acyl chain dynamics, is not the only property of a membrane that is monitored by the fluorescence anisotropy of linear probes such as DPH or its derivatives. The average extent of chain motion, or acyl chain "order", contributes substantially to the observed anisotropy, and, depending on the conditions of the measurement, can overshadow the contribution of chain dynamics. In the remainder of this section, I review attempts to sort out the contributions of chain dynamics and chain order to the observed depolarization of DPH fluorescence. Nearly all treatments of this problem have assumed a single homogeneously rotating species of probe molecules. These treatments will be reviewed first in Section IV.B.2. An alternative treatment of the rotational contribution to fluorescence anisotropy, allowing for multiple rotating species, will be briefly mentioned in Section IV.B.3. of the current section.

*2. Single Rotating Species Treatment*

The first effort to extend the isotropic rotation treatments of Weber[22] and Tao[24] to consider hindered or anisotropic motions was made by Kinosita et al.[26] and specifically applied to DPH by Kawato et al.[27] In order to apply the Smoluchowski treatment of rotational diffusion, these authors limited consideration to fluorophores with either parallel excitation and emission transition dipoles, or axial symmetry around the emission dipole. To a reasonable approximation (see above and Figure 2), DPH and its derivatives should fit into both of these categories. Without making any assumption as to the form of the intermolecular potential that hinders probe rotation, Kinosita et al. demonstrated that two important pieces of information could be obtained from the zero-time and long-time behavior of fluorescence anisotropy decay. From the limiting or long-time anisotropy ($r_\infty$), they defined the "degree of orientational constraint" ($r_\infty/r_0$). From the initial slope of the fluorescence anisotropy decay, the average wobbling diffusion constant ($<D_w>$) was obtained according to Equation 6:

$$r(t \to 0)/r_0 = 1 - 6 <D_w> t \qquad (6)$$

where the average is over all orientations of the fluorophore in a local membrane segment.[26] For the specific case of a rod-shaped molecule wobbling freely in a hard conical potential ("free rotation in a hard cone model"; see Figure 3), an expansion of r(t) in an infinite set of exponentials was obtained by Kinosita et al. To a reasonable approximation, this could be written as:

$$r(t)/r_0 = r_\infty/r_0 + (1 - r_\infty/r_0)e^{-t/\phi} \qquad (7)$$

where $(r_\infty/r_0)^{1/2} = \frac{1}{2} \cos \theta_{max} (1 + \cos \theta_{max})$ and where $\phi$ is defined as the characteristic time with which the initially photoselected fluorophore orientational distribution approaches the average distribution.

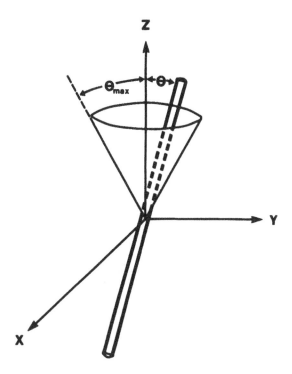

FIGURE 3. The rotational motion of a linear fluorescent probe in terms of the "free rotation in a hard cone" model of Kinosita et al.[26] As described in Figure 1, the dependence of the angular distribution on the other spherical coordinate ($\phi$) and on rotation about the main axis ($\psi$) is removed by symmetry considerations. The maximal angle of rotation away from the director (Z) axis is given as $\theta_{max}$.

Using this formalism to interpret measured fluorescence anisotropy decay curves, Kawato et al.[27] estimated the rotational relaxation time of DPH in dipalmitoylphosphatidylcholine (DPPC) multilamellar vesicles to be roughly 1 nsec, both above and below the lipid phase transition (ca. 41°C). By contrast, the maximum cone angle within which the probe was free to wobble ($\theta_{max}$, defined in terms of $r_\infty$, see Equation 7) increased from roughly 20° below the phase transition to 65 to 70°C at high temperature. In order to explore this larger cone with a roughly constant characteristic time of 1 nsec, the probe wobbling diffusion constant also increased from about 0.05 nsec$^{-1}$ at 30°C to about 0.25 nsec$^{-1}$ at 50°C.

Significant further insight into the interpretation of measured DPH fluorescence anisotropies was provided by Heyn[28] and Jähnig,[29] who recognized, nearly simultaneously, that the model-independent "degree of orientational constraint" $(r_\infty/r_0)$[26] was related to the probe orientational order parameter $(S_2)$:[29]

$$r_\infty/r_0 = S_2^2 = <P_2(\cos\theta)>^2 \tag{8}$$

In Equation 8, the subscript "2" on S indicates the second rank order parameter. This is defined in terms of the orientational-distribution-function-weighted average (indicated by brackets) of the second order Legendre polynomial (see Figure 4) in the cosine of the angle of the probe rotation away from the bilayer normal (see Figure 3). Therefore, to the extent that DPH is aligned in a membrane with the phospholipid acyl chains,[30] this quantity can be taken as reflective of the membrane acyl chain order parameter. Both Heyn[28] and Jähnig[29] noted that the order parameter derived in this way from DPH fluorescence anisotropy agreed

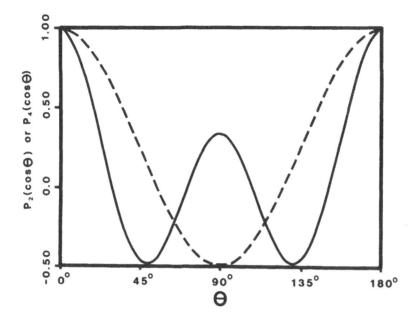

FIGURE 4.   Dependence of the second ($P_2$; dashed line) and fourth ($P_4$; solid line) order Legendre polynomials in cosine $\theta$ on the angle $\theta$, where $P_2$ and $P_4$ are given by: $P_2(\cos \theta)$ = $(3 \cos^2\theta - 1)/2$ and $P_4(\cos \theta) = (35 \cos^4\theta - 30 \cos^2\theta + 3)/8$. $P_2 (\cos \theta)$ has zeros at $\pm 54.7°$, while $P_4 (\cos \theta)$ has zeros at roughly $\pm 30.5°$ and $\pm 70.1°$.

with deuterium NMR order parameters reported for deuterium atoms located on the ninth or tenth carbon atom down the acyl chain. This agreement, although it could be fortuitous, was taken[28] as reinforcing the view that DPH was oriented, on average, parallel to the lipid acyl chains in a membrane and centered at roughly the ninth or tenth carbon atom down the chain. In addition to providing this insight into the interpretation of DPH fluorescence properties, both Heyn and Jähnig recognized that the hindered single-exponential decay law predicted for DPH fluorescence anisotropy (see Equation 7), when combined with single exponential decay of total fluorescence intensity, would lead (see Equations 3 and 4 of Section II) to the following modified Perrin equation for hindered rotation of DPH in a membrane:

$$r - r_\infty = (r_0 - r_\infty)/(1 + \tau/\phi) \tag{9}$$

This equation provides a simple means of dissecting out of the steady-state fluorescence anisotropy (r) the contributions due to probe dynamics ($\phi$) and to hindered rotational freedom ($r_\infty$). Examination of Equation 9 reveals that the steady-state anisotropy is equal to the long-time or limiting fluorescence anisotropy ($r_\infty$) plus a contribution that is largest when the probe rotational correlation time is large relative to the lifetime of the fluorescence excited state. In this limit, the probe can rotate through only a fraction of its available cone before it emits a photon, and the resulting depolarization of fluorescence is a measure of the rate of probe orientational motion. In the limit of a very short rotational correlation time, however, the probe is free to explore its available cone thoroughly in the time between absorption and emission of a photon. In this limit, the steady-state fluorescence anisotropy is approximately equal to the limiting anisotropy and is, therefore, a measure of the orientational order in the membrane, rather than of probe dynamics. As noted by Jähnig,[29] for the case of DPH, the fluorescence excited-state lifetime is roughly eight times the rotational correlation time, leading to Equation 10:

$$r \cong 9/8 \; r_\infty - 1/20 \tag{10}$$

Recognition that DPH fluorescence anisotropy reflected average orientational order of the probe more than its rate of orientational motion stimulated a number of papers[31-36] attempting to extend or refine the basic treatment of hindered rotational motion presented by Kinosita et al.[26] Worthy of note is the development by Lakowicz et al.[31] of expressions for obtaining $r_\infty$ and $\phi$ in Equation 7 from the phase shift and modulation ratio of polarized fluorescence. In stressing a model-independent interpretation of the data, however, these authors erroneously interpreted the angle $\theta$ in Equation 8, a flaw picked up by Lipari and Szabo[32] and later by Engel and Prendergast.[36] Another worthwhile contribution was made by Fulford and Peel,[34] who applied to the expressions of Kinosito et al.[26] a Gaussian, rather than hard cone model, for the orientational distribution function of the probe. This more reasonable approximation of the probe orientational distribution led to qualitatively the same picture as that provided by the hard cone model, but with a different parameter to describe the orientational order (the standard deviation of the angular distribution).

A major advance in the theoretical description of the orientational order of DPH in a bilayer has been the generalized treatment developed by Zannoni and co-workers.[37,38] These authors made no assumption about the nature of the orientational distribution function (i.e., cone model, Gaussian model, etc.) except that it could be expanded in an infinite set of spherical harmonic functions, the Wigner functions. The time-correlation function for probe rotation was then described in terms of a function, called a rotational propagator, that described the time evolution of this generalized orientational distribution function. The form of the rotational propagator then became dependent on the model used to describe the hindered rotation of the probe molecule. The "strong collision model" (effectively equivalent to the "free rotation in a hard cone" model[26]) presumed that the probe experienced sudden changes in its orientation at short time intervals, presumably due to collision with large neighboring molecules. The "diffusion model" presumed that reorientation was due to random solvent collisions, as well as to an ordering torque related to an effective aligning potential. When the aligning potential was approximated by the second order Legendre polynomial in cosine $\theta$ (the Maier-Saupe effective potential[38]), solution of the equation of rotational Brownian diffusion yielded a Gaussian orientational distribution function, reducing the diffusion model to the previously mentioned Gaussian approximation of Fulford and Peel.[34]

For the case of a cylindrically symmetric probe molecule with either the excitation or emission dipole parallel to the molecular long axis (e.g., DPH), Zannoni et al.[38] showed that the Wigner-function expansion of the orientational distribution function reduced to an expansion in Legendre polynomials. The second term of this expansion yielded the second rank order parameter (see Equation 8), while higher order terms yielded higher rank order parameters which provide additional insights into the average orientational order of the probe in a membrane. In the context of the two dynamic models considered, the time dependence of fluorescence anisotropy could be expressed in this treatment in terms of either three (diffusion model) or two (strong collision model) rotational correlation times. For the strong collision model, which should approximate reasonably well the motion of a hydrophobic probe in a membrane, one of these described rotation of the long molecular axis while the other described rotation about this axis. For DPH and its derivatives, to the extent that *both* the excitation and emission dipoles are parallel to the long axis, and that the two types of rotation are uncoupled, this expression reduced to a single exponential describing rotation of the long axis (see Equation 7).

The elegant treatment of Zannoni et al.[37,38] has been generalized by van der Meer, Ameloot, and co-workers[39,40] to accommodate a more complex effective aligning potential, and by Szabo[41] to model, in addition to orientational dynamics, the consequences of excited-state dynamics. The treatment of van der Meer et al.[39] requires special note, in that it offers an alternative explanation for the nonexponential decay of fluorescence anisotropy observed for DPH in membranes.[5,23] These authors expressed the effective aligning potential in terms of a *sum* of the second and fourth order Legendre polynomials in cosine $\theta$, rather than in terms of only the second order term assumed by Zannoni et al.[38] Thus, the treatment of van der Meer et al. yielded both second and fourth rank order parameters ($S_2 = <P_2>$; $S_4 = <P_4>$; see Equation 8 for the meaning of the brackets and Figure 4 for the definition of $P_4$) and allowed, thereby, for a bimodal probe orientational distribution function.

The potential importance of describing probe angular distribution in terms of two Legendre polynomials is illustrated in Figure 4, where the dependence of these functions on the angle $\theta$ is shown. If $<P_2>$, as measured in terms of $r_\infty/r_0$, is positive, we know that the majority of probes in our membrane system are at an angle of less than 54.7° relative to the membrane director. However, we cannot determine from $<P_2>$ alone whether the angular distribution of probes is broad and extends between 0° and some much larger angle (perhaps even greater than 54.7°) or is narrow and sharply distributed about some angle less than 54.7°. On the other hand, if we also had a measure of $<P_4>$ and found it to be negative, we would know that the probes were sharply distributed about an angle between 30.5° and 54.7° rather than broadly distributed over a wide range of orientations. These two possibilities correspond to vastly different situations in terms of molecular order in a membrane (e.g., an $L_\beta$ phase vs. an $L_\alpha$ phase) and yet would be indistinguishable without knowledge of both the second and fourth rank order parameters. Of course, even with knowledge of both order parameters, the distribution may still not be adequately determined. Higher rank terms may be necessary to accomplish an adequate description. Nonetheless, knowledge of two terms will always provide a better indication of the actual orientational distribution function than will knowledge of only one.

Generalization to allow for a bimodal orientational distribution of probe molecules leads to significantly different theoretical predictions relative to simpler models assuming mon-omomodal distributions. With simple models (i.e., the Gaussian or cone models) and with the reasonable assumption that the excitation and emission dipoles of DPH are parallel (see Section IV.C), a monoexponential decay of fluorescence anisotropy is predicted (see Equation 7). Within the context of these simple models, only the second rank order parameter is defined in terms of the preexponential factor in Equation 7 (i.e.,$r_\infty/r_0$; see Equation 8), and the fourth rank order parameter is uniquely determined by the value of the second rank order parameter.[39] Even with the assumption of parallel excitation and emission dipoles, allowance for a bimodal orientational distribution function predicts the decay of fluorescence anisotropy in terms of three exponentials, with $<P_4>$ contained in preexponential factors as an adjustable parameter.[39,40] Thus, assumption of a bimodal orientational distribution allows $S_4$ to be estimated directly in terms of the nonexponential character of the experimentally determined decay of fluorescence anisotropy.

In terms of the treatment of van der Meer et al.,[39] the observed[5,23] multiexponential decay of DPH fluorescence anisotropy was interpreted by Ameloot et al.[40] as indicating a significant contribution from a population of DPH molecules oriented normal to the bilayer director (i.e., in the plane of the bilayer). Of the several models (hard-cone potential, parallel dipoles;[26] soft monomodal potential, parallel dipoles;[32,39,42] hard-cone potential, nonparallel dipoles;[32,39] soft bimodal potential, parallel dipoles;[39] purely mathematical curve fitting[5,23]) considered by Ameloot et al.,[40] only the model allowing for a bimodal potential fits the data as well as the purely mathematical treatment. Applying the bimodal distribution model to the fluorescence anisotropy decay for DPH in DPPC vesicles, Ameloot et al. found that the orientational distribution function was dominated by $<P_2>$ in the low-temperature, ordered state, suggesting an orientation of DPH parallel to the acyl chains. This has previously been proposed on the basis of oriented sample studies of fluorescence anisotropy.[30] However, at temperature above the order-disorder phase transition, the best fit of the decay of DPH fluorescence anisotropy was obtained with a large contribution from $<P_4>$, such that there was roughly an equal probability of finding the probe aligned roughly parallel and perpendicular to the acyl chains.[40] This possibility has been mentioned previously by Lentz et al.[2] and is not inconsistent with the data of Andrich and Vanderkooi.[30]

Very recently, Straume and Litman[43] have applied the bimodal distribution model to analyze polarized phase shift and modulation ratio data obtained with DPH and TMA-DPH incorporated into model membrane vesicles containing either saturated or unsaturated phosphatidylcholine species and varying amounts of cholesterol. With DPH, a bimodal orientational distribution function best fits the data. These authors discussed the resulting orientational distribution functions in terms of the volume available for probe rotation, normalizing this volume, relative to that available to an unhindered, isotropically rotating probe. This fractional volume varied between zero and one, as the degree of ordering experienced by the DPH moiety increased. Not surprisingly, these authors found with DPH that larger fractional volumes, indicative of less ordering, were obtained with higher sample temperature, greater acyl chain unsaturation, and reduced cholesterol content. In contrast, with TMA-DPH, a monomodal distribution function provided an adequate fit to the data. Interestingly, the fractional volume available to DPH decreased much more dramatically with increasing cholesterol content or decreasing temperature, than did the same parameter applied to the motion of TMA-DPH. This led Straume and Litman to suggest that the ordering of the hydrophobic core of a bilayer might be more sensitive to cholesterol and temperature than is the ordering in the interface or headgroup regions. This important conclusion illustrates the potential value of a bimodal orientational analysis as applied to comparing the motion of DPH to that of intrinsically oriented probes such as TMA-DPH or DPHpPC.

### 3. Multiple Rotating Species Treatment

All of the theoretical treatments of probe motion considered to this point make a common fundamental assumption, namely that the depolarization of fluorescence reflects the rotational motion of a single, homogeneously distributed, rotating species. In the context of this assumption, any multiexponential character in the decay of fluorescence anisotropy has been interpreted either in terms of contributions from rotations about different molecular axes[32,39] or in terms of a bimodal orientational distribution function.[40] Taking note of the slight biexponential character often displayed by the decay of total DPH fluorescence,[2,5,44,45] Gratton and co-workers[46] have recently shown that a simple treatment of DPH motion in terms of two rotating species, each with its own hindered rotational behavior (see Equation 7) and fluorescence lifetime, yielded a good fit to the differential polarized phase shifts and modulation ratios observed at multiple modulation frequencies in DPPC vesicles. This result offers an alternative explanation for the multiexponential character of the decay of DPH fluorescence anisotropy. In order to contrast this model with the bimodal distribution model of Ameloot et al.,[40] Gratton and co-workers applied their model to data obtained for both DPHpPC and TMA-DPH.[96] Both of these probes should have their DPH moiety aligned parallel to the bilayer acyl chains, due to the polar portion of the probe molecule being anchored at the membrane interfacial region (see Figure 2). In both instances, the decay of fluorescence anisotropy had a distinctly multiexponential character, which could be accounted for in terms of multiple rotating species. While the data could also be fit by the bimodal distribution model, it is difficult to justify, in terms of the physical properties of these probes, the large $<P_4>$ require for such a fit. One must conclude that a possible reason for the multiexponential character of r(t) may be the existence of multiple rotational environments for DPH in a bilayer. While other factors (e.g., nonparallel absorption and emission dipoles or bimodal orientational distribution functions) may contribute to multiexponential behavior of r(t), it will be difficult to demonstrate these contributions unequivocally in membranes wherein multiple fluorescence lifetimes are observed.

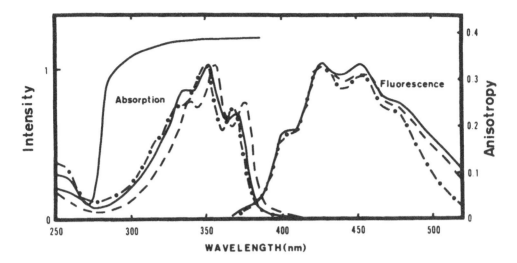

FIGURE 5.   Emission (right) and absorption (left) spectra of DPH in ethanol (solid line), dioxane (dashed line), and *N*-hexane (dot-dash line). Data adapted from Reference 38. The variation of fluorescence anisotropy recorded across the absorption spectrum in polypropylene glycol at $-50°C$ (data adapted from Reference 1) is also shown (solid line).

Brand and co-workers[47] have also recently addressed the issue of rotational heterogeneity of DPH in dimyristoylphosphatidylcholine membranes. Using pulse response techniques, they resolved a spectral component associated with DPH molecules located in an environment that was less mobile than the average environment. This demonstrated at least two rotational species in these membranes. A significant conclusion of these authors was that the different rotational species shared the same total fluorescence decay characteristics (i.e., the same set of lifetimes), a phenomenon referred to as *nonassociation* of $\phi$ and $\tau$. This contrasts with the *associative* model assumed by Gratton et al.[46]

In comparing the results of Gratton and colleagues,[46] Straume and Litman,[43] and Brand and co-workers,[47] it is evident that the physical origin of the multiexponential character of DPH anisotropy decay is still a matter of controversy. This controversy will have to be resolved before the time dependence of DPH fluorescence anisotropy can be useful for detecting structural microheterogeneity within a membrane bilayer.

## C. Properties of DPH-Related Probes

Before any fluorescence probe can be used intelligently to monitor membrane structure or dynamics, its fluorescence characteristics and physical properties in a variety of well-defined model membranes must be established. Of all the probes of membrane fluidity, none has been so thoroughly characterized as DPH and its derivatives. In this section are summarized the properties of these probes, followed by a brief review and description of how they have been used to monitor order within the membrane bilayer.

### 1. Fluorescence Characteristics

The characteristics of DPH fluorescence have been widely studied over the past twenty or so years, both because of interest in this molecule as a membrane probe and because of the insight such information can offer into the photophysics of polyenes. It is because of the latter motivation that the unusual photophysical properties of DPH have been so thoroughly documented.[48,51] The absorption and emission spectra of DPH and DPHpPC dissolved in several solvents are shown in Figures 5 and 6 and illustrate some of these unusual properties. The significant shift in absorption spectrum with solvent polarity along with the

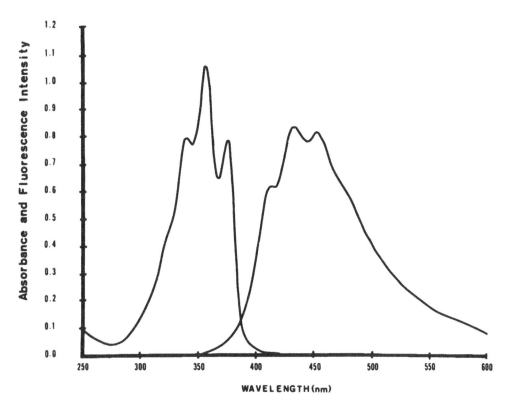

FIGURE 6.    Emission (right) and absorption (left) spectra of DPHpPC in ethanol for comparison with Figure 5.

large peak extinction coefficient (Table 1) suggest a strong transition moment to a highly polarized excited state, consistent with the $\pi \rightarrow \pi^*$ transition expected for a polyene system.[38] On the other hand, the fluorescence spectrum is considerably less sensitive to solvent polarity, and shows very little expected mirror symmetry or overlap with the absorption spectrum. More detailed studies have shown that the position and intensity of only the high-energy band was sensitive to the solvent, to the temperature, and to the nature of the DPH derivative.[52] While there is wide agreement that these observations demonstrate an emitting excited state substantially altered relative to the initially produced excited state, many different models have been proposed to account for this. One model has suggested a molecular conformational shift in the first excited state,[55] perhaps being solvent-dependent and leading to a three-step relaxation mechanism.[49] Most proposals, however, view DPH and similar molecules as having two closely spaced excited states ($^1Ag^*$ and $^1Bu^*$; see a textbook on group theory and molecular spectroscopy[58] for nomenclature).[51-53,56] Theoretical calculations allowing for excited-state mixing have shown that the normal, symmetry-disallowed $^1Ag^*$ state can contribute to the excited-state configuration of polyenes.[57] While most researchers in polyene spectroscopy appear now to acknowledge the significant contribution of the $^1Ag^*$ excited state to the fluorescence of DPH, disagreement focuses on exactly how the contribution comes about. The weight of recent evidence seems to favor simultaneous contributions from these two very closely spaced ($\Delta E \sim 0.4$ to $2.4$ kcal/mol[52]) and vibrationally linked (exchange time on the order of picoseconds[51]) excited states. This picture would seem, at least qualitatively, to account not only for the steady-state spectral properties of DPH, but also for its anomalous fluorescence decay properties.

The real-time decay of DPH fluorescence also shows anomalous behavior. Fluorescence lifetimes for different DPH derivatives have been gleaned from several sources and summarized in Table 1. The decay of DPH (or DPH-derivative) fluorescence has generally been

**Table 1**

**FLUORESCENCE CHARACTERISTICS OF DPH AND DPH DERIVATIVES**

| Fluorescence characteristics | DPH | TMA-DPH | DPHpPC or CE-DPH |
|---|---|---|---|
| Excitation maximum | 355 nm[1], *n*-hexane | 355 nm[10], DMF | 355 nm, ethanol |
| Emission maximum | 425 nm, corrected?[1] *n*-hexane | 427 nm corrected[10], propylene glycol | 434 nm, corrected ethanol |
| Extinction coefficient | 80,600 OD cm$^{-1}$ $M^{-1}$, 355 nm[21] chloroform ~80,000 OD cm$^{-1}$ $M^{-1}$, 355 nm, cyclohexane[a] | 30,200 OD cm$^{-1}$ $M^{-1}$, 355 nm, DMF[a,10], 53,000 OD cm$^{-1}$ $M^{-1}$, 354 nm, methanol[48] | 60,100 OD cm $^{-1}$ $M^{-1}$, 355 nm, chloroform |
| Quantum yield | 0.80[a], cyclohexane, 25°C 0.24[49], ethanol, 26°C | 0.04[48], ethanol | 0.55, ethanol |
| Lifetime (In solution) | 9.55 ± 0.03 nsec[23], paraffin oil, temperature invariant 15.7 nsec[49], *n*-hexane, 25°C 5.6 nsec[49], ethanol, 25°C | 0.6 nsec[10], ethanol | |
| (In membranes) | 7.9 nsec, f[b] = 0.93[5] 4.0 nsec, f = 0.07 DMPC, 37°C 8.3 nsec, f = 0.96[44] 3.0 nsec, f = 0.04 DPPC, 40.5°C 8.4 nsec, f = 1.0[45] DPPC, 50°C | ~ 6.7 nsec[10,c], DMPC, 20°C ~ 4 nsec[10,b], DMPC, 37°C | 7.0 nsec[12,d], DPPC[f], 43°C 6.42 nsec; f[e] = 0.93[12,e] 2.22 nsec, f = 0.07, DPPC[f], 43°C 6.32 nsec, f = 0.95[12,e] 1.52 nsec, f = 0.05 DPPC[f], 20°C |
| Intrinsic fluorescence anisotropy ($r_0$) | 0.39[31] 0.395[27] | 0.39[9] | |

[a]    From Berlman, I. B., *Handbook of Fluorescence Spectra of Aromatic Molecules*, 2nd ed., Academic Press, New York, 1971.

[b]    $f_i$ = fractional intensity of the fluorescence contribution to the i$^{th}$ lifetime ($\tau_i$) = $\alpha_i\tau_i/\Sigma\alpha_i\tau_i$, where $\alpha_i$ = normalized perexponential factor giving the fractional contribution of the i$^{th}$ component to the total decay of fluorescence intensity [$F(t) = \Sigma\alpha_i e^{-t/\tau_i}$].

[c]    Approximate lifetimes, obtained by the phase shift method[31] at an excited-light modulation frequency of 18 MHz. Measurements at other frequencies allowed estimation of a second short lifetime component.

[d]    Data obtained on the SLM 4800 phase fluorometer at 6, 18, and 30 MHz.

[e]    Data obtained on a continuous frequency phase fluorometer [47] in the laboratory of Dr. E. Gratton.

[f]    Abbreviations: DMF, dimethylformamide; DMPC, dimyristoylphosphatidylcholine; DPPC, dipalmitoylphosphatidylcholine.

Data are taken from the literature as indicated or were obtained in the reviewer's laboratory and are reported here for the first time.

reported to be single exponential in organic solvents,[10,22,49] although a serious effort to test a multiexponential model has been made only with paraffin oil as solvent.[23] DPH fluorescence lifetimes are quite sensitive to phenyl ring substitutions on the DPH molecule[48,52] and to temperature.[49] Quite surprisingly, the DPH lifetime in some solvents (e.g., 3-methylpentane and methylcyclohexane) has been observed to *decrease* at low temperature (below − 30°C)[49] and to become distinctly nonexponential at sufficiently low temperature (below − 110°C).[49] Finally, the long DPH fluorescence lifetime commonly observed in organic solvents, along with observed quantum yields, predicts an intrinsic or radiative lifetime five to six times larger [53,55] than that predicted from the observed absorption spectrum using the well-established expressions of Strickler and Berg.[59] All of these anomalies seem to this reviewer to be qualitatively consistent with the dual-excited state ($^1Ag^*$, $^1Bu^*$) model discussed above. Thus, the long decay time would be consistent with emission from the slightly lower energy, but largely symmetry-forbidden, $^1Ag^*$ state. The somewhat shorter lifetimes observed for many DPH derivatives[48,52] might reflect more rapid accessibility of the $^1Ag^*$ state in these molecules. Finally, the more rapid and nonexponential decay at very low temperature[48] could be due to decreased molecular vibrations, thus trapping a small portion of the molecules in the $^1Bu^*$ state, from which they will decay with a very short intrinsic lifetime, as predicted by the Strickler-Berg expressions.

Given the anomalous decay of DPH fluorescence in organic solvents, it is not surprising that the subject of DPH fluorescence decay in membranes has been a controversial one. While some studies have reported a single exponential decay of fluorescence from DPH in the fluid phase of model membranes,[1,27,45,60] others have found multiple exponential decay from such membranes.[4,22,44,61] Recent cooperation between this reviewer's laboratory and the laboratories of Gratton and Szabo revealed that exposure of a membrane sample to light increased the proportion of a second, short lifetime component (2 or 3 nsec) associated with DPH fluorescence decay in the fluid phase.[44,45] Thus, resolution of the issue of whether DPH displays multiple decay times in the fluid phase may rest with development of an instrument capable of making highly accurate measurements while not overexposing the sample to light. In view of the properties of DPH in organic solvents, as outlined above, it seems to this reviewer that the appearance of a short lifetime component in membranes could also be explained in terms of emission from the $^1Bu^*$ state. It may be that exposure to light alters the properties of the phospholipids or the probe in such a way as to stabilize this state slightly with respect to the $^1Ag^*$ state. Whatever the origin of the short lifetime component, its sensitivity to light somewhat complicates efforts (see, e.g., Reference 60) to resolve coexisting membrane domains through the resolution of heterogenous DPH fluorescence decay.[45]

Two other anomalous features of the decay of DPH fluorescence in membranes are worthy of note. First, the average fluorescence lifetimes of both DPH and DPHpPC drop sharply above 0.01 mol fraction,[12,45] of probe in the membrane, and this drop is not accompanied by a comparable drop in fluorescence intensity. This observation makes the use of changes in fluorescence intensity to follow changes in lifetime,[1,2] an inappropriate approximation, especially when changes in the local concentration of probe in the membrane might occur. Second, the average fluorescence decay time of DPH has been observed to decrease somewhat at temperatures sufficiently far below the phospholipid phase transition.[12,45] In the case of DPH, this drop in lifetime at low temperature could be accounted for by an increase in the proportion of the second, short lifetime component observed below the phospholipid phase transition.[45] As for the drop in fluorescence lifetime observed in organic solvents at very low temperature,[49] this may reflect decreased, thermally induced exchange from the $^1Bu^*$ to the $^1Ag^*$ state, although no direct evidence exists to confirm this interpretation.

Whatever the explanation for the unusual photophysical properties of DPH and its derivatives, it seems clear that the configuration of the excited state(s) is such that the absorption and emission dipole moments are roughly colinear and parallel to the long axis of the molecule (see Figure 2). The evidence for this derives from estimates of intrinsic fluorescence anisotropies (see Table 1) and from studies of polarized absorption[54] or emission[30] by oriented samples. Using the values for $r_0$ given in Table 1 and Equation 2 from Section II, we can estimate that the angle, $\alpha$, between the absorption and emission dipole moments of DPH is no greater than 5° to 7°. These observations confirm the assumptions that are routinely made in treating the motion of DPH in a bilayer (see Section IV.B.2).

Before leaving the subject of DPH photophysical anomalies, it is worth noting a practical consequence of these properties. Shinitzky and Barenholz, in their original paper on the use of DPH as a membrane probe,[1] noted that prolonged exposure ($\geq$ 1 min) to exciting light led to reversible bleaching of DPH fluorescence intensity. We have also noted this, especially at temperatures well above the phospholipid phase transition. A recent publication has documented this phenomenon and noted that bleaching was much more rapid in less ordered membranes.[62] Because of this dependence on membrane order, this bleaching phenomenon has been proposed to result from photoinduced *trans-cis* isomerization.[1,62] This proposal is based on the model of an excited-state conformational shift accounting for DPH photophysical anomalies.[49,55] A phenomenon that could be related to photobleaching is the induction of a short lifetime by exposure to light.[44,45] Whatever the reasons for photobleaching and the increase in short lifetime decay, these phenomena must be taken into account when using DPH-containing probes under conditions of extended observation.

## 2. Physical Properties in Membranes
In this section, I will review some critical aspects of the physical behavior of DPH and its derivatives in membranes. The location of these probes in the bilayer will first be considered. Second, the abilities of these probes to exchange between and partition into different membranes will be examined. This discussion will relate directly to the behavior of these probes in partitioning between different local domains in the same membrane. Finally, consideration will be given to the extent to which these probes perturb the membrane structure that they are intended to measure.

### a. Bilayer Location
One of the principal criticisms of DPH as a membrane probe is its uncertain location in the bilayer. The hydrophobic character of DPH all but guarantees that it will partition strongly to the hydrophobic interior of the bilayer;[3] the question is exactly where in the interior is it located and how is it oriented?[2] These questions are addressed in detail in Sections IV.B.2 and IV.B.3. As discussed in those sections, the exact location of DPH in a membrane is still a matter of controversy. A recent publication has addressed this controversy directly using fluorescence resonance energy transfer between DPH and a fluorophore located at the membrane surface.[62a] The conclusion was that the transverse distribution of DPH was governed by a strong tendency to partition preferentially into highly disordered regions of the bilayer. Thus, in egg phosphatidylcholine vesicles, DPH was found primarily close to the center of the bilayer, while in DPPC vesicles, a fairly broad spread about the center was indicated.

The DPH derivatives, DPHpPC and TMA-DPH, avoid ambiguities about the location of the DPH fluorophore in the membrane. Both molecules have hydrophilic headgroups and, as such, should be anchored to the interface region of the bilayer, guaranteeing that the DPH moiety is aligned with the phospholipid acyl chains. Especially in the case of the phospholipid analog, DPHpPC, this should mean that the probe order parameter will reflect the overall molecular order experienced by phospholipids in the membrane.

### b. Partitioning into and Exchange Between Membranes and Membrane Domains
Two key considerations in the choice of a membrane probe are the ease of incorporation into the membrane of interest and, related to this, the tendency of the probe to remain in and report only the properties of that membrane during an experiment. In general, probes that readily enter a membrane after injection into the aqueous medium from an organic solvent will also leave that membrane easily and enter another membrane population (e.g., DPH[2] and TMA-DPH[12]). This property must be considered when designing experiments with such probes. On the other hand, a probe such as DPHpPC, which does not transfer

readily between membrane populations[11,12] is not so easily incorporated into membranes. We[12,64] and others[65] have routinely incorporated DPHpPC into phospholipid vesicles at the time of vesicle preparation by detergent or solvent dilution techniques, but have noted nonreproducible decreases in fluorescence intensity associated with preparation of sonicated vesicles containing DPHpPC. We have not been able to identify the exact cause of this anomalous decrease in fluorescence, but suspect sensitivity of the probe to locally high levels of thermal energy induced by the sonication process or susceptibility to oxidation, as we have observed with DPH.[66] Using a Heat Systems Cup-horn® Sonicator®[62b] and taking care to avoid exposure to oxygen or light, we have recently been able to prepare sonicated vesicles containing DPHpPC with minimal loss of fluorescence intensity.[97] Stubbs et al.[63] have succeeded in incorporating a DPH-containing lipid derived from egg phosphatidylcholine (DPHePC) into preformed sarcoplasmic reticulum membranes by injecting the probe dissolved in an unspecified organic solvent (probably ethanol) into the agitated vesicle suspension. However, these authors noted that uptake of DPHePC by the membranes was less efficient than incorporation of DPH or TMA-DPH under the same conditions. Because of this, unincorporated DPHePC had to be removed by repeated sedimentation and resuspension of the membranes.[63]

Mention should be made of appropriate solvents for preparing stock solutions of the DPH-related probes and for introducing these into membrane suspensions. Not only must the probe be soluble in the solvent of choice, but the solvent should be moderately miscible with water as well, so that it will be completely dispersed upon injection of the DPH stock solution into water. For DPH, tetrahydrofuran,[2] acetonitrile, and acetone[44] have been used for this purpose, although the latter two are preferable to tetrahydrofuran because of the tendency of this solvent to form peroxides during storage. TMA-DPH is soluble in a number of slightly polar organic solvents,[9] including some that are useful for introduction into aqueous solutions (e.g., acetonitrile, methanol). DPHpPC is soluble in chloroform and ethanol,[11] but only the latter is useful for introduction of the probe into aqueous suspensions of membranes.

Once introduced into a membrane suspension via an appropriate organic solvent, the incorporation of DPH and TMA-DPH into membranes is accompanied by a dramatic increase in fluorescence (up to 1000-fold), since the quantum yield of these probes in water is negligible.[1,63] However, DPHePC (and presumably DPHpPC) does show appreciable fluorescence alone in aqueous suspension, probably due to aggregate formation.[63] The rate of incorporation of DPH into synthetic lipid vesicles is much faster above than below the phospholipid order/disorder phase transition.[2] In addition, this reviewer and his co-workers have observed that incorporation into more fluid membranes is faster than incorporation into ordered bilayers, such as those containing cholesterol. Incorporation of DPH into multilamellar vesicles has been shown to be a multiphasic process, probably due to the slow rate of passage through the aqueous regions between bilayers.[2] No data exist for uptake of TMA-DPH or DPHpPC into multilamellar vesicles, but for these there might exist another kinetic barrier, that of flip-flop of the probe from the outer to the inner leaflet of the bilayer. No data exist on the rate of this process.

A fairly unique property of DPH among membrane probes is its tendency to partition equally between different lateral domains in a membrane.[2] This has been demonstrated in the case of membranes of different phases,[2] membranes of different cholesterol content,[68] and membranes containing a transmembrane protein.[69] DPHpPC, on the other hand, partitions preferentially into fluid-phase regions of a mixed-phase membrane ($K_{f/s} = 3.3$).[12] Similarly, TMA-DPH showed a preference for fluid phase vesicles ($K_{f/s} = 3.2$) when it was allowed to equilibrate between fluid and ordered populations of vesicles.[12]

#### c. Perturbation of Bilayer Structure

Because membrane probe molecules are seldom identical to the phospholipid molecules constituting the bilayer matrix, the presence of probe molecules always presents the possibility of perturbing the very structure that is being probed. This perturbation could show up as an alteration in the overall phase structure of the membrane. Alternatively, the perturbation may be local, i.e., only in the near vicinity of the probe, and, therefore, much more subtle and difficult to detect. Both types of perturbing effects have been investigated in the cases of DPHpPC and DPH.

The presence of either DPH or DPHpPC in DPPC vesicles lowered the calorimetrically detected phase transition of this lipid only slightly (0.5°C for DPH,[70] 0.1 to 0.2°C for DPHpPC[12]). This demonstrates that neither probe perturbs substantially the overall phase structure of the membrane bilayer in which it finds itself, at least not at the fairly low concentrations generally used in membrane probe experiments (0.1 to 2 mol %). In the case of DPH, the phase transition as detected by probe fluorescence anisotropy occurred at the same temperature reported by scanning calorimetry,[70] indicating that the probe accurately reflected the average structure of the membrane bilayer. The fluorescence anisotropy of DPHpPC, on the other hand, revealed the phase transition at a temperature 1.1°C lower than revealed by DPH fluorescence anisotropy or scanning calorimetry.[12] High-sensitivity differential scanning calorimetry detected only a very subtle heat capacity peak (1/30th the magnitude of the main transition) at this lower temperature.[12] Fluorescence probes report local structural events as opposed to the highly cooperative, extensive events reported by scanning calorimetry. Thus, these observations suggest that DPHpPC disrupts bilayer order in its near vicinity. Finally, the minor heat capacity peak observed at low temperature[12] offers the possibility that several disrupted regions coalesce below the phospholipid phase transition to form a microdomain rich in DPHpPC. The anomalous drop in DPHpPC lifetime observed below the phospholipid phase transition as well as the preference of DPHpPC for fluid regions of the bilayer are consistent with this possibility.[12]

In the case of TMA-DPH, no comparable evaluations of bilayer-perturbing properties have been reported. However, one might expect this probe to behave in a way similar to DPHpPC because of their common amphipathic structure.

### 3. Probing Membrane Order

As discussed in Section IV.A.2, within the context of the monomodal orientational distribution model, the limiting fluorescence anisotropy of DPH-containing probes yields the second rank order parameter for the probe. This should reflect the average orientation of the probe (and presumably surrounding phospholipids) relative to the bilayer normal. Jähnig[29] pointed out that the limiting anisotropy observed for DPH in DPPC vesicles above their order/disorder phase transition (50°C, ca. 0.035[31]) implied an order parameter (ca. 0.3) consistent with that observed by deuterium NMR[71] at the ninth or tenth carbon of the palmitoyl chain. Our measurements of the limiting anisotropy of DPHpPC yielded a value of 0.50 for the probe orientational order parameter in DPPC vesicles at 50°C.[12] This corresponds to the molecular order parameter obtained by deuterium NMR at carbon number 4 of the DPPC acyl chains.[71] Data for TMA-DPH under similar conditions[9] yield an even larger value (0.57) for the probe order parameter.

It is evident from the structure of DPHpPC that its DPH moiety is located much further down in the hydrocarbon region of the bilayer than carbon number 4. However, the rigid DPH moiety is free to move in the bilayer only to the extent that C–C rotations can occur in the carboxyethyl group that attaches DPH to the phosphoglycerol backbone. Similar arguments would predict that the motion of TMA-DPH would be even more restricted since this probe has no C–C bonds around which rotation can occur (see Figure 2). Therefore, it appears that the rotational freedom experienced by the DPH portion of both DPHpPC and TMA-DPH is determined by molecular order in the uppermost regions of the bilayer. The much smaller limiting anisotropy of free DPH may indicate a location near the center of the bilayer and perhaps an orientation that is not distributed only about the bilayer normal, as suggested by the bimodal distribution model.[39,40]

## IV. NONLINEAR PROBES

### A. Overview and Summary

While DPH and its derivatives have been by far the most popular probes of membrane fluidity, at least two other molecules deserve mention. One of these, perylene, is of interest mainly for historical reasons,[20,21,72] although it is still recommended for estimating fluidity in less highly ordered systems.[3] The other probe class, the anthroyloxy fatty acids, have been moderately popular because of the ability to position the fluorescent group at different positions along a fatty acid chain, and, therefore, at different depths in the membrane bilayer. In this section, I first review the spectral properties of these probes; second, the physical properties of these molecules in the bilayer; and, third, current knowledge of the motion of these probes in the bilayer and how this motion is revealed by probe fluorescence anisotropy and reflects bilayer molecular dynamics and order. It is concluded that the complex motions of these probes are not as easily interpreted in terms of acyl chain dynamics as are the motions of the linear, DPH-related probes. Nonetheless, specific instances are considered for which the fluorescence properties of the nonlinear probes can yield qualitative information about membrane dynamics.

### B. Background and Spectral Properties
#### 1. Anthroyloxy Fatty Acids

The synthesis of anthroyloxy fatty acids as membrane probes was first described by Waggoner and Stryer[16] based on a method published by Parish and Stock.[73] These probes contain an anthroyloxy group attached to a saturated fatty acid chain and are named according to the number of carbon atoms separating the carboxyl group from the anthroyloxy-substituted carbon (see Figure 2). The most commonly used probes are the n-(9-anthroyloxy) stearic acids (n-9AS). The synthesis and properties of a similar, but not identical family of probes was described by Stoffel and Michaelis.[15] In contrast to the anthroyloxy fatty acids, Stoffel and Michaelis' probes had partially unsaturated fatty acid chains with an anthracene (rather than anthroyloxy) group at the end of the chain. The properties of these probes[18] are in many ways similar to those of the much more common anthroyloxy probes and, for this reason, will not be discussed here in detail.

The photophysics of the anthroyloxy fatty acids have been reasonably well-studied. The first property that should be noted is the tendency of these compounds to photobleach if overexposed to light in their absorption band.[74] A second important observation is that the emission spectrum of the anthroyloxy fatty acids is sensitive to the viscosity of nonpolar hydrocarbon solvents in which these probes are dissolved.[75] A shoulder at 418 nm became increasingly more prominent in more viscous solvents and with decreasing temperature or increasing cholesterol in membranes.[75] The behavior of the fluorescence spectrum has been suggested as a sensitive means of estimating the fluidity or viscosity gradient through a lipid bilayer using probes with the anthroyloxy group located at different positions along the fatty acid chain.[75] This viscosity-dependence is thought to be due to structural relaxation of the fluorophore excited state on the time scale of the excited-state lifetime.[76,77] In support of this interpretation, both the fluorescence anisotropy and the excited-state lifetime of the anthroyloxy fatty acids have been reported to vary with the wavelength at which fluorescence was detected.[65,76]

A third noteworthy feature of anthroyloxy fatty acid spectroscopy is the relative orientation of the emission and excitation transition dipoles. Vincent et al.[78] showed that the intrinsic anisotropy ($r_0$) of anthroyloxy fatty acids varied from roughly 0.087 with excitation at 316 nm to 0.325 with excitation at 381 nm. This implied at least two excitation bands with different orientations of their transition dipoles relative to the emission dipole (see Figure 2 and Equation 2). Along with the excited-state geometry changes discussed above, the existence of multiple excitation dipoles further complicates the interpretation of fluorescence anisotropy measurements in terms of bilayer molecular dynamics and order (see Section V.C).

Finally, the fluorescence excited-state lifetimes of several anthroyloxy-substituted fatty acids have been reported.[75,76] Phase shaft and modulation ratio measurements detected for 12-9AS at least two decay components (ca. 2 nsec and 14 to 16 nsec), whose relative proportions varied with emission wavelength, presumably due to the existence of two interconvertible excited-state geometries.[76] Direct fluorescence decay measurements have demonstrated that the average fluorescence lifetime also varied with the position of attachment of the anthroyloxy moiety to the fatty acid chain (from 9.5 nsec at the 2-position to ca. 13.5 nsec at the 12-position).[75]

## 2. Perylene

The excitation and emission spectra of perylene in ethanol were reported by Shinitzky et al.,[21] who first suggested its use as a probe of membrane fluidity. These authors also reported that the intrinsic fluorescence anisotropy ($r_0 \cong 0.37$) was constant over the main, $S_0$-to-$S_1$ absorption band (350 to 450 nm) but decreased to negative values (minimum at $r_0 = -0.15$) at the lower wavelength, $S_0$-to$S_2$ absorption band (maximum ca. 256 nm). These $r_0$ values imply a main excitation dipole nearly parallel to the emission dipole, but a secondary excitation dipole at close to a right angle to the emission dipole (see Figure 2). Finally, the fluorescence decay of perylene in glycerol has been reported to be monoexponential and independent of temperature (4.5 to 4.8 nsec between 10 and 40°C) and excitation wavelength,[79] except in the region where the excitation and emission bands overlap.

## C. Physical Properties in a Bilayer

As for the DPH-related probes, I will consider for the anthroyloxy fatty acids and perylene, first, the location of the probes in the bilayer, second, the partitioning of the probes into and between membranes, and finally, the extent to which these probes perturb membrane structure.

## 1. Anthroyloxy Fatty Acids

Thulborn et al.,[75] Chalpin and Kleinfeld,[80] and Thulborn and Sawyer[81] have all used water-soluble quenching agents to confirm that the anthroyloxy moieties of these probes are located at depths in the bilayer consistent with expectations based on the point of attachment to the fatty acid chains. Uptake of these probes into sonicated egg phosphatidylcholine vesicles has been reported as requiring less than 2 hr, although the influence of lipid composition or vesicle type remains uncertain.[74] Also uncertain at present is the time required for a probe initially incorporated into the outer monolayer of a vesicle lipid bilayer to redistribute to the inner monolayer. Surprisingly, the partitioning from the aqueous phase into the membrane could be modeled best in terms of saturable binding, rather than unsaturable partitioning, although the binding constants obtained varied somewhat with phospholipid vesicle concentration.[74] Hydrophobic rather than electrostatic contributions made the dominant contribution to the free energy of binding.[74] At least one group has reported that the anthroyloxy fatty acids photodimerize and may form domains in monolayers,[82] and this photodimerization phenomenon has also been suggested for membranes below the phospholipid phase transition.[83] Finally, the anthroyloxy fatty acids are widely reported[83-86] to perturb bilayer structure in a substantial fashion, even at the low concentrations normally used in fluorescent probe studies of membrane fluidity.[86] Perturbation of bilayer structure was the least for probes with the anthroyloxy group located near or at the noncarboxyl end of the fatty acid chain.[83,85,86] Similar conclusions were reached concerning the perturbing influence of the unsaturated anthracene-containing probes studied by Stoffel and Michaelis.[18]

Given their tendency to perturb bilayer structure, it is not surprising that the anthroyloxy fatty acids, especially those with the probe near to the carboxyl end of the fatty acid, are not effective probes for reporting the phospholipid phase transition in monolayers[83] or bilayers.[18,20,85] When these probes were used to monitor the phase transition in DPPC vescicles, abrupt changes in fluorescence anisotropy occurred at temperatures below the actual phase transition temperature for the 2- and 12-substituted fatty acids, while the 16-substituted species reported nearly the correct transition temperature.[20] High concentrations of probe (20 mol %) led to a distorted overall phase behavior (detected by calorimetry) similar to that reported by the fluorescence intensity or polarization of the probe.[18,85] Therefore, distortions of the reported phase behavior appear to reflect local perturbations of the bilayer in the vicinity of probe, perhaps due to clustering in the gel phase[82,83] as noted above. Also consistent with the possibility of clustering is the hysteresis in the phase transition reported by probe fluorescence intensity for the related anthracene-containing probes.[18]

### 2. Perylene

Because of its hydrophobic character, perylene is surely located in the hydrocarbon region of the bilayer. However, Bieri and Wallach[87] have interpreted this probe's anomalous increase in fluorecence intensity associated with the order-to-disorder phospholipid phase transition[20] as indicating two regions of probe binding to the bilayer. Quenching experiments with nitroxide spin labels were consistent with location of the probe in the neighborhood of carbon 6 below the phase transition, while a location somewhat deeper into the bilayer was suggested at temperatures above the phase transition.[87] Although no detailed study of the rate of perylene uptake into membranes has been made, Pagano et al.[88] have reported that the probe partitioned rapidly between the different membranes of a fibroblast cell line. Finally, it is impossible to make any definitive statement here regarding perturbation of the bilayer by perylene since no published reports address this issue in any detail.

### D. Probe Motion and Membrane Fluidity

One can imagine that probes lacking the essentially colinear and symmetrically placed excitation and emission dipoles of DPH (see Figure 2) will not yield fluorescence anisotropy values that are as easily interpreted in terms of probe motion. In the case of DPH, a reasonable description of the decay of fluorescence anisotropy could be obtained in terms of a single exponential decay model describing free rotation in a hard cone or hindered rotation in a Gaussian potential (see Section IV.B.2). To the extent that a single exponential is not adequate to describe the decay of DPH fluorescence anisotropy, additional exponential terms have been rationalized either in terms of a bimodal orientational distribution function[40,43] or in terms of multiple fluorescing species.[46] For disk-like probes such as the anthroyloxy group and perylene, the minimal appropriate description, even of free isotropic rotation requires two rotational diffusion coefficients (in-plane and out-of-plane) and, therefore, at least three exponentials to describe the decay of fluorescence anisotropy.[24] To describe hindered rotation of such a probe in a model-independent fashion, Zannoni et al.[38] derived an expression containing six orientational distribution functions, each containing an exponential decay term. Because of such complications, there has been, relative to the situation with DPH, little success in obtaining *detailed* information on phospholipid average structure and dynamics from analysis of either anthroyloxy or perylene fluorescence anisotropy. In the following two sections, I review briefly the attempts that have been made in this area.

### 1. Anthroyloxy Probes

The popularity of the anthroyloxy fatty acid probes has derived in large part from their potential for monitoring fluidity at specific depths into the phospholipid bilayer. In order to realize this potential, it is necessary to develop a theoretical framework for interpreting observed fluorescence anisotropy in terms of some parameters of probe motion. One of the first attempts at this was made by Bashford et al.[89] These authors applied a simple treatment of free isotropic rotation of a disk-like molecule in a manner entirely analogous to the treatment of Shinitzky and co-workers,[21] for the motion of DPH or perylene. It is well-documented in the case of DPH (see Section IV.B.1) that the assumptions of such a treatment are not valid in a membrane bilayer, and it is even less likely that these assumptions would be sound in the case of a probe attached by a covalent bond to a partially oriented fatty acid chain. Tilley et al.[90] in recognizing this, have argued that it is not possible to interpret the dependence on fatty acid chain position of anthroyloxy fluorescence anisotropy in terms of a fluidity gradient. These authors proposed that a detailed analysis of the time-dependence of fluorescence anisotropy in terms of the orientation of the emission dipole relative to the bilayer normal would be required. Such a treatment would have to take into account specific steric consequences of the location of the fluorophore group on the chain and would be so complex and involve so many exponentials as to be nearly intractable. Finally, Matayoshi and Kleinfeld[76] have also pointed out that the excited-state geometry changes characteristic of the anthroyloxy fatty acids (see Section V.B) also greatly complicate an interpretation of anisotropy data in terms of the bilayer fluidity gradient.

In the past few years, there have appeared several analyses of the time-dependence of anthroyloxy fatty acid fluorescence anisotropy.[78,91-94] In general, these have taken advantage of the existence of two absorption bands with excitation dipoles at very different angles to the emission dipole[78] (see Figure 2). Vincent et al.[78] argue that excitation at the short wavelength band (316 nm; dipole approximately aligned with the long molecular axis of anthracene) should enhance the contribution from rotation about the short molecular axis (i.e., out-of-plane rotation about the ester bond; see Figure 2). Conversely, excitation at the 381 nm absorption band should lead to depolarization reflecting mainly the in-plane rotation of the anthracene moiety[78] (see Figure 2). In isotropic media such as propylene glycol or the detergent Primol®342, the rates of both these motions were very similar and a single exponential decay of fluorescence anisotropy was observed following excitation at any wavelength.[78] This presumably reflected the requirement for movement of the entire fatty acid chain in order to observe the in-plane rotation, thus slowing this rotation relative to the out-of-plane rotation.[78]

In DPPC vesicles, however, Vincent and co-workers found that a biexponential description was the simplest one possible at any wavelength, reflecting the different rates of in-plane and out-of-plane rotations. Nonetheless, by taking advantage of the excitation wavelength dependence of the results, these two types of rotational motion could be distinguished.[78] Thus, the in-plane rotational correlation time was roughly 8 nsec for 12-(9-anthroyloxy) stearic acid in DPPC vesicles at 21°C, but the out-of-plane rotational motion was roughly one half as fast (correlation time ~20 nsec) under the same conditions.[78] Quite surprisingly, Vincent et al. found that the out-of-plane rotation of the anthracene group was essentially unhindered ($r_\infty \cong 0$) both above *and* below the DPPC phase transition. By contrast, the in-plane rotation was at least partially hindered at all temperatures within and below the phase transition ($r_\infty \neq 0$) and was unhindered only above the phase transition. These observations have allowed Vincent and colleagues[78,91,94] to propose that the out-of-plane rotational rate of anthroyloxy fatty acid probes should offer a true measure of the fluidity gradient through the bilayer; while the degree of hindered in-plane rotation ($r_\infty$) was suggested as a measure of chain order. We note that the unhindered out-of-plane rotation seems reflective of substantial bilayer perturbation by the anthroyloxy fatty acid probe, thus providing a local "hole" in which the anthroyloxy group could rotate fairly freely. If so, the flexibility gradient suggested by the in-plane rotation would be likely to be distorted. In addition, this treatment is certainly a crude approximation to the complex, six-correlation-function description required for a general description of the motion of the anthroyloxy fatty acid probes.[38] For instance, the results of Vincent et al. themselves suggest that at least one additional, even more rapid motion occurs.[78] Nonetheless, the relatively simple approach proposed by Vincent et al. offers some hope of interpreting at least the time-dependence of anthroyloxy fatty acid fluorescence in terms of parameters of membrane structure and dynamics.

In fairness, I must note that the results of Vincent and co-workers[78,91,94] are not universally accepted. Blatt et al.[92] report essentially unhindered in-plane rotation of a whole series of anthroyloxy fatty acids both above *and* below the phase transition of dimyristoylphosphatidylcholine vesicles. However, the observations of Thulborn and Beddard[93] support the report of hindered 12-9AS motions below the DPPC phase transition. These differences could reflect different properties of the different lipid systems involved or differences in the details of experimental approaches. In either case, further analysis using the approach and observations of Vincent et al. would appear to offer the best possibility of obtaining insight into bilayer structure using the anthroyloxy fatty acid probes.

### 2. Perylene

The earliest attempts at interpreting the fluorescence anisotropy of perylene in terms of membrane fluidity utilized the isotropic rotation model of Shinitzky et al.[21,72] However, Lakowicz and Knutson[95] have shown that the rotation of perylene in synthetic phosphatidylcholine bilayers is hindered ($r_\infty \geqq 0.1$ below the phase transition and $r_\infty \neq 0$ above the phase transition), as might be expected. In addition, Barkley et al.[79] have shown that the rotation of perylene is anisotropic even in an isotropic medium such as glycerol, with the in-plane rotational rate approximately eight to ten times greater than that of the out-of-plane rate. In making this analysis, Barkley et al. made use of the two perpendicular excitation transition moments described by Shinitzky et al.[21] (see Figure 2) in a manner similar to the treatment of anthroyloxy probes by Vincent et al.[78] Unfortunately, no thorough treatment of the motion of perylene in a membrane has been made, although it is clear from the data in an isotropic medium[79] and from the general expression developed by Zannoni et al.[38] that this motion will be complex and difficult to relate in any meaningful way to the motions of acyl chains in a bilayer.

## ACKNOWLEDGMENTS

I thank Drs. M. Straume, B. J. Litman, and Dr. E. Gratton for sharing their data before publication. In addition, gratitude is due to Drs. E. Gratton, M. Straume, L. Loew, and K. Jacobsen as well as to Ms. S. Windes for critical reading of the manuscript.

## REFERENCES

1. **Shinitzky, M. and Barenholz, Y.,** Dynamics of the hydrocarbon layer in liposomes of lecithin and sphingomyelin containing dicetylphosphate, *J. Biol. Chem.*, 249, 2652, 1974.
2. **Lentz, B. R., Barenholz, Y., and Thompson, T. E.,** Fluorescence depolarization studies of phase transitions and fluidity in phospholipid bilayers. I. Single component phosphatidylcholine liposomes, *Biochemistry*, 15, 4521, 1976; II. Two-component phosphatidylcholine liposomes, *Biochemistry*, 15, 4529, 1976.
3. **Shinitzky, M. and Barenholz, Y.,** Fluidity parameters of lipid regions determined by fluorescence polarization, *Biochim. Biophys. Acta*, 515, 367, 1978.

4. **Shinitzky, M.,** Membrane fluidity and cellular function, in *Physiology of Membrane Fluidity,* Shinitzky, M., Ed., CRC Press, Boca Raton, Fla., 1984, 1.
5. **Chen, L. A., Dale, R. E., Roth, S., and Brand, L.,** Nanosecond time-dependent fluorescence depolarization of diphenylhexatriene in dimyristoyllecithin vesicles and the determination of "microviscosity", *J. Biol. Chem.,* 252, 2163, 1977.
6. **Hare, F., Amiel, J., and Lussan, C.,** Is an average viscosity tenable in lipid bilayers and membranes?, *Biochim. Biophys. Acta,* 555, 388, 1979.
7. **Cantor, C. R. and Schimmel, P. R.,** *Biophysical Chemistry, Part II: Techniques for the Study of Biological Structure and Function,* W. H. Freeman, San Francisco, 1980, 433.
8. **Lakowicz, J. R.,** Fluorescence polarization, in *Principles of Fluorescence Spectroscopy,* Plenum Press, New York, 1983, chap. 5; Lakowicz, J. R., Time-dependent decay of fluorescence anisotropy, in *Principles of Fluorescence Spectroscopy,* Plenum Press, New York, 1983, chap. 6.
8a. **Teale, F. W. J.,** Fluorescence depolarization by light scattering in turbid solutions, *Photochem. Photobiol.,* 10, 363, 1969.
8b. **Lentz, B. R., Moore, B. M., and Barrow, D. A.,** Light scattering effects in the measurement of membrane microviscosity with diphenylhexatriene, *Biophys. J.,* 25, 489, 1979.
8c. **Eisinger, J. and Flores, J.,** Fluorometry of turbid and absorbant samples and the membrane fluidity of intact erythrocytes, *Biophys. J.,* 48, 77, 1985.
9. **Sklar, L. A., Hudson, B. S., and Simoni, R. D.,** Parinaric acid as a fluorescent probe for studying phase transitions in model membranes, *Biochemistry,* 16, 819, 1977.
10. **Prendergast, F. G., Haugland, R. P., and Callahan, P. J.,** 1-[4-(Trimethylamino)phenyl]-6-phenylhexa-1,3,5-triene: synthesis, fluorescence properties, and use as a fluorescent probe of lipid bilayers, *Biochemistry,* 20, 7333, 1983.
11. **Morgan, C. G., Thomas, E. W., Moras, T. S., and Yianni, Y. P.,** The use of a phospholipid analogue of diphenylhexatriene to study melittin-induced fusion of small, unilamellar phospholipid vesicles, *Biochim. Biophys. Acta,* 692, 196, 1982.
12. **Parente, R. A. and Lentz, B. R.,** Advantages and limitations of 1-Palmitoyl-2-[[2-[4(6-phenyl-*trans*-1,3,5-hexatrienyl)phenyl]ethyl]carbonyl]-3*sn*-phosphatidylcholine as a fluorescent membrane probe, *Biochemistry,* 24, 6178, 1985.
13. **Morgan, C. G., Thomas, E. W., and Yianni, Y. P.,** The use of fluorescence energy transfer to distinguish between poly(ethyleneglycol)-induced aggregation and fusion of phospholipid vesicles, *Biochim. Biophys. Acta,* 728, 356, 1983.
14. **Hudson, B. and Cavalier, S.,** Studies of membrane dynamics and lipid-protein interactions with parinaric acids, in *Spectroscopic Membrane Probes,* Vol. 1, Loew, L. M., Ed., CRC Press, Boca Raton, Fla., 1988.
15. **Stoffel, W. and Michaelis, G.,** Synthesis and characterization of antracene-labelled fatty acids, phosphatidylcholines, and cholesterol esters, *Hoppe-Seyler's Z. Physiol. Chem.,* 357, 7, 1976.
16. **Waggoner, A. S. and Stryer, L.,** Fluorescent probes of biological membranes, *Proc. Natl. Acad. Sci. U.S.A.,* 67, 579, 1970.
17. **Cadenhead, D. A., Kellner, B. M. J., Jacobson, K., and Papahadjopoulos, D.,** Fluorescent probes in model membranes. I. Anthroyl fatty acid derivatives in monolayers and liposomes of dipalmitoylphosphatidylcholine, *Biochemistry,* 16, 5386, 1977.
18. **Stoffel, W. and Michaelis, G.,** Lipid-lipid and lipid-protein interactions as studied with a novel type of fluorescent fatty acid and phospholipid probes, *Hoppe-Seyler's Z. Physiol. Chem.,* 357, 21, 1976.
19. **Ruby, B. and Gittler, C.,** Microviscosity of the cell membrane, *Biochim. Biophys. Acta,* 288, 231, 1972.
20. **Papahadjopoulos, D., Jacobson, K., Nir, S., and Isac, T.,** Phase transitions in phospholipid vesicles. Fluorescence polarization and permeability measurements concerning the effect of temperature and cholesterol, *Biochim. Biophys. Acta,* 311, 330, 1973.
21. **Shinitzky, M., Dianoux, A.-C., Gitler, C., and Weber, G.,** Microviscosity and order in the mydrocarbon region of micelles and membranes determined with fluorescent probes. I. Synthetic micelles, *Biochemistry,* 10, 2106, 1971.
22. **Weber, G.,** Rotational brownian motion and polarization of the fluorescence of solutions, in *Advances in Protein Chemistry,* Vol. 8, Anson, M. L., Bailey, K., and Edsall, J. T., Eds., Academic Press, New York, 1953, 415.
23. **Dale, R. E., Chen, L. A., and Brand, L.,** Rotational relaxation of the "microviscosity" probe diphenylhexatriene in paraffin oil and egg lecithin vesicles, *J. Biol. Chem.,* 252, 7500, 1972.
24. **Tao, T.,** Time-dependent fluorescence depolarization and brownian rotational diffusion coefficients of macromolecules, *Biopolymers,* 8, 609, 1969.
24a. **Jacobson, K. and Wobschall, D.,** Rotation of fluorescent probes localized within lipid bilayer membranes, *Chem. Phys. Lipids,* 12, 117, 1974.
25. **Hare, F. and Lussan, C.,** Variations in microviscosity values induced by different rotational behavior of fluorescent probes in some aliphatic enviroments, *Biochim. Biophys. Acta,* 167, 262, 1977.

26. **Kinosita, K., Kawato, S., and Ikegami, A.,** A theory of fluorescence polarization decay in membranes, *Biophys. J.,* 20, 289, 1977.
27. **Kawato, S., Kinosita, K., Jr., and Ikegami, A.,** Dynamic structure of lipid bilayers studied by nanosecond fluorescence techniques, *Biochemistry,* 16, 2319, 1977.
28. **Heyn, M. P.,** Determination of lipid order parameters and rotational correlation times from fluorescence depolarization experiments, *FEBS Lett.,* 108, 359, 1979.
29. **Jähnig, F.,** Structural order of lipids and proteins in membranes: evaluation of fluorescence anisotropy data, *Proc. Natl. Acad. Sci. U.S.A.,* 76, 6361, 1979.
30. **Andrich, M. P. and Vanderkooi, J. M.,** Temperature dependence of 1,6-Diphenyl-1,3,5-hexatriene fluorescence in phospholipid artificial membranes, *Biochemistry,* 15, 1257, 1976.
31. **Lakowicz, J. R., Prendergast, F. G., and Hogen, D.,** Differential polarized phase fluorometric investigations of diphenylhexatriene in lipid layers. Quantitation of hindered depolarizing rotations, *Biochemistry,* 18, 508, 1979.
32. **Lipari, G. and Szabo, A.,** Effect of librational motion on fluorescence depolarization and nuclear magnetic resonance relaxation in macromolecules and membranes, *Biophys. J.,* 30, 489, 1980.
33. **Szabo, A.,** Theory of polarized fluorescent emission in uniaxial liquid crystals, *J. Chem. Phys.,* 72, 4620, 1980.
34. **Fulford, A. J. C. and Peel, W. E.,** Lateral pressures in biomembranes estimated from the dynamics of fluorescent probes, *Biochim. Biophys. Acta,* 598, 237, 1980.
35. **Lipari, G. and Szabo, A.,** Padé approximations to correlation functions for restricted rotational diffusion, *J. Chem. Phys.,* 75, 2971, 1981.
36. **Engel, L. W. and Prendergast, F. G.,** Values for and the significance of order parameters and "cone angles" of fluorophore rotation in lipid bilayers, *Biochemistry,* 20, 7338, 1981.
37. **Zannoni, C.,** A theory of fluorescence depolarization in membranes, *Mol. Phys.,* 42, 1303, 1981.
38. **Zannoni, C., Arcioni, A., and Cavatorta, P.,** Fluorescence depolarization in liquid crystals and membrane bilayers, *Chem. Phys. Lipids,* 32, 179, 1983.
39. **van der Meer, W., Pottel, H., Herreman, W., Ameloot, M., Hendrickx, H., and Schroder, H.,** Effect of orientational order on the decay of the fluorescence anisotropy in membrane suspensions. A new approximate solution of the rotational diffusion equation, *Biophys. J.,* 46, 515, 1984.
40. **Ameloot, M., Hendrickx, H., Herreman, W., Pottel, H., Van Cauwelaert, F., and van der Meer, W.,** Effect of orientational order on the decay of the fluorescence anisotropy in membrane suspensions. Experimental verification on unilamellar vesicles and lipid/α-lactalbumin complexes, *Biophys. J.,* 46, 525, 1984.
41. **Szabo, A.,** Theory of fluorescence depolarization in macromolecules and membranes, *J. Chem. Phys.,* 81, 150, 1984.
42. **Kinosita, K., Jr., Ikegami, A., and Kawato, S.,** On the wobbling-in-cone analysis of fluorescence anisotropy decay, *Biophys. J.,* 37, 461, 1982.
43. **Straume, M. and Litman, B. J.,** The influence of cholesterol on the equilibrium and dynamic structural properties of unsaturated lipid bilayers, *Biophys. J.,* 49, 310a, 1986; and personnal communication of a manuscript in preparation.
44. **Parasassi, T., Conti, F., Glaser, M., and Gratton, E.,** Detection of Phospholipid phase separation. A multi-frequency phase fluorimetry study of 1,6-Diphenyl-1,3,5-hexatriene, *J. Biol. Chem.,* 259, 14011, 1984.
45. **Barrow, D. A. and Lentz, B. R.,** Membrane structural domains. Resolution limits using diphenylhexatriene fluorescence decay, *Biophys. J.,* 48, 221, 1985.
46. **Wang, S., Glaser, M., and Gratton, E.,** Frequence domain fluorescence studies of rotational motion of DPH in multilamellar vesicles, *Biophys. J.,* 49, 307a, 1986; and personnal communication of a manuscript in preparation.
47. **Davenport, L., Knutson, J. R., and Brand, L.,** Anisotropy decay associated fluorescence spectra and analysis of rotational heterogeneity. II. 1,6-Diphenyl-1,3,5-hexatriene in lipid bilayers, *Biochemistry,* 25, 1811, 1986.
48. **Cundall, R. B., Johnson, I., Jones, M. W., Thomas, E. W., and Munro, I. H.,** Photophysical properties of DPH derivatives, *Chem. Phys. Lett.,* 64, 39, 1979.
49. **Cehelnik, E. D., Cundall, R. B., Lockwood, J. R., and Palmer, T. F.,** Solvent and temperature effects on the fluorescence of all-*trans*-1,6-diphenyl-1,3,5-hexatriene, *J. Phys. Chem.,* 79, 1369, 1975.
50. **Cehelnik, E. D., Cundall, R. B., Lockwood, J. R., and Palmer, T. F.,** 1,6-Diphenyl-1,3,5-hexatriene as a fluorescence standard, *Chem. Phys. Lett.,* 27, 586, 1974.
51. **Felder, T. C., Choi, K.-J., and Topp, M. R.,** Picosecond observation of fluorescence from Bu states of diphenylpolyenes, *Chem. Phys.,* 64, 175, 1982.
52. **Alford, P. C. and Palmer, T. F.,** Fluorescence of DPH derivatives. Evidence for Emission from $S_2$ and $S_1$ Excited States, *Chem. Phys. Lett.,* 86, 248, 1982.

53. **Birks, J. B., Tripathi, G. N. R., and Lumb, M. D.,** The fluorescence of all-*trans* diphenylpolyenes, *Chem. Phys.*, 33, 185, 1978.

54. **Cehelnik, E. D., Cundall, R. D., Timmonds, C. J., and Bowley, R. M.,** Spectroscopic studies of *trans*-1,6-diphenyl-1,3,5-hexatriene in ordered liquid crystal solutions, *Proc. R. Soc. London, Ser. A,* 335, 387, 1973.

55. **Birks, J. B. and Dyson, D. J.,** The relations between the fluorescence and absorption properties of organic molecules, *Proc. R. Soc. London, Ser. A,* 275, 135, 1963.

56. **Hudson, B. S. and Kohler, B. E.,** A low-lying weak transition in the polyene α, ω-diphenyloctatetraene, *Chem. Phys. Lett.,* 14, 299, 1972.

57. **Schulten, K. and Karplus, M.,** On the origin of a low-lying forbidden transition in polyenes and related molecules, *Chem. Phys. Lett.,* 14, 305, 1972.

58. **Cotton, F. A.,** *Chemical Applications of Group Theory,* Interscience, New York, 1963.

59. **Strickler, S. J. and Berg, R. A.,** Relationship between absorption intensity and fluorescence lifetime of molecules, *J. Chem. Phys.,* 37, 814, 1962.

60. **Klausner, R. D., Kleinfeld, A. M., Hoover, R. L., and Karnovsky, M. J.,** Lipid domains in membranes: evidence derived from structural perturbations induced by free fatty acids and lifetime heterogeneity analysis, *J. Biol. Chem.,* 255, 1286, 1980.

61. **Kinosita, K., Jr. and Ikegami, A.,** Reevaluation of the wobbling dynamics of diphenylhexatriene in phosphatidylcholine and cholesterol/phosphatidylcholine membranes, *Biochim. Biophys. Acta,* 769, 523, 1984.

62. **Duportail, G. and Weinreb, A.,** Photochemical changes of fluorescent probes in membranes and their effect on the observed fluorescence anisotropy values, *Biochim. Biophys. Acta,* 736, 171, 1983.

62a. **Davenport, L., Dale, R. E., Bisby, R. H., and Cundall, R. B.,** Transverse location of the fluorescent probe 1,6-diphenyl-1,3,5-hexatriene in model lipid bilayer membrane systems by resonance excitation energy transfer, *Biochemistry,* 24, 4097, 1985.

62b. **Barrow, D. A. and Lentz, B. R.,** Large vesicle contamination in small, unilamellar vesicles, *Biochim. Biophys. Acta,* 597, 92, 1980.

63. **Stubbs, C. D., Kinosita, K., Jr., Munkonge, F., Quinn, P. J., and Ikegami, A.,** The dynamics of lipid motion in sarcoplasmic reticulum membranes determined by steady-state and time-resolved fluorescence measurements on 1,6-diphenyl-1,3,5-hexatriene and related molecules, *Biochim. Biophys. Acta,* 775, 374, 1984.

64. **Parente, R. A. and Lentz, B. R.,** Fusion and phase separation monitored by lifetime changes of a fluorescent phospholipid probe, *Biochemistry,* 25, 1021, 1986.

65. **Cranney, M. A., Cundall, R. B., Jones, G. R., Richards, J. T., and Thomas, E. W.,** Fluorescence lifetime and quenching studies on some interesting diphenylhexatriene membrane probes, *Biochim. Biophys. Acta,* 735, 418, 1983.

66. **Barrow, D. A. and Lentz, B. R.,** A model for the effect of lipid oxidation on diphenylhexatriene fluorescence in phospholipid vesicles, *Biochim. Biophys. Acta,* 645, 17, 1981.

67. **Haugland, R. P.,** *Molecular Probes Handbook of Fluorescent Probes and Research Chemicals,* Molecular Probes, Junction City, Ore., 1985.

68. **Lentz, B. R., Barrow, D. A., and Hoechli, M.,** Cholesterol-phosphatidylcholine interactions in multilamellar vesicles, *Biochemistry,* 19, 1945, 1980.

69. **Moore, B. M., Lentz, B. R., and Meissner, G.,** Effects of sarcoplasmic reticulum $Ca^{2+}$-ATPase on phospholipid bilayer fluidity: bounding lipid, *Biochemistry,* 17, 5248, 1978.

70. **Lentz, B. R., Freire, E., and Biltonen, R. L.,** Fluorescence and calorimetric studies of phase transitions in phosphatidycholine multilayers: kinetics of the pretransition, *Biochemistry,* 17, 4475, 1978.

71. **Seelig, A. and Seelig, J.,** The dynamic structure of fatty acyl chains in a phospholipid bilayer measured by deuterium magnetic resonance, *Biochemistry,* 13, 4839, 1974.

72. **Cogan, U., Shinitzky, M., Weber, G., and Nishida, T.,** Microviscosity and order in the hydrocarbon region of phospholipid and phospholipid-cholesterol dispersions determined with fluorescent probes, *Biochemistry,* 12, 521, 1973.

73. **Parish, R. C. and Stock, L. M.,** A method for the esterification of hindered acids, phosphatidylcholine and cholesterol esters, *J. Org. Chem.,* 30, 927, 1965.

74. **Haigh, E. A., Thulborn, K. R., Nichol, L. W., and Sawyer, W. J.,** Uptake of n-(9-anthroyloxy) fatty acid fluorescent probes into lipid bilayers, *Aust. J. Biol. Sci.,* 31, 447, 1978.

75. **Thulborn, K. R., Tilley, L. M., Sawyer, W. H., and Freloar, F. E.,** The use of n-(9-anthroyloxy) fatty acids to determine fluidity and polarity gradients in phospholipid bilayers, *Biochim. Biophys. Acta,* 558, 166, 1979.

76. **Matayoshi, E. D. and Kleinfeld, A. M.,** Emission wavelength-dependent decay of the 9-anthroyloxy-fatty acid membrane probes, *Biophys. J.,* 35, 215, 1981.

77. **Werner, T. C., Mathews, T., and Soller, B.,** An investigation of the fluorescence properties of carboxyl-substituted anthracenes, *J. Phys. Chem.,* 80, 533, 1976.

78. **Vincent, M., de Foresta, B., Galley, J., and Alfsen, A.,** Nanosecond fluorescence anisotropy decays of n-(9-anthroyloxyl fatty acids) in dipalmitoylphosphatidylcholine vesicles with regard to isotropic solvents, *Biochemistry,* 21, 708, 1982.

79. **Barkley, M. D., Kowalszyk, A. A., and Brand, L.,** Fluorescence decay studies of anisotropic rotations of small molecules, *J. Chem. Phys.,* 75, 3581, 1981.

80. **Chalpin, D. B. and Kleinfeld, A. M.,** Interaction of fluorescence quenchers with the n-(9-anthroyloxy) fatty acid membrane probes, *Biochim. Biophys. Acta,* 731, 465, 1983.

81. **Thulborn, K. R. and Sawyer, W. H.,** Properties and the locations of a set of fluorescence probes sensitive to the fluidity gradient of the lipid bilayer, *Biochim. Biophys. Acta,* 511, 125, 1978.

82. **Teissie, J., Tocanne, T. F., and Baudras, A.,** Characterization of 9-anthroyl stearic acid photodimerization in monolayers, *Eur. J. Biochem.,* 83, 77, 1978.

83. **McGrathy, A. E., Morgan, C. G., and Radda, G. K.,** Photobleaching: a novel fluorescence method for diffusion studies in lipid systems *Biochim. Biophys. Acta,* 426, 175, 1976.

84. **Cadenhead, D. A., Kellner, B. M. J., and Müller-Landau, F.,** Perturbation of bilayer structure caused by fluorescent and spin probes. Monolayer flims, *Biochim. Biophys. Acta,* 382, 253, 1975.

85. **Cadenhead, D. A., Kellner, B. M. J., Jacobsen, K., and Papahadjopoulos, D.,** Fluorescent probes in model membranes. I. Anthroyl fatty acid derivatives in monolayers and liposomes of dipalmitoylphosphatidylcholine, *Biochemistry,* 16, 5386, 1977.

86. **Ashcroft, R. G., Thulborn, K. R., Smith, J. R., Coster, H. G., and Sawyer, W. H.,** Perturbations to lipid bilayers by spectroscopic probes as determined by dielectric measurements, *Biochim. Biophys. Acta,* 602, 299, 1980.

87. **Bieri, V. G. and Wallach, D. F.,** Fluorescence quenching in lecithin and lecithin/cholesterol liposomes by paramagnetic lipid analogues. Introduction of a new probe approach, *Biochim. Biophys. Acta,* 389, 413, 1975.

88. **Pagano, R. E., Ozato, K., and Ruysschaert, J. M.,** Intracellular distribution of fluorescent probes in mammalian cells, *Biochim. Biophys. Acta,* 465, 661, 1977.

89. **Bashford, C. L., Morgan, C. G., and Radda, G. K.,** Measurement and interpretation of fluorescence polarization in phospholipid dispersions, *Biochim. Biophys. Acta,* 426, 158, 1976.

90. **Tilley, L., Thulborn, K. R., and Sawyer, W. H.,** An assessment of the fluidity gradient of the lipid bilayer as determined by a set of n-(9-anthroyloxy) fatty acids (n = 2,6,9,12,16), *J. Biol. Chem.,* 254, 2592, 1979.

91. **Vincent, M., de Foresta, B., Galley, J., and Alfsen, A.,** Fluorescence anisotropy decays of n-(9-anthroyloxy) fatty acids in dipalmitoylphosphatidylcholine vesicles. Location of the effects of cholesterol addition, *Biochem. Biophys. Res. Commun.,* 107, 914, 1982.

92. **Blatt, E., Sawyer, W. H., and Ghiggino, K. P.,** The rotational motion of n-(9-Anthroyloxy) fatty acids, *Aust. J. Chem.,* 36, 1079, 1983.

93. **Thulborn, K. R. and Beddard, G. S.,** The effects of cholesterol on the time-resolved emission anisotropy of 12-(9-anthroyloxy) stearic acid in dipalmitoylphosphatidylcholine bilayers, *Biochim. Biophys. Acta,* 693, 246, 1983.

94. **Vincent, M. and Gallay, J.,** Time-resolved fluorescence anisotropy study of effects of a *Cis* double bond on structure of lecithin and cholesterol-lecithin bilayers using n-(9-anthroyloxy) fatty acids as probes, *Biochemistry,* 23, 6514, 1984.

95. **Lakowicz, J. R. and Knutson, J. R.,** Hindered depolarizing rotations of perylene in lipid bilayers. Detection by lifetime-resolved fluorescence anisotropy measurements, *Biochemistry,* 19, 905, 1980.

96. **Gratton, E.,** personal communication.

97. **Burgess, S., Windes, S. C., and Lentz, B. R.,** unpublished.

Chapter 3

# STUDIES OF MEMBRANE DYNAMICS AND LIPID-PROTEIN INTERACTIONS WITH PARINARIC ACID

**Bruce Hudson and Sarina A. Cavalier**

## TABLE OF CONTENTS

## I. INTRODUCTION: GENERAL PROPERTIES OF PARINARIC ACID

Parinaric acid, 9,11,13,15-octadecatetraenoic acid, is a naturally occurring 18-carbon conjugated tetraene fatty acid with useful properties as a probe of membrane structure and lipid binding to proteins.[1-10] Parinaric acid was first isolated in 1933 by Tsujimoto and Koyanagi from *Parinari glaberrium*.[11,12] It is also found in a few other plant species.[13] Parinaric acid is mentioned in the 1944 organic chemistry text by the Fiesers.[14] It was this reference that led the senior author on the trail of this fluorescent probe. *Parinari glaberrium* grows in the Fiji and New Solomon Islands of the South Pacific. The plant is known as the makita tree in Fiji where the native Fiji Islanders use the seed oil to caulk their canoes. The acyl chains of the triglyceride oil are about 60% parinaric acid with much of the rest being the conjugated triene eleosteric acid (9,11,13-octadecatrienoic acid). The unsaturated fatty acid triglycerides are food storage for the seed which has the appearance and hull of a small coconut. Several grams of the fatty acid can be isolated[15] from a single seed, but the yield is quite variable. A procedure for isolation from an alternative source, seeds of *Impatiens glandulifera,* has recently been described.[16] A yield of 7 g of the fatty acid from 100 g of dry seeds was reported. This plant is grown as an ornamental and seeds are available from commercial sources. The initial steps of the reported[16] isolation procedure appear to be more convenient than those required for isolation from *Parinari glaberrium*.[15]

The naturally occurring form of the fatty acid has *cis* geometry at the 9-10 and 15-16 double bonds. The all-*trans* isomer is easily formed by iodine-catalyzed photoisomerization.[4] (Visible light is absorbed by iodine to form atomic iodine.) The ready availability of two slightly different structural isomers of this membrane probe permits detailed examination of bilayer structure and dynamics. In particular, the *trans* isomer has what appears to be the unique property that it preferentially accumulates in solid-like gel phase lipid regions, as discussed below.

The structure of parinaric acid was established by classical organic methods,[17-20] UV spectroscopy,[21] and IR spectroscopy.[22] A recent 300-MHz NMR study of parinaric acid confirmed the proposed structures of both isomers.[23] The synthesis of the β or all-*trans* isomer of parinaric acid has recently been reported using two different routes.[24,25] One of these synthetic routes[25] has been used to prepare parinaric acid with the eight vinyl protons replaced by deuterons. This synthetic route could be used to prepare tritiated parinaric acid. The enzymatic synthesis of parinaric acid using an enzyme prepared from *Impatiens balsamina* seeds has been demonstrated.[26] Since the substrate is a *cis*-9 double bond acyl-CoA ester (such as oleoyl-CoA), this method also permits preparation of radioactively labeled parinaric acid. The determination of the crystal structures of both isomers of parinaric acid is in progress.[27]

The carboxyl functionality of parinaric acid permits its use as an acyl chain component of a lipid. Phosphatidylcholines labeled in either the *sn*-1[16] or *sn*-2[28] positions have been prepared. Phosphatidylethanolamine,[28] phosphatidylglycerol,[28] phosphatidylinositol,[29] and sphingomyelin[30] lipids have also been prepared labeled at the *sn*-2 position. Several other parinaroyl derivatives have been prepared including the methyl, cholesterol,[31,32] and hydroxylsuccinimide esters,[28] parinaroylglucosamine[28] and parinaroylglucosamine-6-phosphate.[28] Parinaric acid is commercially available.[33-35]

A major feature of interest in terms of the use of parinaric acid as a lipid analog is the close structural similarity it shows to components of biological membranes and commonly studied model membrane lipid bilayers. This structural similarity is demonstrated by the biosynthetic incorporation of parinaric acid into biological membranes.[6,36,37]

The primary utility of parinaric acid as a fluorescent probe is its potential for detailed characterization of the probe environment at a fundamental level. This potential derives from the structural similarity of this species with normal membrane components and the known location and orientation of the probe. Several unique properties of this probe, including the sensitivity of several spectral characteristics to environment, permit characterization in terms of independent measurable quantities. Parinaric acid is, in fact, not particularly useful as a general fluorescent label because of its relatively low quantum yield and its photochemical and oxidative lability.

In addition to the structural and synthetic papers cited above, there have been about 60 publications dealing with, or using, parinaric acid as a fluorescent probe. These may be roughly divided into those dealing with characterization of probe behavior and those describing specific applications. The applications of parinaric acid fall into four classes: detailed studies of model membrane structure and dynamics, studies of binding to lipophilic proteins, assays of enzymatic or exchange properties of lipids, and studies of biological membranes and other lipid-containing structures. This review discusses the spectroscopic properties of parinaric acid, both in solution and in bilayer environments, and then turns to these applications. A summary of the experimental methods unique to parinaric acid is given at the end.

## II. SPECTROSCOPIC PROPERTIES OF PARINARIC ACID IN SOLUTION

### A. Absorption and Emission Spectra and State Ordering

The tetraene chromophore of parinaric acid provides it with a characteristic strong absorption band in the 320 to 270 nm region. For hexane or ethanol solution the positions of the absorption maxima are at 314, 299, 286, and 275 (shoulder) nm for *trans*-parinaric acid and at 319, 304, 291, and 280 (shoulder) nm for *cis*-parinaric acid.[3] The maximum extinction coefficient is 91,000 $\ell$ mol$^{-1}$ cm$^{-1}$ for *trans*-parinaric acid and 78,000 $\ell$ mol$^{-1}$ cm$^{-1}$ for the *cis* isomer.[3] The structure of the parinaric acid absorption band is due to a vibrational progression in the polyene chain single and double bond vibrations. The relatively sharp structure of this absorption permits the detection of small spectral shifts in the absorption maxima. The absorption spectrum of parinaric acid is, in fact, quite sensitive to its environment.[3] In particular, the excitation energy depends primarily on the high-frequency polarizability of the solvent as determined from its refractive index. Parinaric acid is not optically active and, therefore, its absorption does not exhibit circular dichroism. Circular dichroism is induced in several biopolymer environments, however, as discussed below.

The fluorescence emission of parinaric acid is unstructured and is very widely separated from the strong absorption region, having a maximum near 410 nm. The emission spectrum is essentially the same for both isomers and does not depend significantly on the solvent polarizability or any other properties. This wide separation of the absorption and emission bands, and the differential sensitivity of the absorption and emission spectra to solvent polarizability, are due to the fact that the lowest energy-excited state responsible for the emission of tetraenes is not the state responsible for the strong absorption band but is, instead, due to a forbidden transition at lower energy.[38-41]

The absorption spectrum of parinaric acid makes it a good acceptor in energy transfer experiments where tryptophan is the donor species.[1,2,5,7,8] The distance at which energy transfer is 50% efficient, $R_0$, is 3.0 to 3.5 nm. Eleostearic acid has an absorption maximum near 280 nm giving it an $R_0$ for transfer from tryptophan of about 1.5 to 2.0 nm. The energy transfer properties of parinaric acid have been used in several applications as discussed below.

### B. Fluorescence Quantum Yield and Decay Kinetics

The fact that the lowest energy-excited state of parinaric acid gives rise to a symmetry-forbidden electronic transition to (or from) the ground state has several consequences. One of these is the insensitivity of the emission spectrum to environment discussed above. Another is that the intrinsic or radiative lifetime of parinaric acid is relatively long, about 100 nsec. The intensity of this emission transition is "borrowed" from the nearby strongly allowed absorption transition by vibrations that mix the two neighboring electronic states. This means that the polarization of the emission and strong absorption (to the second excited singlet state) have parallel polarization.

The fluorescence quantum yield of parinaric acid in aqueous or other polar solvents (including glycerol) is very low but becomes considerable when the probe is in a hydrophobic environment. In fact, the quantum yield of the fluorescence is monotonically related to the dielectric constant of the solvent.[42] The reason for this is not at all obvious. It should be noted that in this study[42] variation of the dielectric constant was achieved by changing the composition of water/dioxane mixtures. Properties of this mixed solvent system other than the dielectric constant may determine the quantum yield.

Parinaric acid deuterated in its eight vinylic positions has recently been synthesized.[25] The fluorescence lifetime of this species in ethanol is the same as that for the proto form.[43]

The fluorescence decay observed for parinaric acid in solution is not describable by a single exponential decay.[44-46] Two, or more likely, three components are needed to describe the emission. As an example, the decay observed at 20°C in hexane solution consists of 10% of a 0.8 nsec component, 30% at 2.5 nsec, and 60% at 5.5 nsec.[45] Similar behavior is observed for 1,3,5,7-octatetraene.[45] It must be emphasized that this apparent decomposition of the decay into discrete components may not reflect the true description which may, instead, correspond to a continuum of decay rates reflecting a thermal distribution of environments present in solution. In recent studies it has been demonstrated that variation of the excitation wavelength does not result in changes in the observed decay behavior for solutions.[45] This indicates that, in contrast to the case of lipid bilayers discussed below, the mechanism leading to the complex decay in solution is "homogeneous" in the sense that it is not due to preexisting ground state species. This behavior in solution is probably due to a photochemical process leading to several emitting excited-state conformers that thermally revert to the all *trans* form on returning to the ground state.

**C. Transition Dipole Orientation and Limiting Anisotropy**

The polarization of the absorption of parinaric acid is at least roughly aligned along the polyene chain. This has been determined by measurement of the polarization of the excitation of fluorescence of parinaric acid, oriented in urea inclusion crystals.[47] The value of $r_0$, the anisotropy of emission observed in rigid matrices or extrapolated to zero time, is very close to the theoretical maximum value of 2/5.[9,46] For example, the value reported for propylene glycol at $-55°C$ is 0.391 and is independent of the excitation wavelength.[46] This demonstrates that, as expected, the transition moment for the emission is oriented parallel to that for absorption.

Parinaric acid undergoes an irreversible photochemical dimerization reaction.[48-50] This reaction proceeds both in solution and in bilayers but leads to different products in these two environments.[50] The kinetics of the photoreaction are first-order in illumination intensity and second-order in parinaric acid concentration indicating that a photo-excited species reacts with a ground state molecule. The product of the solution reaction appears to contain a triene chromophore (absorbing near 280 nm). One triene chromophore is generated for two tetraene chromophores lost. Similar spectral results are obtained in the bilayer but the products are chromatographically resolvable. Preliminary NMR[50] data indicate the photoproduct in solution is the result of a photo-cyclization reaction followed by a thermal Diels-Alder reaction. Thermal decomposition of parinaric acid also occurs, particularly in the solid state. The initial products also show absorption spectra characteristic of triene chromophores.[25]

## III. BEHAVIOR OF PARINARIC ACID IN MODEL BILAYER MEMBRANES

### A. Steady-State Fluorescence and Absorption Properties

The behavior of parinaric acid in lipid bilayers was established in a series of papers in the middle and late 1970s.[1,2,4,9] At the thermal gel to liquid crystalline phase transition of one-component model membranes, this probe in any of its forms exhibits a decrease in quantum yield and lifetime and decreased steady-state polarization. The fluid, liquid crystalline phase is characterized by a low quantum yield and lifetime and low anisotropy due to rapid chain motions. The lifetime and quantum yield of parinaric acid are much more sensitive to environment than is observed for diphenylhexatriene (DPH).

Another observation that is relevant to the interpretation of current studies presented below is that parinaric acid exhibits a blue shift in its fluorescence excitation and absorption spectra on going from the higher density gel phase to the lower density liquid crystalline phase. For DPPC vesicles, the fluorescence excitation maximum of *trans*-parinaric acid shifts from 319.8 nm at 40°C to 318.3 nm at 43°C. This shift in the spectral maximum is due to the lower polarizability of the high-temperature phase. For a given chemical environment, the polarizability is related to the density. Calibration of the position of the spectral maxima as a function of solvent polarizability[3] permits a determination of the density change in the bilayer from this spectral observation. The results are in good agreement with direct density change determinations.[4] Thus, the environment of the probe can be characterized in terms of its density by determination of the excitation maximum. Conversely, if the probe exists in more than one environment, variation of the excitation wavelength will change the relative contributions of the heterogeneous environmental components.

### B. Lipid Phase Partitioning Behavior

The behavior of parinaric acid in bilayers is, for the most part, independent of the particular form of the probe, i.e., *cis* or *trans* isomer, free fatty acid, or phospholipid. There is, however, one important difference between the *cis* and *trans* isomers that is observed when two phases of phospholipid coexist, either as two distinct vesicle populations of different composition or as laterally separated phases. The *cis* isomer partitions nearly equally between these two phases, but the *trans* isomer preferentially accumulates in the solid phase with a partition coefficient of 4 to 9.[4,51,52] Differential partitioning behavior is also observed for phospholipids containing a parinaroyl chain.[53-55] This partitioning behavior is in the direction expected on the basis of the shapes of these two probe species and the dominant structures of the acyl chains in the two phases. The preferential partitioning of *trans*-parinaric acid and *trans*-parinaroyl phospholipids into solid phases causes the transition temperatures for the gel to liquid crystalline transition to be observed at slightly higher temperatures (1 to 2°C) for this probe than for *cis*-parinaric acid or by other physical techniques.[1,4,53] More importantly, in studies of mixed lipid systems, the *trans* isomer detects the presence of small amounts of solid phase regions that are undetected with other fluorescent probes.

Another distinction between the *cis* and *trans* forms is that the *trans* isomer exhibits a greater sensitivity in its quantum yield and lifetime to changes in its environment. This behavior, in conjunction with the partitioning behavior discussed above, makes *trans*-parinaric acid uniquely sensitive to the presence of small amounts of "solid-phase" regions in lipid bilayers.

### C. Time-Resolved Fluorescence Behavior

In recent experiments we have used the superior time resolution of a laser-based single-photon counting apparatus to reexamine the behavior of parinaric acid in liquid crystalline phase phospholipids. These studies[45,56-59] have confirmed previous results[4,9,46] that the fluorescence decay, even in the nominally one-component high temperature phase, cannot be described with a single exponential. The best analyses seem to require three exponential terms.[45,56-59] The shortest lifetime component (about 0.8 nsec) has a relatively small amplitude (about 10%). The amplitude of this short lifetime component is essentially constant as a function of temperature. More interesting behavior is observed for the intermediate and long-lifetime components. For dipalmitoylphosphatidylcholine (DPPC) vesicles with a thermal phase transition of 41°C, the long-lifetime component persists at temperatures well above the phase transition and decreases in amplitude rather abruptly at about 48 to 55°C.[58,59] The values of the lifetimes themselves are rather insensitive to temperature. The amplitude of this long-lifetime does not depend on the probe concentration, the particular isomer of parinaric acid used, the method of preparation of the bilayer vesicles, or the thermal history of the sample.[9] It does depend on the excitation wavelength. A model explaining this behavior is presented below.

**D. Fluorescence Anisotropy Decay in Bilayers**

Time-resolved fluorescence anisotropy decay determinations for parinaric acid in model membrane systems have been reported.[9] The major observation is that the time dependence of the anisotropy of parinaric acid is very different in the two bilayer phases. In the solid-like gel phase, the anisotropy is high and essentially constant in time. In the fluid-like liquid crystalline phase, there is a very rapid decay of the anisotropy to a nearly constant value. The magnitude of this residual anisotropy at long times is a measure of the order parameter after this initial, rapid, reorientational motion. The value of the order parameter obtained in this way is in good agreement with values estimated for these conditions from deuterium NMR measurements.[9] This has the interesting implication that there are no slow reorientational processes with significant amplitudes occurring on the time scale between nanoseconds and tens of microseconds.[9,10] Furthermore, the values of the anisotropy extrapolated back to zero time are very close to the theoretical maximum value of 0.4.[9] This demonstrates that no rapid rotational processes are being missed in the data collection or analysis. Thus, the fluorescence measurements appear to be of the proper time scale to detect the significant motions in these model membrane systems. These measurements are able to separate the rate of reorientational motion from the amplitude of this motion. This is not possible with steady-state fluorescence anisotropy determinations or conventional magnetic resonance techniques. An interpretation of the depolarizing motions occurring in bilayers is presented in the material that follows.

## IV. APPLICATIONS OF PARINARIC ACID AND ITS DERIVATIVES

**A. Detailed Studies of Simple Model Membrane Systems**
*1. Heterogeneous Environments in Pure Lipid Phases*

The results presented above concerning the fluorescence decay behavior of parinaric acid in fluid phase model bilayers require an explanation in terms of the dynamic structure of these systems. A consistent hypothesis is that the residual long-lifetime species is due to parinaric acid chains in an environment that is characterized as "solid-like". This association is based initially on the fact that parinaric acid in the low temperature solid gel phase exhibits a long lifetime. Parinaric acid is able to sense the presence of this "phase" or "environment" because of its sensitivity to its local surroundings. We can confirm the nature of the probe environment by measuring the amplitude of this long-lifetime component as a function of the excitation wavelength.[45,58,59] The solid-like environment, because of its higher density, should exhibit a relative red-shift in its excitation maximum as discussed above. Thus, variation of the excitation wavelength should result in enhancement of the long-lifetime component, with a characteristic spectral signature shifted to the red of the overall absorption or fluorescence excitation. This is indeed observed.[45,58,59] The fact that differential excitation is observed demonstrates that these solid-like environments have a characteristic lifetime for their existence that is long compared to the fluorescence lifetime of several nanoseconds.

We also expect that this long-lived species, if it indeed reflects a solid-like environment, should have a high limiting anisotropy compared to the fluid environment with a shorter lifetime. This should result in upward curvature for the anisotropy decay at long times. The reason for this is that the solid-like environments with their expected higher limiting anisotropies also have longer fluorescence lifetimes. Thus, at long times the measured anisotropy will be dominated by this solid-like component.[56-59] The case of parinaric acid in dielaidoylphosphatidylcholine (DEPC) has been studied in detail.[45,59] What is found is that for temperatures in the range of 10 to 15°C above the thermal phase transition temperature (ca. 10°C) where there is a significant amplitude for a long-lived component, the anisotropy decays to a minimum and then rises at long times. As expected, the limiting anisotropy observed depends on the excitation wavelength because selective red excitation increases the contribution from the solid-like phase. Thus the environment sensed by parinaric acid giving rise to a long lifetime for the fluorescence decay has a red-shifted absorption and a high asymptotic anisotropy, all characteristic of an ordered, high-density, solid-like phase.

Addition of cholesterol to phospholipid bilayers induces the presence of a very long-lifetime component (10 to 35 nsec) indicating that this sterol somehow stabilizes a solid-like environment.[45,56,59] The effect on the anisotropy behavior is particularly dramatic. At low (5%) cholesterol levels a large increase in the anisotropy is observed at long times. This is interpreted as being due to an environment with a long lifetime and a very high asymptotic anisotropy. At high cholesterol levels, all of the bilayer is converted into this solid-like phase and the anisotropy decays to a high limiting value. The variation of the amplitudes and lifetimes for the fluorescence decay components with cholesterol concentration is rather complex. The major observation of interest is that the lifetime of the longest component of the decay decreases precipitously from 35 to 11 nsec as the cholesterol content is increased from 5 to 15% and then remains constant up to 25% cholesterol.

Our present interpretation of all of these results is that phospholipid bilayers exhibit behavior similar to that seen for critical systems where long-range correlations give rise to transient fluctuations in the order parameter. This has been predicted on theoretical grounds[60-62] and is consistent with the results of acoustic absorption measurements.[63-66] The lifetime of these critical density fluctuations has been estimated from the resonance behavior of the acoustic absorption to be on the order of 30 nsec for temperatures slightly above the main chain melting transition.[65] This is consistent with the fluorescence results. In particular, a lifetime for these domains in this time regime would not permit lateral equilibration of the probe species between the fluid and solid-like regions. This is consistent with the lack of preferential partitioning of *cis-* and *trans-*parinaric acid between the two regions. We view the bilayer, then, as a flickering structure with transient alignment and packing of the chains trapping the probe species in that environment and then dissipating to reform the fluid domain on a time scale somewhat longer than the fluorescence lifetime, but shorter than the time required to establish lateral spatial equilibrium of the probe. Cholesterol must stabilize these high-density regions. A simple model for this stabilization consistent with very recent monolayer results[67-69] is that cholesterol binds to the interface between the two structural types, orienting its planar surface[70] toward the solid-like domains and its methylated face toward the fluid-like domains. The decrease in the lifetime with increasing cholesterol content is then associated with the breakup of large solid-like domains in order to provide sufficient interfacial regions between the solid and fluid phases for positioning of the cholesterol.[45,59] In order to be effective in satisfying the growth requirement of organisms requiring cholesterol and to exhibit the condensing effects of cholesterol, the steroid must have one planar surface.[70] This is explicable in the present model in that the interfacial region between solid-like and fluid-like domains can best be stabilized by a structure with one flat and one "bumpy" surface.

## 2. Acyl Chain Dynamics

The motion of parinaric acid in these model bilayers leading to rapid depolarization of the fluorescence could be of two kinds: the entire acyl chain could move as a unit in a rigid fashion or there could be excitation of *gauche* conformations in the methylene segment of the upper part of the probe. Excitation or migration of "kinks", correlated *gauche+*, *trans*, *gauche−* configurations, does not significantly reorient the tetraene chromophore and therefore does not lead to depolarization. The conjugated section of the polyene is rather rigid and internal excitation of this segment is unlikely, at least in the ground electronic state. Rigid reorientation of the entire probe chain is expected to be collective in nature because of the large angle swept out. In this case the motions of the bilayer chains might resemble the behavior of a wheat field. *Trans-gauche* excitation in the upper part of the probe chain might result from statistically uncorrelated motions. This excitation depends on the flexibility of the methylene segment and the amplitude of this kind of motion depends on the number of methylene segments. An experiment is in progress where an analog of parinaric acid is being synthesized which has fewer methylene segments between the carboxyl group and the tetraene chromophore. If the acyl chain motion is of the rigid body variety, then this probe should have the same behavior as that observed for parinaric acid. If *trans-gauche* excitations are required for depolarization, then this analog should have a decreased amplitude for its depolarization, i.e., a high anisotropy at long times.

It was argued above and in earlier publications[9,10] that the agreement observed between the order parameter extracted from the limiting anisotropy in these fluorescence experiments and deuterium NMR order parameters indicates that there are no significant amplitude reorientational motions with time scales in the region intermediate between the fluorescence (10 nsec) and deuterium NMR (10 μsec) time regions. This conclusion is somewhat compromised by the fact that the deuterium NMR measurements refer to saturated acyl chains rather than the polyene chromophore used for the fluorescence experiments. That is the reason for synthesis of deuterated parinaric acid[25] preparatory to deuterium NMR experiments on this species.

## 3. The Effect of Protein Components on Anisotropy Decay

The effect of incorporated proteins on the chain dynamics of model membranes as measured by parinaric acid has been established in two papers.[8,10] In each case the incorporated protein was the coat protein of the M13 bacteriophage. In the life cycle of the virus the coat protein exists in the *E. coli* membrane prior to final phage assembly. This protein has a very hydrophobic central sequence and is believed to exist in the bilayer as a simple transmembrane α helix. The presence of the protein tends to abolish the thermal transition of one-component phospholipid bilayers. Energy transfer measurements from the single tryptophan of the M13 coat protein to the random distribution[71] of parinaric acid in the bilayer have been used to characterize the nature of the lipid surrounding the protein.[8] The major conclusion is that the protein disorders the bilayer in its vicinity relative to the solid gel phase. This conclusion is based on the avoidance of the vicinity of the protein by the *trans* isomer, consistent with a relative lack of straight chain structures. On the other hand, the protein orders the bilayer relative to the fluid-like liquid crystalline state as revealed by an increase in the limiting anisotropy in the presence of the protein.[10] This overall behavior is consistent with the loss of the phase transition upon incorporation of the protein, since the difference between the acyl chain structure at low and high temperatures in the vicinity of the protein is diminished.

An upward curvature is observed for the anisotropy decay of parinaric acid in the presence of the M13 coat protein.[10] The interpretation of this rising anisotropy at long times is similar to that given above for phospholipid bilayers themselves. Here the behavior is based on the combination of facts that the protein induces an increase in the fluorescence lifetime relative to that seen for the fluid bilayer and increases the anisotropy (reduces motional freedom) for chains in the vicinity of the protein. At short times all probe species contribute to the fluorescence, and therefore to the anisotropy, in proportion to their amount; at longer times the short-lived species distant from the protein become less important and do not contribute as much to the anisotropy. The longer lifetime for the probe species adjacent to the protein means that at long times the observed anisotropy is that due to these "perturbed" chains. Since the anisotropy is higher for the chains near the protein than that for the bulk lipid, the observed anisotropy rises at long times. A smaller effect is seen for the *cis* isomer than the *trans* isomer due to the smaller change in its lifetime with change in environment. This behavior can be analyzed to provide an estimate for the number of perturbed chains.[10] It is found that the number of perturbed chains is roughly 12, the number of chains immediately adjacent to the protein.

## B. Studies of Binding to Proteins
### 1. Binding to Serum Albumins and α-Fetoprotein
When parinaric acid binds to hydrophobic sites of soluble proteins there is a considerable (about 50-fold) increase in the fluorescence quantum yield. Furthermore, in the bound form, the tetraene chromophore is able to participate in energy transfer from protein tryptophan residues and may transfer energy to other ligands. These properties of parinaric acid have been used to characterize the fatty acid binding sites of serum albumins[5,7,72] and human α-fetoprotein.[73] Binding affinities and number of sites have been determined, competition studies have demonstrated that the affinities for parinaric acid are comparable to those of other fatty acids and energy transfer from tryptophan to bound parinarate and from bound parinarate to bilirubin have been used to obtain distance information. This distance information has been correlated with a domain binding site model based on sequence information.[7] The distances obtained for α-fetoprotein are similar to those for human serum albumin indicating strong three-dimensional homology.[73] The proximity and relative orientation of pairs of fatty acids bound to HSA has been determined by measurement of the quenching of bound parinarate fluorescence by bound nitroxide-labeled fatty acids. The effect of non-enzymatic glycosylation on the affinity of human serum albumin for fatty acids has been determined using parinaric acid.[74] This reaction specifically modifies lysine 157. A very large (20-fold) decrease in affinity for the first binding site was observed.[74]

Parinaric acid exhibits a strong, highly structured, induced circular dichroism (CD) when bound to bovine and human serum albumin[5,7] as well as that from other species.[75] This induced CD depends strongly on the species source of the albumin. At low ratios of fatty acid to protein the induced CD for HSA and BSA is positive and resembles the absorption spectrum although for HSA there is a significant shift of the long wavelength bands as if there was a negative component at low energies. For rabbit serum albumin the CD spectrum is negative and has much greater maximum to minimum ratios than does the absorption spectrum.[75] When more than one fatty acid is bound there is a strong exciton interaction resulting in a conservative CD pattern. This indicates that the first two fatty acids bind very close together in the complex.

### 2. Binding to Phosphatidylcholine Transfer Protein
Static and time-resolved fluorescence experiments employing parinaric acid have been used to examine the binding of lipids to the intracellular protein phosphatidylcholine transfer protein (PC-TP).[76] This protein is of particular interest because of its probable role in membrane biogenesis and because of the lipid specificity exhibited by this and related proteins. This study[76] employed phosphatidylcholines with a *cis*-parinaroyl chain in either the *sn*-1, *sn*-2, or both *sn*-1 and *sn*-2 positions. By monitoring the fluorescent properties of the parinaroyl chains in different environments, Wirtz and co-workers determined the ori-

entation of the labeled phosphatidylcholine molecule bound to PC-TP and characterized the binding site of the protein. Changes in the fluorescence intensity were used to monitor intravesicular movement of lipids, binding of the fluorescently labeled lipid to PC-TP, and quenching of PC-TP tryptophan fluorescence by the parinaroyl-phosphatidylcholine. Vesicles of pure parinaroyl-phosphatidylcholine are self-quenching and the binding of lipid molecules to PC-TP results in enhanced fluorescence. Subsequent transfer into unlabeled phosphatidylcholine vesicles catalyzed by this protein results in a further increase in fluorescence intensity of up to five times relative to that observed for the lipid bound to the protein. Wirtz and co-workers have developed a linear relationship between the labeled phosphatidylcholine concentration in vesicles and the fluorescent intensity. A comparison of the fluorescence intensity for premixed vesicles of labeled and unlabeled lipids with that of equilibrated vesicles after transfer showed that 60% of the parinaroyl-phosphatidylcholine could be transferred. This can be accounted for by assuming that PC-TP will only transfer lipids from the outer monolayer of a phospholipid bilayer. The diparinaroyl-substituted lipid showed only 30% of the fluorescence intensity of the singly substituted phospholipid in nonlabeled bilayers, indicating that the two neighboring chains interact in such a way as to quench themselves.

A linear relationship was also proposed relating the amount of labeled lipid per mole of PC-TP and the PC-TP tryptophan fluorescence intensity. The fluorescent decay data of the tryptophans of PC-TP, parinaroyl-phosphatidylcholine and protein-lipid complex show that in the lipid-protein complex, the parinaroyl lipid affects only the shortest lifetime component of the three lifetime components of PC-TP. This shortest lifetime component contributes 50% of the fluorescent decay. In the phospholipid/protein complex this lifetime component exhibits a decrease in lifetime by one half. PC-TP has five tryptophan residues and therefore a detailed model of the quenching mechanism was not developed.

Time-resolved anisotropy measurements have been used to further characterize the phosphatidylcholine/ PC-TP complex. The correlation times of *sn*-2 and *sn*-1 substituted lipids are distinctly different: 11 nsec and 26 nsec, respectively. The initial anisotropy is high (0.3) in both cases. The high initial anisotropy, relatively long, and strongly different correlation times suggest a rigid binding of the phospholipid acyl chains to the protein and a difference in orientation of the two fatty acyl chains when bound to PC-TP. In the diparinaroyl substituted phosphatidylcholine PC-TP complex, the rotational correlation time in this case is 15 nsec, in agreement with the harmonic mean value for the singly labeled phospholipids. This observation, and the lack of quenching of one parinaroyl chain by the other, indicates that the two acyl chains are spatially separated in the complex and behave independently of each other.

These time-resolved fluorescence measurements have led to important conclusions on the nature of the interaction of the lipid with the protein. The single rotational correlation times for both analogs of phosphatidylcholine indicate immobilization of the acyl chains when bound. The large difference in correlation values indicate a distinct binding site for each acyl chain and a nonparallel orientation of the two chains. The *sn*-1 chain with its longer correlation time appears to be oriented parallel to the long symmetry axis of the supposed β sheet within PC-TP.[77] The *sn*-2 chain with its lower correlation value implies binding at a 60 to 90 degree angle to the *sn*-1 chain.

### 3. Binding to Other Proteins

The fluorescence enhancement observed for parinaric acid upon binding to proteins has been developed into a general method for characterizing protein surface hydrophobicity.[78] The basic method used is the determination of the initial slope of the increase in fluorescence with protein concentration. This is effectively a measure of the number of strong hydrophobic binding sites. A good correlation was observed between this rapid and simple method and methods using hydrophobic partitioning chromatography.

The hydrophobic binding sites of the surface of myosin and its fragments have been characterized using parinaric acid fluorescence.[79] In unpublished preliminary studies we have characterized the light-dependent binding of parinaric acid to phytochrome.[80] This protein is responsible for regulating plant growth and development in response to changes in the spectral distribution of skylight. It was found that illumination with far red light exposed a hydrophobic binding site that was not present following illumination with red light. It is believed that the initial step following illumination of this protein with far red light is binding to a membrane.

### C. Assay Methods Using Parinaric Acid

Several of the applications of parinaric acid are analytical methods and assays for chemical or physical transformations of lipid components. These methods use one or more of the spectral properties of parinaric acid described above and may depend on the carboxyl functionality and/or the close structural similarity of the chain with natural lipid components for their effectiveness.

The competitive binding of endogenous fatty acids and parinaric acid to serum albumin have been developed into a method to assay the concentration of free fatty acids in serum.[81] Here, for technical generality, the absorption shift associated with binding of the conjugated fatty acid to albumin is monitored. This method requires only 30 $\mu\ell$ of serum or plasma.

The bimolecular nature of the photochemical dimerization of parinaroyl-labeled lipids in bilayers has been used to assay membrane "fusion".[48] Fusion is defined here as the intimate comixing of lipid components from two different vesicle populations. This intimate mixing results in a dilution of the parinaroyl-labeled phospholipids originally present in part of the vesicle population with unlabeled vesicles. The resulting lateral dilution results in a decrease in the photochemical dimerization rate.

An assay for the hydrolysis of lecithins by phospholipase $A_2$ (PLA$_2$) from snake venom was developed utilizing the fluorescent properties of the all-*trans* isomer of parinaric acid.[82] Synthetic 2-*trans*-parinoyl-phosphatidylcholine was incorporated into vesicles. In the presence of albumin and pH and salt conditions correct for PLA$_2$ activity, the transfer of the all-*trans* parinoyl chain from the phospholipid to the hydrophobic binding site of albumin was monitored by measuring the increase in polarization of the parinaric acid fluorescence. When bound to albumin, the fluorescent lipid is in a restricted hydrophobic binding site and its polarization value increases relative to that observed for the probe in the lipid bilayers used. It was shown that pure parinaric acid (nanomolar amounts) incubated with nonfluorescent vesicles has a low polarization value (ca. 0.17). Introduction of albumin to a solution of parinaric acid and vesicles results in an increase in polarization to a value close to 0.31. Addition of EDTA to chelate Ca ions required for PLA$_2$ activity abolishes the increase in polarization. Experiments allowing the hydrolysis of the fluorescent phospholipid in the absence of albumin showed a decrease in depolarization. (This indicates that the free fatty acid has a lower steady-state polarization than the phospholipid. This is utilized in another assay described below.) When albumin is introduced, the polarization immediately increases, indicating that the accumulated free parinaric acid is transferred from the bilayer to the protein. This fluorescent assay method of PLA$_2$ activity is less time-consuming, is continuous, and is as sensitive as radioassays. The calculated sensitivity of the assay is 0.3 nmol of substrate hydrolyzed per minute.

This same assay method for deacylation has been applied to the thrombin-induced aggregation of platelets.[83] In this case the parinaric acid label was introduced into platelet membranes by biosynthetic incorporation. Lipid analysis demonstrated incorporation into triglyceride, phosphatidylethanolamine, phosphatidylcholine, phosphatidylserine, and phosphatidylinositol. Hydrolysis by snake venom phospholipase $A_2$ showed that 68% of the parinaric acid was at the *sn*-2 position. Thrombin-induced aggregation resulted in a release of parinaric acid due to activation of endogenous phospholipase activity.

A fluorimetric assay method has also been developed for characterizing the activity of the phospholipid transfer proteins.[84] The basis of this method is the self-quenching of parinaric acid-labeled phospholipids when they are present at high concentrations in lipid bilayers. Transfer to unlabeled vesicles eliminates this quenching and thus results in an increase in fluorescence intensity. This transfer assay has been applied to the phosphatidylcholine transfer protein (PC-TP), phosphatidylinositol transfer protein (PI-TP), and a nonspecific phospholipid transfer protein.[84] By monitoring the change in fluorescence intensity, studies were conducted on the transfer efficiency of each protein and the effect that membrane charge and substrate specificity have on such a transfer process. Preliminary experiments showed that there is transfer of 2-parinoyl-phosphatidylcholine from the outer membrane of pure labeled vesicles into unlabeled vesicles as registered by the increase of fluorescence intensity in the presence of PC-TP. Monitoring the increase in fluorescence intensity, the efficiency of transfer was also studied in the presence of negatively charged membranes. Of the three proteins, the phosphatidylcholine transfer protein was the most sensitive to membrane charge. The rate of transfer, proportional to the rate of increase of fluorescence intensity, was stimulated at low concentrations of negatively charged lipid into the acceptor vesicles. Both of the other proteins remained relatively unaffected by negatively charged membranes.

This fluorimetric assay was also useful in determining the substrate specificities of each protein. Here the acceptor vesicles were varied in structure. It is assumed that there is a strictly coupled exchange between phosphatidylcholine and analogs giving rise to the fluorescence increase. Three types of modifications of the phosphatidylcholine substrate were used: diester, diether, and dialkyl substituted at the linkage to the head group. Once again, PC-TP was the most sensitive to the substitutions. The fluorescence intensity increased most with the diester analog of PC, indicating an increased rate of transfer. PI-TP, and the nonspecific transfer protein exhibited essentially no effect on transfer rates. Background fluorescence intensities of the analogs in vesicle form proved to be independent of the type of analog used. This assay for phospholipid exchange has also been applied to phosphatidylinositol transfer protein using synthetic-labeled phosphatidylinositol.[29] This simple assay method provides important information about transfer protein specificity necessary for the further characterization of the mechanism of action of these proteins.

The change in polarization of *cis-* parinaric acid was used as a method to monitor activity of the membrane-bound enzyme acylCoA acyltransferase in rat brain synaptosomes.[85] This study was prompted by the observation that incubation of synaptosomes with *cis*-parinaric acid resulted in a long-term increase in fluorescence polarization with time. This time dependence was not observed with the *trans* isomer. This was developed into an assay method for the incorporation of parinaric acid into phospholipids enabling an estimate of the speed and efficiency of the acyltransferase reaction. By monitoring the rise in polarization with time, the effect of cofactors of the enzymatic reaction was determined. It was found that with a combination of Mg cations, ATP, and CoA, the reaction proceeded most rapidly. Heating the synaptosome preparation decreased the rate of this fluorescence change indicating that the reaction was indeed enzymatic. These results discounted the possibility of an increase in polarization due to the physical rearrangement of the lipid bilayer caused by addition of the fatty acid. Phospholipid analysis showed that 70% of the *cis*-parinaric acid was incorporated into phosphatidylcholine, 20% into phosphatidylethanolamine and trace amounts into phosphatidylinositol and phosphatidylserine. Oleic acid can also be incorporated into phospholipids by this enzyme and blocks the incorporation of *cis-* parinaric acid. It is interesting to note that this enzyme is active on the *cis* but not the *trans* isomer of parinaric acid. The authors note that the availability of radioactive parinaric acid would permit them to make this assay quantitative.

A comparison of the fluorescence of parinaric acid in synaptosome membranes with vesicles prepared from extracted lipids indicated that a more rigid lipid environment exists around the protein. The initial (and time-independent) polarization of the fluorescence of *trans*-parinaric acid in the membranes was higher than that for the *cis* isomer. This is consistent with the general observation of preference of the *trans* isomer for more solid-like rigid phases. The reason for the increase of the polarization of *cis*-parinaric acid upon enzymatic incorporation is not so obvious but could be due to a difference in the partition coefficient for the free fatty acid and the phospholipid so that there is a relative preference for the lipid to be in solid-like domains.

## D. Studies of Biological Membranes and Lipoprotein Particles

There have been numerous applications of parinaric acid fluorescence to the study of the structure of biological membranes and lipid-protein complexes. Here we will only provide illustrative examples of spectral and interpretation methods and discuss those studies utilizing a novel technique.

Tecoma et al.[6] have developed procedures for the biosynthetic incorporation of parinaric acid into *E. coli* cells auxotrophic for fatty acids. The growth conditions involve inclusion of oleic or elaidic acid in the medium so that the cells become enriched in these fatty acids and thus develop a more nearly homogeneous fatty acid composition. Typical conditions are (a) growth at 35°C in 25μg/mℓ oleic acid and 50 μg/mℓ parinaric acid, or (b) growth at 39°C in 40 μg/mℓ elaidic acid and 40 to 60 μg/mℓ parinaric acid. The cells undergo 2 to 5 doublings under these conditions. *cis*-Parinaric acid is more extensively incorporated than *trans*-parinaric acid. Incorporation levels up to 3% were achieved; phosphatidylethanolamine, phosphatidylglycerol, and cardiolipin were labeled. The washed cells showed absorption and fluorescence excitation spectra characteristic of the structured absorption of parinaric acid. Broad thermal phase transitions were observed by measuring the fluorescence intensity of this endogeneous probe. The transition temperatures ranged from 15 to 40°C, depending on the composition of the cell membrane which could be as much as 62% elaidic acid. The behavior of the cells was very similar to that of vesicles made from extracted lipids and membrane preparations. *cis*- and *trans*-Parinaric acid showed comparable behavior and the results obtained with exogenous parinaric acid were very similar to those observed for the biosynthetically incorporated species. Significant quenching (60%) of the fluorescence of tryptophan residues of membrane proteins was observed.

In a study of the effect of membrane deenergization by colicin K, Tecoma and Wu[36] showed that the fluorescent behavior in response to deenergization depends on the form of the probe used. When free *cis*-parinaric acid was added to respiring bacteria, dissipation of the energized state of the membrane resulted in a dramatic increase in fluorescence; *trans*-parinaric was much less sensitive. Biosynthetically incorporated *cis*-parinaric acid showed no change in fluorescence intensity upon deenergization. It was shown that the effect observed for free *cis*-parinaric acid is due to a large change in the fraction of the probe bound with about 77% of the probe binding to the energy-poisoned *E. coli*, but only 44% binding to the actively respiring form.

A study of the intracytoplasmic membranes of the photosynthetic bacterium *Rhodopseudomonas sphaeroides* has been reported.[47] Parinaric acid was used in this study in part because the absorption properties of another commonly used probe, DPH, and the photosynthetic chromophore of these membranes results in very efficient quenching of DPH due to energy transfer. DPH/lipid ratios of 1/5 to 1/10 are required in order to obtain adequate signal and at these concentrations there is extensive perturbation of the membrane properties. Parinaric acid gave adequate signals at probe to lipid ratios of 1/200. It should be noted, however, that these studies were performed with carotenoid-minus mutants of the bacterium. Strains containing the normal carotenoid complement did not give rise to adequate signals with parinaric acid.

This study[47] included a determination of the effect of the protein component of these photosynthetic membranes on the rotational mobility of the parinaric acid probe. This was achieved by comparison of intact membranes with those formed by extracted lipids and included variation of the protein component by growth at various light levels. These membranes are very rich in protein, 70 to 80% by weight. It was found that the protein had a significant effect on the polarization of the parinaric acid fluorescence. It is necessary to exclude that this is due to energy transfer quenching by the protein component. This was done by determining that the fluorescence lifetime is not affected by the photosynthetic protein content. The increased polarization in the presence of protein is therefore again due to a more rigid environment.

This study[47] of membrane dynamics utilized an interpretation of the fluorescence anisotropy measurements in terms of a rotational rate derived from the Perrin equation. The use of this equation is based on the assumption of isotropic rotational diffusion. This is clearly not valid for the case of lipid bilayers.[9] In particular, the significant residual anisotropy observed for bilayers means that a change in the static anisotropy can be due to either a change in the rate or extent of the acyl chain motion. This latter quantity, reflected in the residual anisotropy or order parameter, seems to be more sensitive to changes in temperature and membrane composition than does the rotational rate.

A study of the effect of membrane structure on cellular differentiation using parinaric acid has been reported.[86] The hypothesis being tested here is that cellular differentiation resulting in altered gene expression is mediated by some prior change in the structure of the plasma membrane. A proerythroblastic cell line that can be caused to differentiate by addition of butyrate or hemin shows characteristic-enhanced agglutinability and induction of hemoglobin synthesis. The steady-state polarization and fluorescence lifetime decay behavior of *cis-* and *trans-*parinaric acid were used to characterize changes in the membrane structure during the differentiation process. The dependence of the lifetime of both parinaric acid isomers and the analysis of the decay into two exponential components was used as the primary characterization method. Phase/amplitude fluorometry methods were used with variable frequency excitation provided by synchrotron radiation. Six modulation frequencies were used up to 51 MHz. Cellular differentiation resulted in a decrease in the steady-state anisotropy of both isomers of parinaric and for DPH, well before significant hemoglobin production. In the nondifferentiated cells the fluorescence decay of *cis-*parinaric acid was described by 4% of a component with a lifetime of 19.2 nsec and 96% with a lifetime of 2.9 nsec. The corresponding values for *trans-*parinaric acid were 10% with a lifetime of 11.8 nsec and 90% with a lifetime of 2.2 nsec. This differential behavior for the two isomers is consistent with the association of the long lifetime value with a solid-like gel phase. For differentiated cells the corresponding values were 18%, 7.8 nsec and 82%, 2.0 nsec for *cis-*parinaric acid, and 18%, 6.9 nsec and 82%, 1.23 nsec for *trans-*parinaric acid. There is clearly a significant effect associated with differentiation with an increase in the amplitude of the long-lifetime component for each probe. The associated change in the lifetime values for the decay components makes it difficult to interpret the steady-state anisotropy measurements in terms of motional behavior.

In a series of papers[86-90] Sklar, Dratz, and co-workers have used the fluorescence of parinaric acid and some of its derivatives to study the organization of the retinal rod outer segment membranes. These membranes are interesting in several respects. The major lipid, phosphatidylcholine, consists of at least 18% with two disaturated acyl chains, primarily palmitic acid, 24% with two polyunsaturated acyl chains, predominantly docosahexaenoic acid (22:6) and 58% with one chain of each type. The phosphatidylcholine and cholesterol of the bilayer are thought to be located primarily in the inner monolayer, while the phosphatidylserine and phosphatidylethanolamine are localized in the outer monolayer. The phosphatidylserine composition is roughly 40% species with one polyunsaturated chain and 60% two polyunsaturated chains; the phosphatidylethanolamine composition is about 75%

and 24% of these species. These three classes of lipids comprise about 98% of the total lipid. The phosphatidylcholine lipids exhibit a phase separation near 30°C believed to be segregation of the disaturated chain PC from the PC containing one or more polyunsaturated chains. The addition of cholesterol (10%) broadens this phase separation considerably.[89] Synthetic mixtures or extracted lipids representing the PE/PS rich outer monolayer also undergo a partial phase separation at low temperature. Addition of calcium ions results in elevation of this transition temperature.[87] In order to study the effect of calcium on bilayer phase transitions it is necessary to use the methyl ester of parinaric acid, rather than the free fatty acid, because of precipitation of the salt of the fatty acid.[87] These probes have essentially identical spectroscopic behavior.[51,88] The measurement of fluorescence polarization for ROS membranes and bilayers with compositions representing the individual monolayers has been applied to the analysis of membrane physical asymmetry.[89] Some of the experimental methods used in these studies have been described in some detail in a review.[90] Energy transfer from rhodopsin tryptophan residues to parinaric acid and from parinaric acid to retinal were also employed in these studies.[88] These studies suggest that opsin and rhodopsin are excluded from the laterally separated solid domains formed by the relatively saturated phospholipids below 7°C.

The circular dichroism observed for cholesteryl-*cis*-parinarate in the inner core of low-density lipoprotein particles has been used to study the structure and thermal transitions of this liquid crystalline structure.[31] The thermal behavior of the CD indicated phase changes at temperatures in good agreement with calorimetric measurements. The fluorescence method provides much greater sensitivity.

In a recent study[91] Alfsen and co-workers have used *cis*- and *trans*-parinaric acid to study the effect of the clathrin protein coat of coated and uncoated vesicles from bovine brain. The organelles are formed in endo-and exocytosis and membrane cycling. The decay of the fluorescence of both *cis*- and *trans*-parinaric acid in these structures is clearly triple exponential. For example, at 18°C the uncoated vesicles had lifetimes of 35, 13, and 3.5 nsec with amplitudes of about 20%, 50%, and 30%. This behavior clearly illustrates the great sensitivity of parinaric acid to its environment. Time-resolved anisotropy measurements of parinaric isomers in these structures clearly showed that the uncoated vesicles had a much higher limiting anisotropy at long times, indicating a considerable enhancement in rigidity.

## V. EXPERIMENTAL METHODS USING PARINARIC ACID

### A. Spectroscopic Methods

The experimental methods used for fluorescence intensity, polarization, lifetime, and anisotropy decay measurements with parinaric acid or its derivatives are substantially the same as those used for any other fluorescent probe. Several features of parinaric acid spectroscopy related to these experiments should be noted. First, the absorption bands, although in the UV, are significantly red-shifted from that of tryptophan. The absorption happens to be in the region of the harmonic of rhodamine 6G, a very efficient laser dye. This permits ready excitation using dye laser sources.[56-59] The fluorescence emission of parinaric acid is in a spectral region where photomultipliers are quite sensitive.

Because of the oxidative and photochemical lability of parinaric acid, experiments are best performed with samples deoxygenated with nitrogen or argon and with low illumination intensities. Photochemical degradation in the region illuminated by the excitation beam can lead to complex and erratic time-dependent fluorescence intensities due to convection. This is avoided by stirring the sample or by mixing with a stream of nitrogen or argon gas. The major photochemical degradation reaction is second order in parinaric acid concentration.[48] This reaction is therefore greatly slowed down by using low labeling levels. BHT may also be added to vesicle preparations to inhibit free radical reactions. At a BHT/lipid ratio of 1/100 this additive does not seem to significantly perturb the bilayer behavior.

**Table 1**
**PARTITION COEFFICIENTS, $K_p$, FOR *CIS*- AND**
***TRANS*-PARINARIC ACID BETWEEN LIPIDS AND**
**BUFFER**

| Lipid | Temp. | *cis*-Parinaric Acid | *trans*-Parinaric acid |
|---|---|---|---|
| DPPC[a] | 22°C; solid | $5.3 \times 10^5$ | $5 \times 10^6$ |
| PDPC[b] | 22°C; fluid | $9 \times 10^5$ | $1.7 \times 10^6$ |
| DMPC[c] | 10°C; solid | $\sim 10^6$ | $2.6 \times 10^6$ |
| DMPC | 30°C; fluid | $\sim 10^6$ | $9.4 \times 10^5$ |

[a]  Dipalmitoylphosphatidylcholine.
[b]  1-Palmitoyl-2-docosahexaeneoyl-phosphatidylcholine.
[c]  Dimyristoylphosphatidylcholine.

## B. Sample Preparation

### 1. Labeling with Free Fatty Acids

The preparation of samples for parinaric acid experiments depends primarily on the particular form of the probe. The free fatty acids or their methyl esters may be simply added to vesicle or membrane preparations or cell suspensions as concentrated ethanolic solutions. Incorporation into membranes is nearly instantaneous. The partition coefficient for the distribution of parinaric acid between aqueous and membrane phases is defined as

$$K_p = \frac{\text{mol parinaric acid bound/mol lipid}}{\text{mol parinaric free/mol water}}$$

where "free" and "bound" refer to the aqueous and lipid phases, respectively. Centrifugation studies[4] determined that $K_p$ for *cis*- parinaric acid between DPPC vesicles and buffer at 22°C is about $5 \times 10^5$. For bovine retinal rod outer segment disc membranes, absorption measurements have established that $K_p$ for either *cis*- or *trans*-parinaric acid is about $2 \times 10^6$. The partition coefficients for both isomers in solid and fluid phase lipids are given in the following table (Table 1).[8,51,52] These values of the partition coefficients are similar to those observed for other fatty acids. Note that for *trans*-parinaric acid the value of $K_p$ is higher by about a factor of three for solid phase lipids while for *cis*-parinaric acid the partition coefficients are nearly equal. This behavior reflects the preference of the *trans* isomer for solid vs. fluid phase lipids and the relative indifference of the *cis* isomer.

Labeling levels of 0.1 to 1% are usually used in these studies. Labeling at greater than 2% broadens the thermal phase transition of model lipid bilayers.[2]

### 2. Incorporation of Parinaroyl Phospholipids

Phospholipids labeled with parinaroyl chains may be incorporated into lipid vesicles by dissolution of labeled and unlabeled lipids in an organic solvent as the first step in vesicle preparation. In our work, vesicles are prepared by injection of an ethanolic solution (free from water) into stirred warm buffer. Other methods of vesicle preparation may be used, but sonication of lipids containing the parinaroyl chain results in destruction of parineric acid.

The incorporation of phospholipids labeled with a parinaroyl chain into membranes or synthetic lipid vesicles can be accomplished with phospholipid exchange protein.[29,84] It has been reported[53] that parinaroyl-labeled phospholipids incorporate into phosphlipid dispersions and membrane preparations in the presence of 50% ethylene glycol.

Biosynthetic incorporation methods have been described above.[6,36,37,83,85] The particular growth conditions necessary for optimum incorporation need to be worked out for each species.

## C. Analysis of Bilayer Phase Transitions

The variation of the fluorescence intensity of parinaric acid or parinaroyl phospholipids as a function of temperature, either in solution or in bilayers away from the phase transition, is roughly linear when plotted as log intensity vs. $1/T$.[1] The slope of such a plot is essentially independent of the solvent[3] and is the same for bilayers well above and well below the order-disorder phase transition. A much steeper dependence is observed in the region of the phase transition. If this limiting slope of the plot of log $I(1/T)$ is defined as $L(1/T)$, then a plot of log $I(1/T)/L(1/T)$ exhibits constant values at high and low temperature and a sharp change from one value to the other at the phase transition.[1] The midpoints of such curves are in excellent agreement with $T_m$ values obtained by other methods.[1,4] Phase diagrams for two-component lipid mixtures constructed using this method are also in excellent agreement with other methods.[4]

The steady-state polarization of the fluorescence of parinaric acid is essentially constant above and below the thermal phase transition for bilayers formed from single-component saturated acyl chain lipids.[4] This permits ready data analysis in terms of the association of sharp decreases in the polarization at particular temperatures with phase changes of the bilayer. Anisotropy measurements also have the advantage, relative to intensity measurements, that they are independent of the total fluorophore concentration and aqueous/lipid partitioning effects. The interpretation of steady-state anisotropy values in terms of motional freedom or bilayer order is ambiguous even if lifetime data are available for the same system. This is because both the rate and amplitude of the reorientational motion influence the steady-state measurement.

The interpretation of fluorescence lifetime data for parinaric acid is complicated by the nonexponential behavior observed in even simple systems. This behavior has to be understood in more detail for model cases before application to membranes can lead to conclusive results. On the other hand, lifetime information can sometimes be useful in a qualitative way and is necessary input for anisotropy measurements. The interpretation of fluorescence anisotropy decay data is again complicated by the sensitivity of parinaric acid lifetime to environment. The principles necessary for interpretation of complex decays are available, but detailed tests of such models have not yet been made. Again, qualitative interpretation in terms of effects of temperature or membrane components on motional rate and amplitude are possible, but in complex systems quantitative analysis is not yet possible.

## REFERENCES

1. **Sklar, L. A., Hudson, B., and Simoni, R. D.,** Conjugated polyene fatty acids as membrane probes: preliminary characterization, *Proc. Natl. Acad. Sci. U.S.A.*, 72, 1649, 1975.
2. **Sklar, L. A., Hudson, B. S., and Simoni, R. D.,** Conjugated polyene fatty acids as fluorescent membrane probes: model system studies, *J. Supramol. Structure*, 4, 449, 1976.
3. **Sklar, L. A., Hudson, B., Petersen, M., and Diamond, J.,** Conjugated polyene fatty acids as fluorescent probes: spectroscopic characterization, *Biochemistry*, 16, 813, 1977.
4. **Sklar, L. A., Hudson, B., and Simoni, R. D.,** Conjugated polyene fatty acids as fluorescent probes: synthetic phospholipid membrane studies, *Biochemistry*, 16, 189, 1977.
5. **Sklar, L. A., Hudson, B., and Simoni, R. D.,** Conjugated polyene fatty acids as fluorescent probes: binding to bovine serum albumin, *Biochemistry*, 18, 1707, 1979.
6. **Tecoma, E., Sklar, L. A., Simoni, R. D., and Hudson, B.,** Conjugated polyene fatty acids as fluorescent probes: biosynthetic incorporation of parinaric acids by *Escherichia coli* and studies of phase transitions, *Biochemistry*, 16, 829, 1977.

7. **Berde, C. B., Hudson, B., Simoni, R. D., and Sklar, L. A.,** Human serum albumin: spectroscopic studies of binding and proximity relationships for fatty acids and bilirubin, *J. Biol. Chem.*, 254, 391, 1978.
8. **Kimelman, D., Tecoma, E. S., Wolber, P. K., Hudson, B., Wickner, W. T., and Simoni, R. D.,** Protein-lipid interactions: studies of the M13 coat protein in dimyristoylphosphatidylcholine vesicles using parinaric acid, *Biochemistry*, 18, 5874, 1979.
9. **Wolber, P. K. and Hudson, B.,** Fluorescence lifetime and time resolved polarization anisotropy studies of acyl chain order and dynamics in lipid bilayers, *Biochemistry*, 20, 2800, 1981.
10. **Wolber, P. K. and Hudson, B.,** Bilayer acyl chain dynamics and lipid protein interactions: the effect of the M13 bacteriophage coat protein on the decay of the fluorescence anisotropy of parinaric acid, *Biophys. J.*, 37, 253, 1982.
11. **Tsujimoto, N. M. and Koyanagi, H.,** New unsaturated acid in the kernel oil of "akarittom", *Parinarium laurinum* I, *J. Soc. Chem. Ind. Jpn.*, 36, 110B, 1933.
12. **Tsujimoto, N. M. and Koyanagi, H.,** New unsaturated acid in the kernel oil of "akarittom", *Parinarium laurinum* II, *J. Soc. Chem. Ind. Jpn.*, 36, 637B, 1933.
13. **Hopkins, C. Y.,** Fatty acids with conjugated unsaturation, *Top. Lipid Chem.*, 3, 37, 1972.
14. **Fieser, L. F. and Fieser, M.,** *Organic Chemistry*, D. C. Heath, Lexington, Mass., 1944, 383, 385.
15. **Sklar, L. A., Hudson, B., and Simoni, R. D.,** Parinaric acid from *Parinarium glaberrimum*, *Methods in Enzymol.*, 72, 479, 1981.
16. **Schmitz, B. and Egge, H.,** 1-*trans*-Parinaroyl phospholipids: synthesis and fast atom bombardment/electron impact mass spectrometric characterization, *Chem. Phys. Lipids*, 34, 139, 1984.
17. **Farmer, E. H. and Sunderland, E.,** Unsaturated acids of natural oils. II. The highly unsaturated acid of the kernels of *Parinarium laurinum*, *J. Chem. Soc.*, 759, 1935.
18. **Bagby, M. O., Smith, C. R., and Wolff, I. A.,** Stereochemistry of alpha-parinaric acid from *Impatiens edgeworthii* seed oil, *Lipids*, 45, 263, 1966.
19. **Gunstone, F. D. and Subbarao, R.,** New tropical seed oils. I. Conjugated trienoic and tetraenoic acids and their oxo derivatives in seed oils of *Chrysobalanus icaco* and *Parinarium laurinum*, *Chem. Phys. Lipids*, 1, 349, 1967.
20. **Takagi, T.,** Steric configurations of parinaric and punicic acids, *J. Am. Oil Chem. Soc.*, 43, 249, 1966.
21. **Kaufmann, H. P., Baltes, J., and Funke, S.,** Field of fats (LXI) constitution of parinaric acid, *Fette Seifen*, 45, 302, 1938.
22. **Kaufmann, H. P. and Sud, R. K.,** Zur Stereochemie der vierfach Konjugiert ungesattigten Parinarsauren, *Chem. Ber.*, 92, 2797, 1959.
23. **Smith, R. M. and Croft, K. D.,** A 300 MHz nmr study of the conjugated tetraenes alpha- and beta-parinaric acid, *J. Chem. Res. Synop.*, 2, 41, 1981.
24. **Hayashi, T. and Oishi, T.,** A new, general and stereo-selective synthesis of long chain tetraenoic acids exemplified by beta-parinaric acid, *Chem. Lett.*, p. 413, 1985.
25. **Goerger, M. M., and Hudson, B.,** Synthesis of all-transparinaric acid-$d_8$ specifically deuterated at all vinyl positions, *J. Org. Chem.*, in press.
26. **Noda, M., Ohga, K., Nakagawa, Y., and Ichihara, K.,** Biosynthesis of parinaric acid (9,11,13,15-Octadecatetraenoic acid), *Dev. Plant Biol.*, 6, 215, 1980.
27. **Craven, B.,** personal communication, 1985.
28. **Tsai, A., Hudson, B., and Simoni, R. D.,** Preparation of parinaric acid derivatives, *Methods in Enzymol.*, 72, 483, 1981.
29. **Somerharju, P. and Wirtz, K. W. A.,** Semisynthesis and properties of a fluorescent phosphatidylinositol analog containing a *cis*-parinaroyl moiety, *Chem. Phys. Lipids*, 30, 81, 1982.
30. **Cohen, R., Barenholtz, Y., Gatt, S., and Dagan, A.,** Preparation and characterization of well-defined D-erythro sphingomyelins, *Chem. Phys. Lipids*, 35, 371, 1984.
31. **Sklar, L. A., Craig, I. F., and Pownall, H. J.,** Induced circular dichroism of incorporated fluorescent cholesterol esters and polar lipids as a probe of human serum low density lipoprotein structure and melting, *J. Biol. Chem.*, 256, 4286, 1981.
32. **Patel, K. M., Sklar, L. A., Currie, R., and Pownall, H. J., Morrisett, J. D., and Sparrow, J. T.,** Synthesis of saturated, unsaturated, spin-labeled and fluorescent cholesterol esters: acylation of cholesterol using fatty acid anhydride and 4-pyrrolidinopyridine, *Lipids*, 14, 816, 1979.
33. **Haugland, R. P.,** *Molecular Probes: Handbook of Fluorescent Probes and Research Chemicals*, Molecular Probes, Inc., Eugene, Ore., 1985, 48.
34. Sigma Chemical Company, P.O. Box 14508, St. Louis, Mo. 63178.
35. Calbiochem, Behring Diagnostics, P.O. Box 12087, San Diego, Calif. 92112.
36. **Tecoma, E. S. and Wu, D.,** Membrane deenergization by colicin K affects fluorescence of exogenously added but not biosynthetically esterified parinaric acid probes in *Escherichia coli*, *J. Bacteriol.*, 142, 931, 1980.
37. **Rintoul, D. A. and Simoni, R. D.,** Incorporation of a naturally occurring fluorescent fatty acid into lipids of cultured mammalian cells, *J. Biol. Chem.*, 252, 7916, 1977.

38. **Hudson, B. and Kohler, B. E.,** A low-lying weak transition in the polyene alpha, omega-diphenyloctatetraene, *Chem. Phys. Lett.,* 14, 299, 1972.
39. **Hudson, B. and Kohler, B. E.,** Polyene spectroscopy: the lowest energy excited singlet state of diphenyloctatetraene and other linear polyenes, *J. Chem. Phys.,* 59, 4984, 1973.
40. **Hudson, B. and Kohler, B. E.,** Linear polyene electronic structure and spectroscopy, *Annu. Rev. Phys. Chem.,* 25, 437, 1974.
41. **Hudson, B., Kohler, B. E., and Schulten, B. E.,** Linear polyene electronic structure and potential surfaces, in *Excited States,* Lim, E. C., Ed., Academic Press, New York, 1982, 1.
42. **Schroeder, F., Holland, J. F., and Vagelos, P. R.,** Use of beta-parinaric acid, a novel fluorometric probe, to determine characteristic temperatures of membranes and membrane lipids from cultured animal cells, *J. Biol. Chem.,* 251, 6739, 1976.
43. **Johnson, I. D. and Hudson, B.,** unpublished results, 1985.
44. **Parasassi, T., Conti, F., and Gratton, E.,** Study of heterogeneous emission of parinaric acid isomers using multifrequency phase fluorometry, *Biochemistry,* 23, 5660, 1984.
45. **Ruggiero, A. and Hudson, B.,** Critical density fluctuations in lipid bilayers detected by fluorescence lifetime heterogeneity, *Biophys. J.,* in press.
46. **Fraley, R. T., Jameson, D. M., and Kaplan, S.,** The use of the fluorescent probe alpha-parinaric acid to determine the physical state of the intracytoplasmic membranes of the photosynthetic bacterium *Rhodopseudomonas sphaeroides, Biochim. Biophys. Acta,* 511, 52, 1978.
47. **Daikh, D. D. and Hudson, B.,** unpublished, 1983; Hudson, B. S., Dou, X.-M., Nester, T., Harris, D., Johnson, I., and Ruggiero, A., Fluorescence studies of bilayers and proteins: critical behavior and genetic engineering, *Opt. Eng.,* in press.
48. **Morgan, C. G., Hudson, B., and Wolber, P. K.,** The photochemical dimerization of parinaric acid in lipid bilayers, *Proc. Natl. Acad. Sci., U.S.A.,* 77, 26, 1980.
49. **Williamson, H. W., Morgan, C. G., Fuller, S., and Hudson, B.,** Melittin induces fusion of phospholipid vesicles, *Biochim. Biophys. Acta,* 732, 668, 1983.
50. **Lynch, R. L. and Hudson, B.,** unpublished results, 1983.
51. **Sklar, L. A., Miljanich, G. P., and Dratz, A. D.,** Phospholipid lateral phase separation and the partition of *cis*-parinaric acid and *trans*-parinaric acids among aqueous, fluid, and solid lipid phases, *Biochemistry,* 18, 1707, 1979.
52. **Sklar, L. A.,** The partition of *cis*-parinaric acid and *trans*-parinaric acids among aqueous, fluid lipid and solid lipid phases, *Mol. Cell. Biochem.,* 32, 169, 1980.
53. **Pugh, E. L., Kates, M., and Szabo, A. G.,** Studies on fluorescence polarization of 1-acyl-2-*cis*- or *trans*-parinaroyl sn-3-glycerophosphorylcholines in model systems and microsomal membranes, *Chem. Phys. Lipids,* 30, 55, 1982.
54. **Welti, R.,** Partition of parinaroyl phospholipids in mixed head group systems, *Biochemistry,* 21, 5690, 1982.
55. **Welti, R. and Silbert, D. F.,** Partition of parinaroyl phospholipid probes between solid and fluid phosphatidylcholine phases, *Biochemistry,* 21, 5685, 1982.
56. **Hudson, B. S., Ludescher, R. D., Ruggiero, A., Harris D. L., and Johnson, I. D.,** Fluorescence anisotropy decay determinations of rapid reorientational motion: complexities in the interpretation of bilayer acyl-chain and protein tryptophan dynamics, Comments, *Mol. Cell. Biophys.,* 4, 171. 1987.
57. **Ludescher, R. D., Peting, L., Hudson, S., and Hudson, B.,** Time-resolved fluorescence anisotropy for systems with lifetime and dynamic heterogeneity, *Biophys. Chem.,* in press.
58. **Ruggiero, A., Ludescher, R. D., and Hudson, B.,** Multiple environments in lipid bilayers: effects on fluorescence lifetimes and polarization anisotropy decays, *Biophys. J.,* 45, 336a, 1984.
59. **Hudson, B., Harris, D. L., Ludescher, R. D., Ruggiero, A., Cooney-Freed, A., and Cavalier, S.,** Fluorescence probe studies of proteins and membranes, in *Fluorescence in the Biological Sciences,* Taylor, D. L., Waggoner, A. S., Lanni, F., Murphy, R. F., and Birge, R., Eds., Alan R. Liss, New York, 1986, p. 159.
60. **Jahnig, F.,** Critical effects from lipid-protein interaction in membranes. I. Theoretical description, *Biophys. J.,* 36, 329, 1981.
61. **Jahnig, F.,** Critical effects from lipid-protein interaction in membranes. II. Interpretation of experimental results, *Biophys. J.,* 36, 347, 1981.
62. **Jahnig, F.,** The ordered-fluid transition in lipid bilayers, *Mol. Cryst. Liq. Cryst.,* 63, 157, 1981.
63. **Sakanishi, A., Mitaku, S. and Ikegami, A.,** Stabilizing effect of cholesterol on phosphatidylcholine vesicles observed by ultrasonic velocity measurement, *Biochemistry,* 79, 2636, 1979.
64. **Mitaku, S. and Okano, K.,** Ultrasonic measurements of two-component lipid bilayer suspensions, *Biophys. Chem.,* 14, 147, 1981.
65. **Mitaku, S. and Date, T.,** Anomalies of Nanosecond Ultrasonic Relaxation in the Lipid Bilayer Transition, *Biochim. Biophys. Acta,* 688, 411, 1982.

66. **Mitaku, S., Jippo, T., and Kataoka, R.,** Thermodynamic properties of the lipid bilayer transition: pseudocritical phenomena, *Biophys. J.,* 42, 137, 1983.
67. **Weis, R. W. and McConnell, H. M.,** Two-dimensional chiral crystals of phospholipid, *Nature (London),* 310, 47, 1984.
68. **Weis, R. W. and McConnell, H. M.,** Cholesterol stabilizes the liquid crystalline interface in phospholipid monolayers, *Biophys. J.,* 47, 44a, 1985.
69. **Fischer, A., Losche, M., Mohwald, H., and Sackmann, E.,** On the nature of the lipid monolayer phase transition, *J. Phys. (Paris) Lett.,* 45L, 785, 1984.
70. **Bloch, K. R.,** Sterol Structure and Membrane Function, *CRC Crit. Rev. Biochem.,* 14, 47, 1983.
71. **Wolber, P. K. and Hudson, B.,** An analytic solution to the Förster energy transfer problem in two dimensions, *Biophys. J.,* 28, 197, 1979.
72. **Hsia, J. C. and Kwan, N. H.,** Human serum albumin: binding specificity and allosteric effect of parinarate and stearate, *J. Biol. Chem.,* 256, 2242, 1981.
73. **Berde, C. B., Nagai, M., and Deutsch, H. F.,** Human alpha-fetoprotein: fluorescence studies on binding and proximity relationships for fatty acids and bilirubin, *J. Biol. Chem.,* 254, 12609, 1979.
74. **Shaklai, N., Garlick, R. L., and Bunn, H. F.,** Nonenzymatic glycosylation of human serum albumin alters its conformation and function, *J. Biol. Chem.,* 259, 3812, 1984.
75. **Ludescher, R. D. and Hudson, B.,** unpublished.
76. **Berkhout, T. A., Visser, A. J. W. G., and Wirtz, K. W. A.,** Static and time-resolved fluorescence studies of fluorescent phosphatidylcholine bound to phosphatidylcholine transfer protein of bovine liver, *Biochemistry,* 23, 1505, 1984.
77. **Westerma, J., Wirtz, K. W. A., Berkhout, T., van Deenen, L. L. M., Radhakrishnan, R., and Khorana, H. G.,** Identification of the lipid-binding site of phosphatidylcholine-transfer protein with phosphatidylcholine analogs containing photoreactivable carbene precursors, *Eur. J. Biochem.,* 132, 441, 1983.
78. **Kato, A. and Nakai, S.,** Hydrophobicity determined by a fluorescence probe method and its correlation with surface properties of proteins, *Biochim. Biophys. Acta,* 624, 13, 1980.
79. **Bordejdo, J.,** Mapping of hydrophobic sites on the surface of myosin and fragments, *Biochemistry,* 22, 1182, 1983.
80. **Ludescher, R. D., Georgovich, G., and Hudson, B.,** unpublished results.
81. **Berde, C. D., Kerner, J. R., and Johnson, J. D.,** Use of the conjugated polyene fatty acid parinaric acid in assaying fatty acids in serum of plasma, *Clin. Chem. (Winston-Salem, N.C.),* 26, 1173, 1980.
82. **Wolf, C., Sagaert, L., and Bereziat, G.,** A sensitive assay of phospholipase using the fluorescent probe 2-parinaroyl-lecithin, *Biochem. Biophys. Res. Commun.,* 99, 275, 1981.
83. **Wolf, C., Sagaert, L., Bereziat, G., and Polonovski, J.,** The deacylation of rat platelet phospholipids during thrombin-induced aggregation studied by a fluorescence method, *FEBS Lett.,* 135, 285, 1981.
84. **Somerharju, P., Brockerhoff, H., and Wirtz, K. W. A.,** A new fluorometric method to measure protein-catalyzed phospholipid transfer using 1-acyl-2-parinaroylphosphatidylcholine, *Biochim. Biophys. Acta,* 649, 521, 1981.
85. **Harris, W. E. and Stahl, W. L.,** Incorporation of *cis*-parinaric acid, a fluorescent fatty acid, into synaptosomal phospholipids by an acyl-CoA acyltransferase, *Biochim. Biophys. Acta,* 736, 79, 1983.
86. **Miljanich, G. J., Sklar, L. A., White, D. L., and Dratz, E. A.,** Disaturated and dipolyunsaturated phospholipids in the bovine retinal rod outer disk membrane, *Biochim. Biophys. Acta,* 522, 294, 1979.
87. **Sklar, L. A., Miljanich, G. J., and Dratz, E. A.,** A comparison of the effects of calcium on the structure of bovine retinal rod outer segment membranes, phospholipids, and bovine brain phosphatidylserine, *J. Biol. Chem.,* 254, 9592, 1979.
88. **Sklar, L. A., Miljanich, G. P., Burstein, S. L., and Dratz, E. A.,** Thermal lateral phase separations in bovine retinal rod outer segment membranes and phospholipids as evidenced by parinaric acid fluorescence polarization and energy transfer, *J. Biol. Chem.,* 254, 9583, 1979.
89. **Sklar, L. A. and Dratz, E. A.,** Analysis of membrane bilayer asymmetry using parinaric acid fluorescent probes, *FEBS Lett.,* 118, 308, 1980.
90. **Sklar, L. A. and Dratz, E. A.,** Analysis of rod outer segment disk membrane phospholipid organization using parinaric acid fluorescent probes, *Methods Enzymol.,* 81, 685, 1982.
91. **Alfsen, A., de Paillerets, C., Prasad, K., Nandi, P. K., Lippoldt, R. E., and Edelhoch, H.,** Organization and dynamics of lipids in bovine brain coated and uncoated vesicles, *Eur. Biophys. J.,* 11, 129, 1984.

Chapter 4

# TERTIARY STRUCTURE OF MEMBRANE PROTEINS DETERMINED BY FLUORESCENCE RESONANCE ENERGY TRANSFER

**Alan Kleinfeld**

## TABLE OF CONTENTS

# I. INTRODUCTION

The goal of structural biology is the elucidation of cell function at the molecular level. Membrane protein structure represents a particularly challenging, and as yet unsolved, aspect of the general problem. Indeed, until recently, there were very few primary sequences known for membrane proteins. During the past few years, advances in nucleic acid sequencing techniques have yielded the primary structure of a number of important membrane proteins.[7,12,34,42,48,50,61]

Information about the secondary structure of membrane proteins has been derived from circular dichroism and IR spectroscopies. Most studies suggest that the transmembrane segments have appreciable $\alpha$ helical structures with helices oriented perpendicular to the membrane surface.[41,44,51,58] On the other hand, secondary structure predictions, low resolution diffraction, and some spectroscopy studies indicate that it is also possible for $\beta$ sheet to occur within the lipid bilayer, also in the perpendicular configuration.[27,38,43,59]

Only one protein (actually a complex of four proteins), the photoreaction center from the purple membrane of *Rhodopseudomonas viridis,* has been solved to high resolution (3 Å) by X-ray diffraction.[11] Although the crystal structure is known, there is no direct evidence for the orientation of the protein with respect to the membrane. Nevertheless, a quite plausible configuration has been suggested and this image shows the transmembrane section to be fairly helical in nature (see Figure 1). The helices, while reasonably parallel to the acyl chains over most of their length, bow outward at the surface of the membrane. It remains to be seen whether the five hydrophobic segments identified in the primary sequence of *R. viridis* are in fact part of the helical structure of the putative transmembrane region.

Although the extramembranous portion of several other membrane-associated proteins have been solved to high resolution,[47,71] structural information for proteins which have a large fraction of their mass intimately associated with the lipid bilayer is at too low a level of resolution to identify individual amino acids.[4,12a,23,29,43,59] The seminal study of bacteriorhodopsin (bR) by Henderson and Unwin,[23] for example, which has done much to shape our views of membrane proteins, was obtained at a resolution of 7 Å in the plane and 14 Å perpendicular to the plane of the membrane, whereas 2 to 3 Å is necessary to resolve amino acids. (More recently this resolution has been improved to 6.5 and 12 Å, respectively.)[39] At this resolution one can infer a general outline of the protein shape and secondary structure. The model of Figure 2 indicates that the protein is composed of 7 rods, arranged roughly as a cylinder of radius 15 Å. Approximately 70% of the cylinder, whose length is about 40 Å, is contained within the membrane. Analysis of the electron density, circular dichroism, and primary sequence suggest that the 7 rods are $\alpha$ helices and that the segments connecting the loops are extramembranous and nonhelical.[16,23,28,41,52,59] Plausible models,

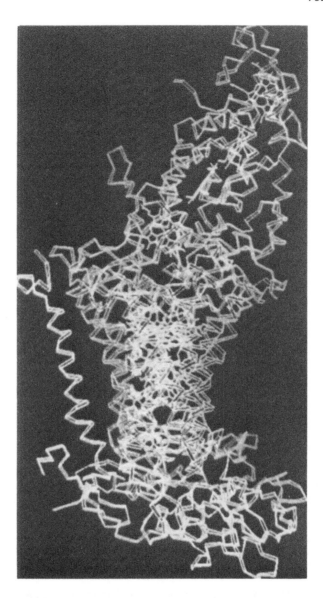

FIGURE 1. Structure of the photosynthetic reaction center of *Rhodopseudomonas viridis* obtained by X-ray crystallography at 3 Å resolution. This view shows the protein complex in the orientation it is likely to have in the membrane. The central helical region is the putative transmembrane portion. (Reprinted by permission from *Nature*, Vol. 318, p. 618. Copyright (c) 1985 Macmillan Journals Limited.)

based upon the α helical configuration, have been proposed for the spatial arrangement of the primary sequence through the membrane.[1,15,25,52,64] To extend this kind of modeling, methods are needed to localize discrete regions of the protein, and thereby define spatial distribution of specific amino acids.

A number of experimental methods have been used to determine the location of particular portions of membrane-bound proteins and to relate these locations to the membrane architecture. These approaches complement diffraction methods that are as yet unable to distinguish individual residues. They also offer the possibility of monitoring at least part of the protein structure *in situ* and possibly in different physiological states. Chemical labeling and

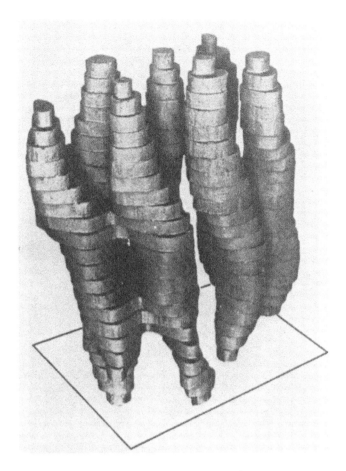

FIGURE 2.   Structure of bacteriorhodopsin at 7 Å resolution. In this study, the naturally occurring two-dimensional crystals of the protein within the membrane were investigated. Therefore, the orientation of the protein (as shown) is known to be perpendicular to the surface of the membrane. This view shows the protein to be a single peptide composed of α helical segments. (Reprinted by permission from *Nature*, Vol. 257, p. 23. Copyright (c) 1975.)

proteolysis with selected proteases have been used to determine the number of peptide loops across the membrane, as well as the location and general topology of the extramembranous segments (band 3,[6] bacteriorhodopsin,[15] and histocompatibility antigen[8]). Regions within the bilayer have been mapped out using ferritin labeling,[23a] photoactivatible reagents attached to fatty acids,[55,67] fluorescence quenching of tryptophan with spin-labeled fatty acids,[40,40a] and with brominated phospholipids.[45]

The primary drawback to obtaining high-resolution information by these methods is the difficulty in determining the spatial dependencies of the interactions. Resonance energy transfer (RET), in contrast, is well-understood and can, therefore, be used to localize membrane protein regions to high resolution. RET methods applied to the study of membrane protein structure are generally of two types. In the first, the protein is labeled with covalent probes and therefore transfer is measured between discrete donor-acceptor pairs. In the second, RET is measured between an intrinsic donor or acceptor and a complementary chromophore which forms a noncovalent association with the membrane, usually a lipid probe. It is this latter method which is the primary subject of this review. A brief discussion of the first approach will be given with emphasis on the theoretical description of single donor-acceptor transfer.

## II. RET BETWEEN COVALENT DONOR-ACCEPTOR LABELS

### A. Single Donor-Acceptor Pair Theory

RET for single donor-acceptor pairs has been described in detail elsewhere;[20,66] here the essential features of the theory are outlined with emphasis on those which are relevant to the determination of membrane protein structure. According to Förster's theory, energy is transferred from an excited donor molecule to an acceptor molecule in its ground state by means of an interaction between the electric dipoles of the two molecules. In the limit that this interaction is too weak to alter the structure of either molecule, the transfer rate can be expressed as

$$k_{DA}^T = 1/\tau_D (R_0/R_{DA})^6 \tag{1}$$

in which $R_{DA}$ is the donor-acceptor separation. The critical transfer radius $R_0$ is given by:

$$R_0 = (8.8 \times 10^{-25} \kappa^2 Q_D J/n^4)^{1/6} \text{ cm} \tag{2}$$

in which $Q_D$ is the quantum yield of the donor in the absence of acceptor, and n is the index of refraction of the medium between donor and acceptor. The overlap integral J is defined as:

$$J = \int \epsilon_A(\lambda) \, f_D(\lambda) \, \lambda^4 d\lambda \tag{3}$$

in which $\epsilon_A(\lambda)$ is the acceptor molar extinction coefficient density at wavelength $\lambda$, and $f_D(\lambda)$ is the corrected fluorescence intensity of the donor, normalized so that the integral of $f_D(\lambda)$ over its emission band equals $Q_D$. The orientation factor $\kappa^2$ between a particular donor and a particular acceptor is given by:

$$\kappa^2 = (\sin\xi_D \sin\xi_A \cos\gamma - 2\cos\xi_D \cos\xi_A)^2 \tag{4}$$

in which $\xi_D$ and $\xi_A$ are the angles between the $\vec{R}_{DA}$ and the donor and acceptor dipole directions, respectively, and $\gamma$ is the azimuthal angle between donor and acceptor dipoles.

For a single donor-acceptor pair the rate equation describing the transfer from the donor D to an acceptor A is as follows:

$$dP_D(t)/dt = -(k_D + k_{DA}^T)P_D(t) \tag{5}$$

In this Equation, $P_D(t)$ dt is the probability that the donor is excited at a time between t and $t + dt$ following a pulse of exciting light, $k_D$ is the donor decay rate in the absence of energy transfer and $k_{DA}^T$ is the resonance energy transfer rate from donor to acceptor.

Since the solution to this rate equation is the simple exponential:

$$P_D(t) = P_D(0)e^{-t(k_D + k_{DA}^T)} \tag{6}$$

$R_{DA}$ may be determined by measuring the rate of decay (rate = 1/lifetime) of the donor, in the presence $(k_D + k_{DA}^T)$, and absence $(k_D)$, of acceptor. Distance information may also be obtained by measuring the time-integrated fluorescence intensity of the donor:

$$I_D = \int_0^\infty P_D(0)e^{-ik}dt \qquad (7)$$

The degree of quenching of donor fluorescence intensity (or quantum yield) is given by:

$$k_D/(k_D + k_{DA}^T) \qquad (8)$$

The transfer efficiency, T, is the ratio of the RET rate to the total donor decay rate, and can be related to $R_{DA}$ by:

$$T = k_{DA}/(k_D + k_{DA}^T) = 1 - I_D/I_D^\circ = 1/(1 + R_{DA}/R_0)^6) \qquad (9)$$

Similar information may be obtained by measuring the sensitized emission of the acceptor when the acceptor is fluorescent.

The donor-acceptor separation can be determined by measuring T and expressing Equation 9 as $R_{DA} = R_0(1/T - 1)^{1/6}$. In order that $R_{DA}$ correctly reflects the donor-acceptor separation, it is necessary that the observed quenching of the donor fluorescence be due entirely to the Förster mechanism. This can be verified in *single* donor configurations when the acceptor is also fluorescent. In such cases, the transfer efficiency determined from donor quenching and the sensitized emission of the acceptor are equal.

The major source of uncertainty in the determination of $R_{DA}$ for single donor-acceptor pairs is due to the generally unknown orientation of the donor and acceptor dipoles, the orientation factor $\kappa^2$. In the extreme case in which $\kappa^2$ can vary between 0 and 4 the determination of $R_{DA}$ is completely uncertain. In practice, however, a number of factors (see below) ameliorate these extrema. Moreover, limits can be placed on the uncertainty in $\kappa^2$ by measurements of the donor, acceptor, and transfer depolarization factors.[10,35]

## B. RET Studies with Covalent Probes

### 1. The Anion Exchange Protein

The anion exchange protein from human erythrocytes, band 3, is 95 kdaltons of which 40 kdaltons is located on the cytoplasm side of the red cell membrane and contains three sulfhydryl groups.[34] Anion exchange can be specifically inhibited by stilbenesulfonates added to the outside of the membrane. The location of the anion exchange inhibitor site has been determined by RET methods.[54] In their study, Rao et al. labeled the cytoplasmic SH and the stilbenesulfonate binding sites with donor-acceptor pairs and measured RET to determine the separation between these two groups. Their results indicate that the SH and stilbenesulfonate sites are separated by 34 to 42 Å. The major source of uncertainty in this determination is the uncertainty in the orientation factor $\kappa^2$, due, primarily, to the immobility of the stilbenesulfonates. In spite of this large uncertainty these results indicate that the inhibitor site must be buried within the membrane since the cytoplasmic SH groups are outside the membrane and the membrane thickness is about 45 Å.

### 2. Rhodopsin

Neori and Montal[49] used covalent fluorescent labels to obtain information about the structure of rhodopsin from rod outer segments in the bleached and unbleached state. Labeling was accomplished with one of four fluorescent probes, two sulfhydryl and two free amino reagents. The labeled rhodopsin was purified and measurements were carried out on detergent micelles or in proteoliposomes formed by detergent dialysis. For each label the native 11-*cis*-retinal serves as an energy transfer acceptor with $R_0(2/3)$ values between 31 and 42 Å. Upon bleaching, the 11-*cis* retinal is converted to all-*trans* and is transferred from the protein interior to a random location on the surface of the lipid bilayer. For sufficiently diluted

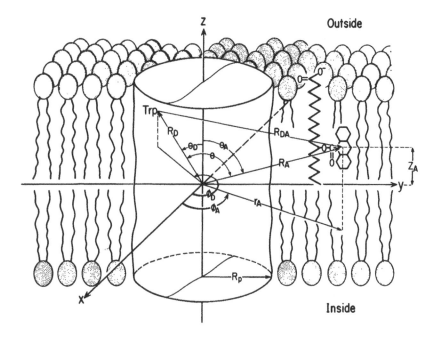

FIGURE 3.   Membrane protein model and the Trp-AO geometry. The origin of the coordinate system is the intersection of the protein Z axis and the midplane (xy) of the lipid bilayer. $R_D$, $\theta_D$, $\phi_D$, and $R_A$, $\theta_A$, $\phi_A$ are the spherical coordinates of the donor (Trp) and acceptor (AO), respectively. $Z_A$ is the z coordinate of a particular acceptor, and $r_A$ is the projection of $R_A$ on the xy plane. (From Kleinfeld, A. M., *Biochemistry*, 24, 1874, 1985. With permission.)

protein densities, the surface retinal will then be far enough away, on average, so that RET is negligible. RET, therefore, was measured by comparing donor intensities in the bleached and unbleached samples. In each case the transfer efficiencies yielded retinal to probe distances of greater than 30 Å, consistent with an elongated cylinder model for rhodopsin. In addition, Neori and Montal also found that the accessibility of the amino reagents to fluorescence quenchers was appreciably reduced by bleaching, suggesting that bleaching induces a major conformational change in certain, probably surface, regions of the protein.

## III. RET BETWEEN PROTEIN AND LIPID PROBES

### A. Introduction

The general problem of RET between membrane proteins and lipid probes is illustrated in Figure 3 for RET from the protein of Trp. Several kinds of lipid probes have been used to study RET in membranes. The discussion presented here, although general, will tend to focus on the properties of n-(9-anthroyloxy) fatty acid (n-AO) probes. These probes insert spontaneously into the membrane with their carboxyl group anchored at the lipid water interface and with the AO moiety buried a distance $Z_A$ from the midplane of the bilayer. RET occurs between Trp and the ensemble of AO probes randomly distributed in the plane of the membrane. The critical transfer radius is about 22 Å and, therefore, is ideal for probing structures with dimensions between 10 and 40 Å. Moreover, since the AO moiety can be positioned at a number of locations along the acyl chain, it is generally possible to determine two Trp coordinates and the protein radius.[30,31]

### B. Multidonor-Multiacceptor Theory

We will now discuss the theory of RET which is appropriate to the problem of a protein containing multiple donors and an array of acceptors randomly distributed within the plane(s) of the membrane. The "Master" or general form of the rate equation appropriate to this problem is a set of $N_D$ coupled differential equations represented by:

$$\frac{dP_i(t)}{dt} = -P_i(t)/\tau_i - \sum_{k=i}^{N_A} k_{ik}^T P_i(t) - \sum_{j \neq i}^{N_D-1} k_{ij}^T P_i(t) + \sum_{j=i}^{N_D-1} k_{ji}^T P_j(t) \tag{10}$$

In this equation, the index i refers to the ith donor, j the jth donor, and k the kth acceptor. The first term, just as in Equation 5, represents the spontaneous decay of the ith donor in the absence of other donors or acceptors. The second term represents the transfer from the ith donor to the ensemble of $N_A$ acceptors and is a generalization of the second term in Equation 5 for a single acceptor. In most instances, back transfer from a heteroacceptor is negligible and is ignored here. The third and fourth terms, respectively, represent transfer from the ith donor to the other $N_D - 1$ donors and from the other $N_D - 1$ donors back to the ith donor. (The molecules involved in homotransfer are referred to as "donors" since they serve as donors in the heterocomponent of the transfer.) The transfer rates $k_{ik}^T$ and $k_{ij}^T$ are given by Equation 1, with $R_0$ and $R_{ij}$ corresponding to the particular donor-acceptor or donor-donor pairs. It is the interaction between donors which couples this set of equations.

### C. Single Donor-Multiacceptor Theory

As will be demonstrated below, in the case of the Trp-AO system, the effects of Trp-Trp coupling are negligible. The third and fourth terms in Equation 11 can, therefore, be neglected and one obtains:

$$\frac{dP_i(t)}{dt} = -P_i(t)/\tau_i - \sum_{k=i}^{N_A} k_{ik} P_i(t) \tag{11}$$

The solution to these equations, which describe the rate of decay from a set of uncoupled donors, has been discussed previously for an axially symmetric geometry.[14,21,35,60,72] In the case of a multidonor protein in which the donors are not constrained to be axially symmetric, the geometry of Figure 3 is appropriate and the solution for this problem which has been discussed in Kleinfeld[30] is given by:

$$P_i(t) = P_i(0)\exp[-t(1/\tau_i + \sum_{k=1}^{N_A} k_{ik}^T)] \tag{12}$$

For the geometry of Figure 3, the donor-acceptor separation which appears in $k_{ik}^T$ is given by:

$$R_{ik}^2 = r_k^2 + Z_k^2 + R_i^2 - 2R_i\{Z_k\cos\theta_i + r_k\sin\theta_i\cos\omega\} \tag{13}$$

in which $R_i$ is the radial and $\theta_i$ and $\phi_i$ are the angular coordinates of a particular tryptophan residue, and $\omega = \phi_k - \phi_i$.

$P_i(t)$ (Equation 12) is the probability density that donor i will be excited at time t, when the $N_A$ acceptors are in a particular configuration. The ensemble average of Equation 12 over all possible acceptor configurations is given by:

$$<P_i(t)> = \int P_i(t)W(R_{ik})dR_{ik} \tag{14}$$

in which the integration is over all possible acceptor positions and $W(R_{ik})dR_{ik}$ is the probability that there is an acceptor between $R_{ik}$ and $R_{ik} + dR_{ik}$. Using Equation 14 for $R_{ik}$ it follows that:

$$W(R_{ik})dR_{ik} = (r_k - r_i \sin\theta_i \cos\omega)/A_T \qquad (15)$$

in which $A_T$ is the total area available to the acceptors. The ensemble average can therefore be expressed (see, e.g., Fung and Stryer)[21] as:

$$\langle P_i(t) \rangle = \exp(-t/\tau_i)\exp[-\sigma L(t)] \qquad (16)$$

in which:

$$L(t) = \int_0^{2\pi} \int_{Rm}^{\infty} [1 - \exp(-t/\tau_i(R_0/R_{ik})^6)](r_k - R_i \sin\theta_i \cos\omega)dr_j d\phi_j \qquad (17)$$

in which $\sigma$ is the two-dimensional number density of acceptor molecules, and Rm is the radial integration parameter (at the depth $Z_k$). The radial integration parameter is equal to the protein radius (Rp) plus the acceptor radius. Equations 12 to 17 represent the solution to the problem of RET between a single donor located at coordinates ($R_D$), and an ensemble of acceptors randomly distributed in the plane of the membrane.

The transfer efficiency can be determined from the steady-state intensities, $I_D$ and $I_D^\circ$, measured in the presence and absence of an acceptor, respectively. In terms of these intensities, T is given by:

$$T_i = 1 - I_i/I_i^\circ \qquad (18)$$

The intensities are obtained, as in Equation 7, by integrating Equation 16 over time.

## D. Uncoupled Multidonor-Multiacceptor Theory
### 1. The Time Domain

To extend the theory to multiple, albeit uncoupled donors, the single donor solution is averaged over all independent donors (e.g., the $N_T$ Trp per protein). The decay of a set of $N_T$ independent donors, therefore, is given by:

$$P(t) = \sum_{i=1}^{N_T} f_i P_i(t) \qquad (19)$$

in which Equation 16 is used for $P_i(t)$, $f_i$ is the intial relative intensity of the ith tryptophan, and the sum is over all tryptophan within the protein. The decay of a set of tryptophan donors, even in the absence of an ensemble of acceptors, may be quite complex (note that for multiacceptor RET, Equation 16, even a single donor will not decay with a single rate). Not only will the individual tryptophan have different lifetimes reflecting their different locations within the protein, but a single donor itself may decay nonexponentially.[56] For these reasons, the time domain has not been used in the study of multitryptophan RET. Future developments may warrant more careful consideration of the time domain. Distributions that are equivalent as far as transfer efficiencies are concerned may be resolved by following the alteration in decay heterogeneity due to RET (see, e.g., Rao et al.[54]).

### 2. Transfer Efficiencies for Multiple Donors
### a. Donor Quenching

The average transfer efficiency determined from donor quenching can be shown[30,62] to be a quantum-yield weighted average of the individual $T_i$:

$$T = \sum_{i=1}^{N_T} \alpha_i T_i \qquad (20)$$

in which:

$$\alpha_i = Q_i / \sum_{i=1}^{N_T} Q_i \qquad (21)$$

is the fractional quantum yield of the ith Trp.

**b. Sensitized Emission**

The observed acceptor fluorescence intensity (I) following excitation at the donor wavelength is the sum of four separate contributions:[17a,31]

$$I = I_T + I_D + I_B + I_R \qquad (22)$$

in which $I_T$ is due to nonradiative transfer, $I_D$ to direct emission, $I_B$ to scattering and Trp background, and $I_R$ to radiative migration. This latter contribution can be eliminated by appropriate choice of cuvette size and donor-acceptor concentrations.[31] The intensity due to nonradiative transfer can be expressed as:

$$I_T = Q_A \sum_{i=1}^{N_T} [D^*]_i T_i \qquad (23)$$

in which $Q_A$ is the acceptor quantum yield, $[D^*]_i$ the concentration of excited donors with properties (quantum yield, location, orientation factor) such that, it will undergo RET with transfer efficiency $T_i$. The $[D^*]_i$ values are

$$[D^*]_i = B\epsilon_D[D]/N_T \qquad (24)$$

in which B is a normalization constant related to the lamp intensity and the geometry of the fluorometer (see Appendix) and [D] is the concentration of ground-state donors (all $N_T$ donors are assumed to be identical in the ground state). Thus $I_T$ is

$$I_T = BQ_A\epsilon_D([D]/N_T) \sum_{i=1}^{N_T} T_i \qquad (25)$$

Furthermore, since $I_D$ and $I_B$ can be determined experimentally, the observed fluorescence intensity can be related to that due to nonradiative transfer as $I_{obs} - I_B - I_D = I_T$, it follows that the average transfer efficiency determined from nonradiative transfer is

$$<T> = \sum_{i=1}^{N_T} T_i/N_T = I_T/(BQ_A\epsilon_D * [D]) \qquad (26)$$

**c. Comparison of Trp Quenching and Sensitized Emission**

An important feature of multidonor transfer is that the "average" T determined from donor quenching (DQ) is not, in general, equal to the value determined from sensitized emission (SE) (Equations 20 and 26). This effect is an intrinsic property of RET in multidonor configurations. These expressions predict that only when all $Q_i$ are equal will $<T^{DQ}> = <T^{SE}>$. In general, $<T^{DQ}>$ is greater than $<T^{SE}>$, since the intensity of sensitized emission is proportional to, and Trp quenching is independent of, the Trp fluorescence intensity. As an example, consider two Trp, one having $Q = O$ and the other transferring with an efficiency $= T_1$. In this case $<T^{DQ}> = T_1$, while $<T^{SE}> = T_1/2$. Thus, a measure of the quantum yield heterogeneity can be obtained by comparing $<T^{DQ}>$ and $<T^{SE}>$.

### 3. The Orientation Factor

As mentioned above, there are a number of factors which ameliorate the uncertainty due to the orientation factor in the Trp-AO system (see also Kleinfeld).[30]

1.  In general, the lack of donor-acceptor correlation leads to a considerable reduction in the $\kappa^2$ sensitivity of the ensemble-averaged energy transfer expressions.
2.  Fluorescence anisotropy measurements of the AO probes in membranes indicate relatively little motional restriction.[29,36,69]
3.  Mixed absorption dipoles of the acceptor in the region overlapped by tryptophan emission ($\sim$300nm to 380nm) also yield more isotropic $\kappa^2$ values.[22]

In the case of the AO acceptors the absorption band is a mixture of orthogonal dipoles.[46] Taken together, the mixed polarization absorption bands and the lack of appreciable rotational constraints, suggest that the acceptor is well-described by an unrestricted isotropic rotor.

Donor orientational constraints, especially in the case of multitryptophan, are more difficult to assess. The effect of donor orientational constraint in the Trp-AO system has been evaluated for the geometry of Figure 3 (Kleinfeld,[30] and Appendix to this paper). These calculations demonstrate that the maximum uncertainty (a completely immobilized tryptophan) in T is less than 25% about the value obtained for isotropic donor-acceptor pairs ($\kappa^2 = 2/3$). For tryptophan with sufficient motional freedom to undergo diffusive motion within a cone of 30° half width, the uncertainty in the efficiency reduces to 10%, about the same order as the experimental uncertainties.

Most experimental studies of Trp mobility have been performed on water-soluble proteins where most results are consistent with a considerable degree of orientational mobility. This occurs even in the case when the hydrophobic Trp residue is buried in the interior of the protein.[26,37] In the case of membrane proteins the hydrophobic residues such as Trp may be oriented outside the core so that contact is made with the hydrophobic region of the lipid acyl chains, in the sense of an inside-out protein as suggested by Engelman and Zacci.[17] Indeed, measurements of Trp mobility in membrane proteins are generally consistent with considerable rotational freedom (cone angles > 30°).[5,18,29]

### 4. Trp-Trp Coupling

As was discussed above, the rate equation appropriate in the general case of a multiTrp protein is the coupled set represented by Equation 10. The solution to Equation 10 depends not only on the acceptor parameters (as is the case for a single donor), but also on the donor coordinates. Thus, if Trp-Trp coupling were significant the determination of the Trp spatial distribution would require an iterative procedure. Moreover, since Equation 10 can only be solved numerically it is important to *estimate* the effect of Trp-Trp coupling on the Trp-AO RET efficiencies in order to determine if Equation 10 must be used in the analysis instead of the much simpler Equation 11.

As we shall see, the effect of Trp-Trp coupling is quite small. This is a direct consequence of the difference in $R_0$ for Trp to AO transfer as compared to Trp-Trp transfer. $R_0$ values for $Q_{Trp} = 0.2$ are about 7 Å for Trp-Trp[13] and 22 Å for Trp-AO.[30] Thus the effect of Trp-Trp coupling will be maximized when the Trp separation is about 7 Å and the RET rate is significantly greater for one Trp than any of the others. Consider for simplicity the two-Trp model illustrated in Figure 4. The transfer rate of the Trp labeled A is greater than the one

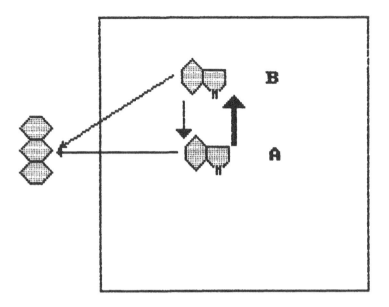

FIGURE 4.   Model for the effect of Trp-Trp coupling on Trp-AO RET. The coordinates of Trp A and B are (10,0) and (10,7), respectively. Calculations were performed with a protein of radius 20 Å and the AO acceptor is at $Z_A = 0$.

at B, either because of its location and/or because of its quantum yield. At first sight, it might seem that severe quenching of $Trp_A$ by AO could, through A to B coupling, serve as a sink (an RET acceptor) for $Trp_B$, or other Trp in the protein. Severe quenching of $Trp_A$ will not, however, increase the rate of energy flow from B to A since the $Trp_A$ ground state (absorbance) is not modified by the interaction with AO. Quenching of $Trp_A$ will reduce the *return* transfer from A to B, and thereby reduce the quantum yield of $Trp_B$. However, since the quantum yield of $Trp_B$ is reduced, this *indirect* quenching of $Trp_B$ by AO is compensated for by a reduction in the *direct* quenching of $Trp_B$ by AO.

To gauge the effects of this cancellation we estimate, with and without Trp coupling, the T values between Trp and AO in the geometry of Figure 4. We choose a quantum yield for $Trp_A$, when there is no coupling, of 0.6 and for $Trp_B$ a value of 0.16 (we have chosen to illustrate the problem using $Q_A > Q_B$ since the effects are entirely negligible for $Q_A \leq Q_B$). Efficiencies of transfer to AO calculated using Equation 18, with these quantum yields, are 0.65 for $Trp_A$ and 0.37 for $Trp_B$. When the coupling is turned on, transfer from A to B decreases the quantum yield of A to 0.5 and increases B to 0.2, according to,

$$Q_{coupling\ on} = Q_{coupling\ off} [1 - (T_{A->B} - T_{B->A})] \qquad (27)$$

Their respective Trp→AO efficiencies, with coupling on, are 0.62 and 0.41. Thus, the result of coupling is to reduce $T_{A \to AO}$ by 3% and to increase $T_{B \to AO}$ by 8%. The average values (Equation 20 change by less than 4%. We conclude, therefore, that the influence of Trp-Trp coupling can be neglected in the Trp-AO problem.

*5. Quantum Yield Heterogeneity*

Quantum yields of Trp in membrane proteins exhibit considerable heterogeneity; observed values have ranged from 0 to greater than 0.5.[5,18,31] Trp with a zero quantum yield is clearly invisible, insofar as RET is concerned. It is, therefore, important to at least be able to place limits on the range of Trp quantum yields in a particular protein. Comparison of $<T^{DQ}>$ and $<T^{SE}>$, as discussed above, allows constraints to be placed on the distribution of quantum yields {Q}, within a protein. In addition, the *magnitude* of the observed T values together with the measured *average* quantum yield (<Q>) of the Trp of the protein imposes further constraints on the possible distribution of quantum yields (designated as {Q}). For example, if {Q} is heterogeneous, $T^{DQ}$ values calculated assuming all $Q_i$ are equal will be too small to yield satisfactory agreement with the measured values. To obtain agreement requires (since <Q> is fixed) some quantum yields to be increased at the expense of others, thereby achieving heterogeneity. This approach was used in the studies of cytochrome $b_5$[18,31] and gramicidin.[5]

Experimental verification of the constraints on {Q} can also be obtained using chemical or genetic modifications (site-directed mutagenesis) of the protein. *N*-bromosuccinimide (NBS) is a Trp-specific reagent which has been used successfully to specifically modify Trp residues in water-soluble proteins[19,24,53,63] and more recently in membrane proteins within the lipid bilayer.[5,29] In the gramicidin studies, e.g., NBS treatment clearly showed that two of the four Trp in each monomer were nonfluorescent, in excellent agreement with the $T^{DQ}$ and $T^{SE}$ difference, and the restrictions imposed by the absolute magnitude of the T values.[5] In addition to providing information about {Q}, modification of the protein can help determine the Trp spatial distribution, by controlling the number of fluorescent Trp. In principle, the number of residues can be reduced to one and, therefore, the spatial resolution as well as the proper assignment of Q, and other spectroscopic properties can be made with the same resolution as for a single Trp.

### 6. AO Probe Characterization
#### a. Bilayer Depth and General Characteristics

In order to derive accurate spatial information from the Trp-AO energy transfer measurements it is necessary to fully understand how the probe is distributed and what effect it has on the membrane. Information about the depth of the probes in the bilayer (summarized in Kleinfeld)[30] indicates that the position of the AO moiety is located to within ± 2 Å of the location expected for an extended fatty acid acyl chain. This result, as we argued in Kleinfeld and Lukacovic,[31] is strongly supported by our finding that the location of the fluorescent Trp in cytochrome $b_5$ is in excellent agreement with earlier studies.

Since the linear dimensions of the anthracene ring system is nearly 9 Å, it may not be obvious that spatial resolution of 2 Å can be achieved. The donor-acceptor interaction can, however, be represented as a multipole-multipole expansion which is a function of the $R_{DA}$ (the separation between the charge centers [the average electronic position] and the moment of the interaction [dipole, quadrupole, etc.]). Thus even though the electrons involved in the excitation are spread over considerable area, the interaction is governed by a unique center to center distance which determines the transfer probability. As long as the moments are independent of the position of the AO moiety along the acyl chain, the unique center will be the same in each probe. Furthermore, most studies suggest that the dipole approximation is sufficiently valid since T values are well-described by the Förster rate (Equation 22) using a single parameter set over a wide acceptor density range.[18,21]

Other properties of the AO probes also indicate that they act as ~ point particles. The dependence with bilayer depth of the quantum yields, lifetimes, and anisotropies exhibit well-defined gradients. Thus, these properties are not averaged out by the 9 Å dimensions of the AO moiety and, as far as electronic properties are concerned, the probes act as particles with dimensions less than 3 Å.[31] The absolute $Z_A$ values relative to each AO are more difficult to assess but are probably accurate to 3 Å.[30]

The AOffa concentration dependence of the energy transfer patterns measured in all the systems we have studied are well-described by the theory. This suggests, therefore, that the probes are randomly distributed about the protein and do not perturb its structure. Measurements of the energy transfer depolarization in egg PC vesicles and red cell ghosts are

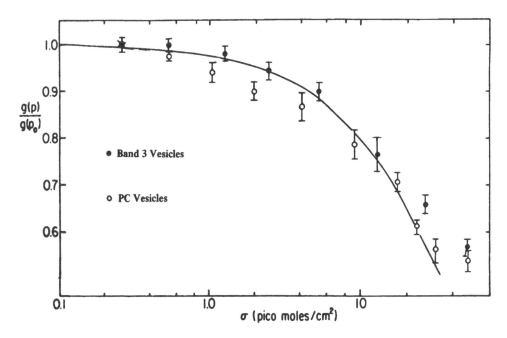

FIGURE 5.   Homo transfer of 12-AS in phosphatidylcholine Vesicles. RET between 12-AS added to SUV of egg PC was determined (data point are circles with error bars) by measuring the steady-state polarization of the AO probe. With increasing AO density in the vesicle the polarization was observed to decrease, although the lifetime (measured by the phase-modulation method of Matayoshi and Kleinfeld, 1981) was constant for probe densities less than 10 mol %. The solid curve was calculated assuming a random surface distribution of the AO probes using the theory of Knox (1968). Good agreement between theory and experiment lends support to the assumption of a random distribution of the probes on the surface of the membrane.

also consistent with a uniform distribution of the probes within the plane of the membrane (Figure 5).

### b. Characterization of the Transbilayer Distribution

The transmembrane distribution of the AOffa probes is an essential element of the Trp imaging method. Although studies in red cell membranes indicate that the probes remain in the outer bilayer leaflet (the side to which they are added),[14,60] we have recently found that in small unilamellar vesicles (SUV) the probes distribute rapidly between both bilayer leaflets.[65] In this study, RET was measured between the *n*-AO probes to 5,6carboxyfluorescein (CF) trapped in the inner aqueous compartment of the vesicles. The measured efficiencies were compared with values calculated for RET between a two-dimensional array of donors (AO) on the surface of a shell and a uniform distribution of acceptors (CF) within the vesicle. Measured and calculated values were in good agreement only if the AO probes were assumed to be uniformly distributed in both hemileaflets of the bilayer.

## IV. SIMULATIONS

### A. Introduction

As the above discussion emphasizes, RET for Trp-AO depends upon a number of both geometric and spectroscopic parameters. These include the Trp spatial distribution (the y and z coordinates), the protein radius $R_p$, the spatial distribution of the AO probes, the quantum yield of the Trp, and the orientation factor. The special nature of the probes and membrane geometry suggest that the orientation factor can to a good approximation be assumed to be isotropic ($\kappa^2 = 2/3$). Methods also exist, as we have discussed above, for

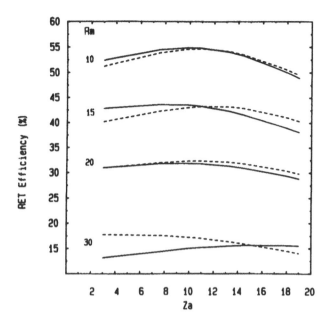

FIGURE 6.  Simulated Two-Trp Patterns. Coordinates used to generate the solid curves were, (0,0) and (Rm,20) and those corresponding to the dashed curve were (0,20) and (Rm,0). AOffa was distributed uniformly on both sides of the bilayer.

determining the AO distribution. The Trp quantum yields can also be estimated by independent means. It should, therefore, be possible to obtain information about the Trp spatial distribution and the protein radius from Trp to AO RET.

## B. RET Patterns

As discussed in Kleinfeld,[30] it is convenient to represent the RET efficiencies by a T vs. $Z_A$ plot, which is called an RET pattern. With experimental uncertainties on the order of 10%, RET patterns for different Trp distributions can readily be distinguished from one another.[30,31] Figures 6 and 7 show how the RET patterns for two Trp proteins depend upon y, z, and Rm. These patterns were evaluated assuming equal quantum yields for both Trp and a uniform transmembrane distribution for the AO probes. It is apparent from these figures that variations in each of the three parameters (y, z, and Rm) change both the shape and magnitude of the patterns. Whether these alterations can be distinguished experimentally will depend upon the particular case. In general, alterations in each of these three parameters are nondegenerate. Intuition suggests that translations in the z direction should produce T changes which are orthogonal to changes produced by both y and Rm. What may be less apparent is that changes in y and Rm are also uncoupled, at least to some degree. This is because RET averaging, corresponding to shifts in the y direction, is not symmetric around the protein axis, whereas an Rm variation clearly is symmetric. The examples shown in Figures 6 and 7 demonstrate that much smaller changes in T are produced by changes in y than by the same changes in Rm. These figures also demonstrate that alterations in both y and z produce changes in the shapes of the patterns that cannot be reproduced by shifts in Rm.

Figure 8 illustrates the expected pattern variation for a 5-Trp-containing protein for which the AO probes are confined to the outside hemileaflet of the bilayer. The important feature of this figure is the large change in both shape and magnitude of the RET patterns that accompany changes in the Trp spatial distribution. These variations are much greater than

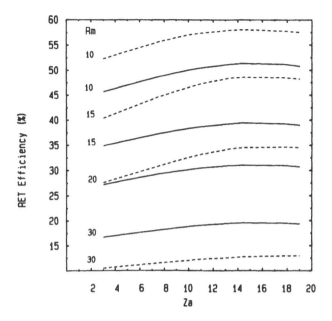

FIGURE 7.   Simulated Two-Trp Patterns. Coordinates used to generate the solid curves were, (Rm,15) and (Rm, −15) and those corresponding to the dashed curve were (0,15) and (0, −15). AOffa was distributed uniformly on both sides of the bilayer.

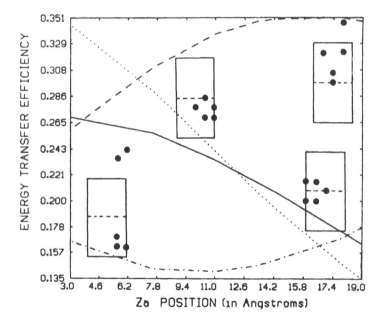

FIGURE 8.   Simulated patterns corresponding to five Trp-containing proteins (from Kleinfeld, 1985). These patterns were calculated assuming a protein radius of 20 Å and equal quantum yields for the five Trp. The Trp distributions used for this simulation are shown in the projections of the protein cylinder which accompany each pattern. These distributions were chosen to illustrate patterns expected for distributions that are skewed to one side or the other of the bilayer, symmetric about the origin, and bimodal, with two of the Trp located 10 to 15 Å outside the bilayer. (Reprinted with permission from Kleinfeld, A. M., *Biochemistry*, 24, 1874, 1985. Copyright 1985 American Chemical Society.)

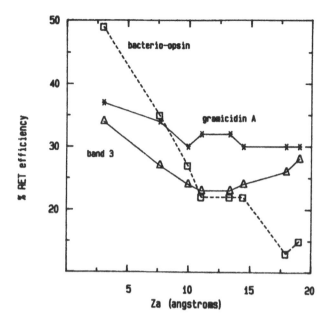

FIGURE 9. Measured RET patterns for three proteins. References and experimental methods are given in the text. All three measurements were obtained with AO probe density equal to 2 mol% of the vesicle phospholipid. The data shown are typical of each protein. These results clearly demonstrate that different proteins give rise to patterns that are distinguishable, both in shape and magnitude.

the expected experimental uncertainties of about 10%. Although the results of experimental studies will be discussed in the following pages, it is worthwhile here to emphasize that the different patterns obtained in simulations are also observed using actual proteins. As Figure 9 shows, three of the proteins studied using the Trp-AO imaging technique yield quite different patterns. These results were obtained using roughly 2% AOffa relative to the total phospholipid in each case. It is seen that the gramicidin pattern is relatively flat, the bacterioopsin almost linear, and the Band 3 concaved upward. In addition to the shapes of the patterns there is also a considerable variation in the RET magnitude displayed by the different proteins.

## C. Monte Carlo Analysis and Trp Density Maps

The preceding discussion, as well as previous work,[30] demonstrates that the Trp-AO RET patterns contain information about the spatial distribution of the Trp and the radius of the protein. Generation of an RET pattern from a particular distribution of Trp is, with the help of Equations 16, 20, and 26, relatively straightforward. Inverting this problem, determination of the Trp distribution from the measured RET pattern, is not quite so simple. Indeed, the problem is underdetermined. In general, there are more parameters to be determined than there are independent measurements of the RET efficiencies. The solution, therefore, will not always be unique.

A Monte Carlo procedure was developed to determine the set of acceptable solutions.[30] The analysis is carried out in several stages. First, single Trp transfer efficiencies are calculated with a resolution of $2 \times 2$ Å at each point within a $30 \times 80$ Å projection of the protein cylinder and are stored in a data base. For a protein with $N_T$ fluorescent Trp, $N_T$ coordinates are randomly chosen within the $30 \times 80$ Å projection of the protein and $<T>$ is formed from the efficiencies corresponding to each of the $N_T$ coordinate pairs. These calculated $<T>$ values are compared with the measured patterns by evaluating the mean square ($R^2$) differences. The patterns and their $N_T$ coordinate pairs are ranked according to $R^2$. The best 100 coordinate sets, out of the typically more than $10^5$ tested, are used to evaluate the frequency distribution and these are plotted as a Trp density map.[30]

The results of simulations carried with the Monte Carlo analysis[30] indicate that for a single Trp the location is specified uniquely with an expected uncertainty of about 2 Å. As the number of Trp increases, the uncertainty in specifying locations of individual Trp increases until about 5 Trp when, in the general case, it is no longer possible to identify individual Trp. Nevertheless, these simulations show that at the very least, the gross features of the distribution are faithfully reproduced and in every case the first two moments of the distribution (the mean and variance) are correctly predicted. While even this degree of resolution represents a considerable advance over what is otherwise available, it is likely that further improvement in resolution will be achieved by modifying the protein chemically (NBS) or genetically (site-directed mutagenesis) to specifically reduce the number of fluorescent Trp.

### D. Deconvolution of the y, z, and Rm Dependencies

In order to demonstrate the sensitivity of the RET to the geometrical parameters, the Monte Carlo method was used to analyze a simulated two-Trp pattern. The results of this analysis are shown in Figure 10 where the caption indicates the parameters of the simulation. The simulated RET pattern was obtained for a protein of radius 20 Å and the analysis was carried out by randomly placing two Trp within an area of 80 Å length and a radius between 10 and 30 Å. In Figure 11 the variation in the minimum $R^2$ value obtained for each trial radius is plotted against Rm. This latter plot demonstrates the extreme sensitivity to Rm that is possible. Clearly if Rm is too large or too small then no combination of y or z values will yield the magnitude or shape of the observed RET values. Figure 10 demonstrates that only at the correct Rm value does the Trp density map yield the correct Trp locations. At Rm values smaller than 20 Å, the optimal distributions correspond to three instead of two groups and at larger Rm values a single group concentrated at the periphery is required to yield the best fit. If the search was confined to a protein of 40 Å length then the $R^2$ values corresponding to Rm <20 would be much larger. This suggests that if additional structural information is known then an even more accurate determination can be made of the Trp spatial distribution.

## V. RET STUDIES USING LIPID PROBES

### A. Cytochrome b$_5$

*1. Surface Probes*

The study of Fleming et al.[18] is the first one in which the location of a Trp residue in a membrane protein was determined using lipid probes. In their study cytochrome b$_5$ was added to unilamellar vesicles composed of dimyristoylphosphatidylcholine (DMPC) (95%) and dimyristoylphosphatidylethanolamine (DMPE) (5%), into which the protein inserts spontaneously from the aqueous phase. To these vesicles was added one of two different fluorescent lipid analogs; dansyldodecylamine (DDA) and *N*-(trinitrophenyl)DMPE, in both of which the acceptor chromophore is located near the lipid-water interface. RET between the tryptophanyl residues of the whole protein or the nonpolar peptide (the portion left after trypsin cleavage of the hydrophilic segment) and these lipid probes, was analyzed using the method of Koppel et al.[35] Their results demonstrated that only one of the three Trp within the membrane was fluorescent and this fluorescent Trp was located 20 to 22 Å from the outer surface of the bilayer.

*2. The AOffa*

Trp imaging studies of whole cytochrome b$_5$ and the nonpolar peptide in DMPC have

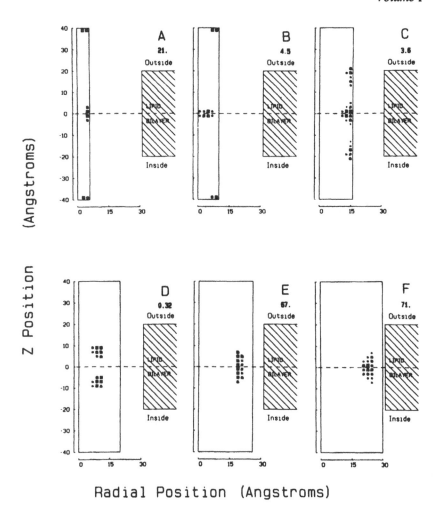

FIGURE 10. Monte Carlo-derived Trp density maps for a simulated two-Trp protein. The radius of the simulated protein was 20 Å and the Trp positions were (10,10) and (10, − 5). Equal Trp quantum yields (0.2) were used. The analysis was performed with radial parameters between 6 and 30 Å. The number in the right-hand corner of each frame is the $R^2$ value for that fit. As is seen, the analysis achieves a minimum $R^2$ and returns the original Trp positions only when the correct radial parameter is used. The patterns obtained for Rm <20 demonstrates that for too small a radius, the RET efficiency is too large for any position within the protein.

also been carried out with the AOffa.[31] RET measured at six different probe densities and for both whole $b_5$ and nonpolar peptide exhibited the same trend, an increase of T with increasing probe depth. Each RET pattern (results with a single probe density) was analyzed by the Monte Carlo analysis and optimal fits were obtained with a single Trp. The fits for every pattern gave the same Trp position, about 20 Å below the surface. Thus, the findings of both cytochrome $b_5$ studies, showing only a single Trp is fluorescent and the Trp is about 20 Å from the surface in *both* whole and nonpolar peptide, are in excellent agreement. In addition, in the study of Kleinfeld and Lukacovic the location of the heme moiety of the whole protein was determined by RET *from* the AOffa and from the Trp to the heme. These results, which help orient the portion of the protein whose crystal structure is known,[47] indicate that the heme moiety is at least 15 Å from the surface of the membrane.

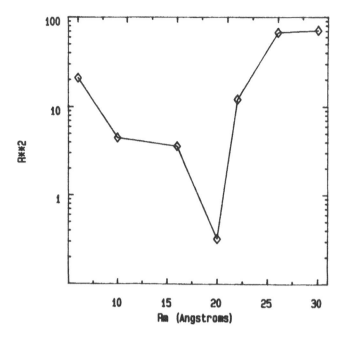

FIGURE 11.   $R^2$ vs. Rm variation for the simulation of Figure 9. This plot clearly emphasizes the nature of the convergence in the Monte Carlo analysis. If the search is confined to a protein of 40 Å length instead of 80 Å, the $R^2$ values for the Rm <20 would increase by more than tenfold.

## B. Gramicidin

Gramicidin, a 15-amino acid polypeptide, has been the subject of intensive investigation since it is believed to represent the simplest form of a cation channel.[2] A good deal is known about its structure and it therefore represents a reasonable model of a multiTrp protein, with known structure. A Trp imaging study has been carried out using gramicidin D, a mixture of 85% Å that has a Trp at Position 11, and 15% B + C that have a phenylalanine or tyrosine at this position.[5] Gramicidin was incorporated into DMPC SUV at lipid to peptide ratios of between 30 and 200. All measurements were performed at about 34°C, well-above the DMPC phase transition. The quantum yield heterogeneity was investigated using N-bromosuccinimide and it was found that two of the four Trp/monomer are not fluorescent. This heterogeneity can be explained within the framework of the Urry[68] model which predicts close pairing of the Trp at Positions 9 and 15. A similar configuration in cytochrome $b_5$ could account for the quenched Trp in that protein.

The results of the energy transfer experiments all exhibit the same general behavior: a slight increase between 2 and 9 and a more significant increase for AO attachment sites >10 (see Figure 9). The RET patterns are best fit with the Trp density map shown in Figure 12. No information concerning the structure or dimer nature of gramicidin has been used in this analysis. The average position of the two Trp groups is ±14 Å about the bilayer center. According to Urry's model of gramicidin (β helix with 6.3 residues/turn) a symmetric dimer in which the C termini meet at the center predicts a mean Trp location of 4 Å about the center, while the N-N model predicts 14 Å. Thus, these results are in excellent agreement with the preferred N-N configuration of the peptide,[70] and suggest that for the first time it is possible to reliably image membrane protein structure in the physiological state.

## C. Bacteriorhodopsin

Bacteriorhodopsin (bR) has been studied by Trp-AO RET in bleached and unbleached purple membrane of *Halobacterium halobium*.[9] RET efficiencies in the unbleached mem-

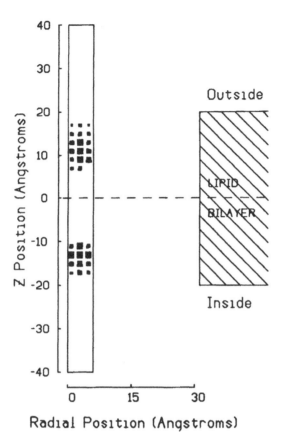

FIGURE 12.   Trp density for gramicidin. This is the Trp density map corresponding to the optimal fit for the Monte Carlo analysis using two Trp per monomer and a radial parameter of 6 Å. No assumptions concerning the dimer nature of the protein were imposed on the search. The results clearly show the peptide as a symmetric dimer with Trp at ± 14 Å around the center of the bilayer, as predicted by the head-to-head model. (Reproduced from the *Biophysical Journal*, 1986, Vol. 49, p. 122 by copyright permission of the Biophysical Society.)

branes were observed to *decrease* with the deeper-lying AO moieties and were essentially independent of probe position in the bleached membranes. These results were interpreted to mean that in the unbleached membrane, Trp at the surface dominated the transfer because the buried residues are quenched by retinal. In the bleached membrane the buried Trp are visible and contribute to the observed RET. A number of important issues remain to be addressed in this study. Although retinal certainly quenches the Trp fluorescence, its large $R_0$ (>30 Å) may preclude the establishment of a Trp fluorescence gradient, due solely to retinal RET. In addition, because of the crystalline nature of the purple membrane, assumptions concerning rotational isotropy are not valid. The RET pattern, therefore, could reflect a strong component, due to orientational, as opposed to positional, differences.

For these and a number of other reasons we have measured Trp-AO RET in liposomes reconstituted with bacterioopsin (bO).[29] The preparation used in this reconstitution was essentially that of Bayley et al.[3] who showed that the protein is virtually 100% asymmetrically oriented. NBS quenching studies indicate that about four of the eight Trp are responsible for most of the fluorescence. Typical results from the Trp-AO transfer measurements are shown in Figure 8 and indicate that T *increases* almost linearly with AO position. Analyzing

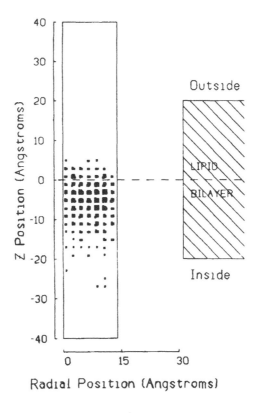

A

FIGURE 13.   Comparaison of the results of a Trp imaging
study of bO (Kleinfeld et al., 1986) and the model of Engelman
et al.[15] for bR. RET patterns used to obtain the bO map (A)
were similar to those shown in Figure 9. For bR the projections
of the eight Trp were evaluated using the model of Engelman
et al. (1982) (B) and this pattern was then analyzed using the
Monte Carlo method to obtain the density map shown in (C).
There is good agreement between the observed and model-
derived density maps, especially since the measurements suggest considerable quantum yield heterogeneity, while the model
density was evaluated assuming equal Trp quantum yields.

these results using a protein of 15 Å radius and with four Trp, each with a quantum yield
of 0.3, results in the Trp density map shown in Figure 13A. A simulation of the model of
Engelman et al.[15] results in a similar map, suggesting good agreement with this model
(Figures 13B and 13C).

## D. Band 3

Preliminary tryptophan imaging studies of Band 3 were carried out using the AOffa.[32] In
this work, red cell ghost were treated to enrich the content of Band 3 to about 80% of the
total protein using the method of Wolosin et al.[73] Typical RET patterns are shown in Figure
9 and exhibit a rather unique curvature. The results of this study suggested that Trp exists
in two well-separated groups, one buried within the outer hemileaflet of the membrane and

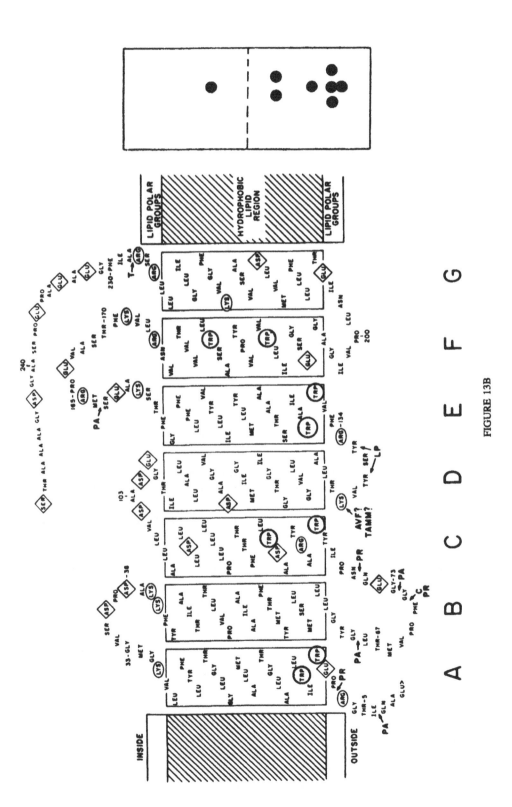

FIGURE 13B

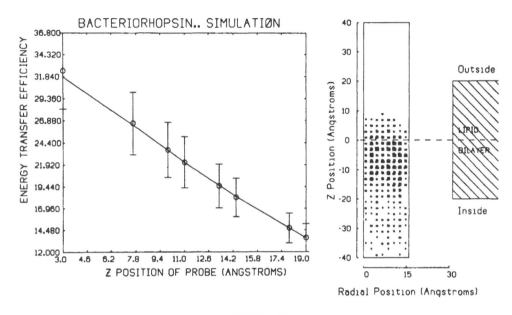

FIGURE 13C

the other extending a considerable distance into the cytoplasm. This is an especially intriguing result in light of the recently elucidated primary sequence of Band 3[34] which shows the Trp grouped into two well-separated regions.

## VI. CONCLUSIONS

Although relatively few RET studies of membrane protein tertiary structure have been carried out, it is clear that this approach offers the possibility of obtaining unique information of high quality. Even without refinement, the Trp-AO method allows the determination of the general outline of the Trp spatial distribution. Probes which are glycosylated at the carboxyl terminus and, therefore, do not flip across the bilayer are under development. Combining results using these and unmodified AO probes, together with Trp modification and lifetime methods will allow a complete resolution of the distribution in proteins containing many Trp. Certainly one of the most exciting aspects of the lipid probe investigations is the possibility of observing the protein under physiological conditions. Thus, it may be possible to map at high resolution, conformational changes associated with actual protein function.

## APPENDIX: NORMALIZATION CONSTANT FOR SENSITIZED EMISSION

In this Appendix, expressions will be derived relating RET efficiencies obtained from sensitized emission of the acceptor, to measured fluorescence intensities. The particular focus of this Appendix will be the evaluation of the normalization constant B, that appears in Equation 26 for the multidonor averaged transfer efficiency.

Fluorescence intensities are usually observed at 90° to the excitation light, that, in the case considered here, is assumed to be a thin parallel beam. It is further assumed that fluorescence is detected only from a short segment, $\Delta x$, along this beam, and that this detection region is centered in a cuvette of Length $2 \times \ell$. If the intensity of the exciting beam at the front surface of the cuvette is designated by $I^{Bm}(0)$, the beam intensity after it traverses a distance $\ell - \Delta x/2$ by $I^{Bm}(1)$, and the beam intensity at a distance $\ell + \Delta x/2$ by $I^{Bm}(2)$, then the intensity lost to excitation of donor (D) molecules in the observation region $\Delta x$ is

$$I^*(D) = I^{Bm}(1) - I^{Bm}(2) = I^{Bm}(0)(10\ OD_1 - 10\ OD_2) \tag{1A}$$

The optical densities corresponding to positions 2 and 1 are

$$OD_1 = \epsilon_D[D](\ell - \Delta x/2) \tag{2A}$$

$$OD_2 = \epsilon_D[D](\ell + \Delta x/2) \tag{3A}$$

in which $\epsilon_D$ is the molar extinction coefficient of the donor at the donor excitation wavelength ($\lambda_D$). Designating the optical density for the excitation path length as $OD(\lambda_D) = 1\epsilon_D[D]$, Equation 1A can be expressed as

$$I^*(D) = I^{Bm}(0)10^{OD(\lambda_D)}(1 - e^{-\epsilon_D[D]\Delta x}) \tag{4A}$$

Since $\epsilon_D[D]\Delta x \ll 1$, the following expression is obtained for the intensity of excited donors:

$$I^*(D) = 2.3\Delta x \epsilon_D[D]I^{Bm}(0)10^{OD(\lambda_D)} \tag{5A}$$

The observed donor fluorescence intensity is, therefore:

$$I^F(D) = KQ_D I^*(D) \tag{6A}$$

in which K is the fraction of fluorescence intensity emitted from the region $\Delta x$ that is intercepted by the detector. The absolute values of the quantities $I^{Bm}(0)$, $\Delta x$, and K are difficult to determine with precision and, therefore, it is necessary to express the transfer efficiencies in terms of ratios of the observed intensities.

If the donor population is heterogeneous ($N_T$ different Trp per protein) and we designate by $I^*(i)$ the intensity of excited donors with properties i (e.g., quantum yield $Q_i$), then the fluorescence intensity of acceptors excited by RET is

$$I^F(A) = Q_A \sum_{i=1}^{N_T} I^*(i)T_i \tag{7A}$$

in which $T_i$ is the RET efficiency from the ith donor, as used in Equations 20 and 26. Substituting Equations 5A and 6A into Equation 7A we obtain:

$$I^F(A) = B\ \epsilon_D[D]Q_A\left(\sum_{i=1}^{N_T} T_i\right)/N_T \tag{8A}$$

(also Equation 25) in which the normalization constant is given by:

$$B = 2.3K\Delta x I^{Bm}(0)10^{-OD(\lambda_D)} \tag{9A}$$

In order to minimize experimental uncertainties it is important to evaluate B in terms of fluorescence intensities that are measured with the same conditions as the spectra used to determine the RET efficiencies. There are three candidates, each with its own advantages and disadvantages:

1. The donor intensity measured in the absence of acceptor
2. The intensity of the acceptor measured at an excitation wavelength where the donor is not excited
3. The intensity of a standard such as quinine sulfate.

First, the donor intensity measured at excitation wavelength $\lambda_D$. Using the above definition of B, the fluorescence intensity of a heterogeneous donor population can be written as:

$$I^F(D) = B\epsilon_D[D]<Q> \tag{10A}$$

with

$$<Q> = \sum_{i=1}^{N_T} Q_i/N_T \tag{11A}$$

Thus, B can be completely expressed in terms of donor parameters according to:

$$B^{(1)} = \frac{I^F(D)}{\epsilon_D[D]<Q>} \tag{12A}$$

In terms of $B^{(1)}$, therefore, the RET efficiency (Equation 26) becomes:

$$<T> = \frac{I_T<Q>}{I^F(D)Q_A} \tag{13A}$$

All the quantities appearing in Equation 13A can be determined with relative ease. $<T>$ determined in this way depends upon the relative quantum yield of the protein ($<Q>$) and the acceptor.

Second, the acceptor intensity excited at $\lambda_A$. Defining the ratio of the excitation intensities at $\lambda_D$ and $\lambda_A$ as:

$$R^E(A) = I_{\lambda_A}^{Bm}(0)/I_{\lambda_D}^{Bm}(0) \tag{14A}$$

and using the definition of B, the fluorescence intensity of the acceptor can be written as:

$$I^F(A) = B\,R^E(A)10^{\Delta OD(A)}\epsilon_A Q_A[A] \tag{15A}$$

in which $\Delta OD(A)=OD(\lambda_D)-OD(\lambda_A)$. The normalization constant can, therefore, be expressed in terms of the acceptor intensity according to:

$$B^{(2)} = \frac{I^F(A)10^{-\Delta OD(A)}}{R^E(A)Q_A\epsilon_A[A]} \tag{16A}$$

In terms of $B^{(2)}$ Equation 26 is

$$<T> = \frac{I_T\,R^E(A)\epsilon_A[A]}{I^F(\lambda_A)10^{-\Delta OD(A)}\epsilon_D[D]} \tag{17A}$$

In this form $<T>$ is independent of quantum yields and $N_T$. $R^E$ can be determined with considerable accuracy by measuring the absorption and fluorescence intensities of a single fluorophore at the two exciting wavelengths.

Third, the fluorescence intensity of a standard such as quinine sulfate measured at an excitation wavelength of $\lambda_{QS}$. Using arguments similar to those for $B^{(2)}$, the normalization constant can be expressed as:

$$B^{(3)} = \frac{I^F(QS)10^{-\Delta OD(QS)}}{R^E(QS)Q_{QS}\epsilon_{QS}[QS]} \qquad (18A)$$

in which $\Delta OD(QS) = OD(\lambda_D) - OD(\lambda_{QS})$ and $R^E(QS) = I^{Bm}_{\lambda_{QS}} / I^{Bm}_{\lambda_A}(0)$. Thus, in terms of $B^{(3)}$ the RET efficiency (Equation 26) is

$$<T> = \frac{I_T\, R^E(QS)Q_{QS}\epsilon_{QS}[QS]}{I^F(QS)10^{-\Delta OD(QS)}Q_A\epsilon_D[D]} \qquad (19A)$$

Here again, quantum yields appear as a ratio which can be determined with considerable accuracy.

## ACKNOWLEDGMENTS

This work was supported by a grant from the National Science Foundation (PCM-830268). This work was done during the tenure of an Established Investigatorship of the American Heart Association and with funds contributed in part by the Massachusetts affiliate (82-174).

## REFERENCES

1. **Agard, D. A. and Stroud, R. M.**, Linking regions between helicies in bacteriorhodopsin revealed, *Biophys. J.*, 37, 589, 1982.
2. **Andersen, O. S.**, Gramicidin channels, *Annu. Rev. Physiol.*, 46, 531, 1984.
3. **Bayley, H., Hojeberg, B., Huang, K., Khorana, H. G., Liao, M., Lind, C., and London, E.**, Delipidation, renaturation, and reconstitution of bacteriorhodopsin, *Methods Enzymol.*, 88, 74, 1982.
4. **Benga, G. and Holmes, R. P.**, Interactions between components in biological membranes and their implications for membrane function, *Prog. Biophys.*, 43, 195, 1984.
5. **Boni, L. T., Connolly, A. J., and Kleinfeld, A. M.**, Transmembrane distribution of gramicidin by tryptophan energy transfer, *Biophys. J.*, 49, 122, 1986.
6. **Brock, C. J., Tanner, M. J. A., and Kemph, C.**, The human erythrocyte anion-transport protein, *Biophys. J.*, 213, 577, 1983.
7. **Buchel, D. E., Gronenborg, B., and Muller-Hill, B.**, Sequence of the lactose permease gene, *Nature (London)*, 283, 541, 1980.
8. **Cardoza, J. D., Kleinfeld, A. M., Stallcup, K. C., and Mescher, M. F.**, Hairpin configuration of H-2K$^k$ in liposomes formed by detergent dialysis, *Biochemistry*, 23, 4401, 1984.
9. **Chatelier, R. C., Rogers, P. J., Ghiggino, K. P., and Sawyer, W. H.**, The transverse location of tryptophan residues in the purple membranes of *Halobacterium halobium* studies by fluorescence, *Biochim. Biophys. Acta*, 776, 75, 1984.
10. **Dale, R. E., Eisinger, J., and Blumberg, W. E.**, The orientational freedom of molecular probes. The orientation factor in intramolecular energy transfer, *Biophys. J.*, 26, 161, 1979.
11. **Deisenhoffer, J., Epp, O., Miki, K., Huber, R., and Michel, H.**, Structure of the protein subunits in the photosynthetic reaction centre of *Rhodopseudomonas viridis* at 3A resolution, *Nature (London)*, 318, 618, 1985.
12. **Dunn, R., McCoy, J., Simsek, M., Majumdar, A., Chang, S. H., RajBhandary, U. L., and Khorana, H. G.**, The bacteriorhodopsin gene, *Proc. Natl. Acad. Sci. U.S.A.*, 78, 6744, 1981.
12a. **Eisenberg, D.**, Three-dimensional structure of membrane and surface proteins, *Annu. Rev. Biochem.*, 53, 595, 1984.
13. **Eisinger, J., Feuer, B., and Lamolla, A. A.**, Intramolecular singlet excitation transfer. Applications to polypeptides, *Biochemistry*, 8, 3908, 1969.

14. **Eisinger, J. and Flores, J.,** The relative locations of intramembrane fluorescent probes and of the cytosol hemoglobin in erythrocytes studied by transverse resonance energy transfer, *Biophys. J.,* 37, 6, 1982.

15. **Engelman, D. M., Goldman, A., and Steitz, T. A.,** The identification of helical segments in the polypeptide chain of bacteriorhodopsin, *Methods Enzymol.,* 88, 81, 1982.

16. **Engelman, D. M., Henderson, R., McLachlan, A. D., and Wallace, B. A.,** Path of the polypeptide in bacteriorhodopsin, *Proc. Natl. Acad. Sci. U.S.A.,* 77, 2023, 1980.

17. **Engelman, D. M. and Zacci, G.,** Bacteriorhodopsin is an inside-out protein, *Proc. Natl. Acad. Sci. U.S.A.,* 77, 5894, 1980.

17a. **Fairclough, R. H. and Cantor, C. R.,** The use of singlet-singlet energy transfer to study macromolecular assemblies, *Methods Enzymol.,* 28, 347, 1978.

18. **Fleming, P. J., Koppel, D. E., Lan, A. L. Y., and Strittmatter, P.,** Intramembrane position of the fluorescent tryptophanyl residue in membrane-bound cytochrome b5, *Biochemistry,* 18, 5458, 1979.

19. **Fleming, P. J. and Strittmatter, P.,** The nonpolar peptide segment of cytochrome b5, *J. Biol. Chem.,* 253, 8198, 1978.

20. **Förster, Th.,** Zwischenmolekulare energi-wandering and Fluoreszenz, *Ann. Phys.,* 2, 55, 1948.

21. **Fung, B. K. K. and Stryer, L.,** Surface density determination in membranes by fluorescent energy, *Biochemistry,* 17, 5241, 1978.

22. **Haas, E., Katchalski-Katzir, E., and Steinberg, I. Z.,** Effect of orientation of donor and acceptor on the probability of energy transfer involving transition of mixed polarization, *Biochemistry,* 17, 5064, 1978.

23. **Henderson, R. and Unwin, P. N. T.,** Three-dimensional model of purple membrane obtained by electron microscopy, *Nature (London),* 257, 23, 1975.

23a. **Henderson, R., Jubb, J. S., and Whytock, S.,** Specific labelling of the protein and lipid on the extracellular surface of purple membrane, *J. Mol. Biol.,* 123, 259, 1978.

24. **Holmgren, A.,** Selective N-bromosuccinimide oxidation of the nonfluorescent tryptophan-31 in the active center of thioredoxin from *Escherichia coli, Biochemistry,* 20, 3204, 1981.

25. **Huang, K.-S., Liao, M.-J., Gupta, C. M., Royal, N., Beimann, K., and Khorana, H. G.,** The site of attachment of retinal in bacteriorhodopsin, *J. Biol. Chem.,* 257, 8596, 1982.

26. **Ichiye, T. and Karplus, M.,** Fluorescence depolarization of tryptophan residues in proteins: a molecular dynamics study, *Biochemistry,* 22, 2884, 1983.

27. **Jap, B. K., Maestre, M. F., Hayward, S. B., and Glaeser, R. M.,** Peptide-chain secondary structure of bacteriorhodopsin, *Biophys. J.,* 43, 81, 1983.

28. **Khorana, H. G., Gerber, G. E., Herlihy, W. C., Gray, C. P., Anderegg, R. J., Nihei, K., and Biemann, K.,** Amino acid sequence of bacteriorhodopsin, *Proc. Natl. Acad. Sci. U.S.A.,* 76, 5046, 1979.

29. **Kleinfeld, A. M.,** Current views of membrane structure, in *Current Topics in Membranes and Transport,* Vol. 22, Klausner, R. D., Kempf, C., and Van Renswoude, J., Eds., Academic Press, New York, 1986, chap. 1.

30. **Kleinfeld, A. M.,** Tryptophan imaging of membrane proteins, *Biochemistry,* 24, 1874, 1985.

31. **Kleinfeld, A. M. and Lukacovic, M. L.,** Energy transfer study of cytochrome b5 using the anthroyloxy fatty acid membrane probes, *Biochemistry,* 24, 1883, 1985.

32. **Kleinfeld, A. M., Matayoshi, E. D., and Solomon, A. K.,** Energy transfer determination of the tryptophan distribution of band 3: a new approach to the study of membrane proteins, *J. Supramol. Struct.,* 4, 163, 1980.

33. **Knox, R. S.,** Theory of polarization quenching by excitation transfer, *Physica,* 39, 361, 1968.

34. **Kopito, R. R. and Lodish, H. F.,** Primary structure and transmembrane orientation of the murine anion exchange protein, *Nature (London),* 316, 234, 1985.

35. **Koppel, D. E., Fleming, P. J., and Strittmatter, P.,** Intramembrane positions of membrane-bound chromophores determined by excitation energy transfer, *Biochemistry,* 18, 5450, 1979.

36. **Kutchai, H., Chandler, L. H., and Zavoico, G. B.,** Effects of cholesterol on acyl chain dynamics in multilamellar vesicles of various phosphatidylcholines, *Biochim. Biophys. Acta,* 736, 137, 1984.

37. **Lakowicz, J. R., Maliwal, B. P., Cherek, H., and Balter, A.,** Rotational freedom of tryptophan residues in proteins and peptides, *Biochemistry,* 22, 1741, 1983.

38. **Lee, D. C., Hayward, J. A., Restall, C. J., and Chapman, D.,** Second-derivative infrared spectroscopic studies of the secondary structures of bacteriorhodopsin abd $Ca^{2+}$ − ATPase, *Biochemistry,* 24, 4364, 1985.

39. **Liefer, D. and Henderson, R.,** Three-dimensional structure of orthorhombic purple membrane at 6.5 Å resolution, *J. Mol. Biol.,* 163, 451, 1983.

40. **London, E. and Feigenson, G. W.,** Fluorescence quenching in model membranes, *Biochemistry,* 20, 1932, 1981.

40a. **London, E. and Feigenson, G. W.,** Determination of the local lipid environment of the calcium adenosinetriphosphatase form sarcoplasmic reticulum, *Biochemistry,* 20, 1939, 1981.

41. **Long, M. M., Urry, D. W., and Stoeckenius, W.,** Circular dichroism of biological membranes: purple membrane of halobacterium halobium, *Biochem. Biophys. Res. Commun.,* 75, 725, 1977.

42. **MacLennan, D. H., Brandl, C. J., Korczak, B., and Green, N. M.,** Amino-acid sequence of a $Ca^{2+}$ + $Mg^{2+}$-dependent ATPase from rabbit muscle sarcoplasmic reticulum, deduced from its complementary CNA sequence, *Nature (London)*, 316, 696, 1985.
43. **Makowski, L., Caspar, D. L. D., Phillips, W. C., Baker, T. S., and Goodenough, D. A.,** Gap junction structures. VI. Variation and conservation in connexion conformation and packing, *Biophys. J.*, 45, 208, 1984.
44. **Mao, D. and Wallace, B. A.,** Differential light scattering and absorption flattening optical effects are minimal in the circular dichroism spectra of small unilamellar vesicles, *Biochemistry*, 23, 2667, 1984.
45. **Markello, T., Zlotnick, A., Everett, J., Tennyson, J., and Holloway, P. W.,** Determination of the topography of cytochrome $b_5$ in lipid vesicles by fluorescence quenching, *Biochemistry*, 24, 2895, 1985.
46. **Matayoshi, E. D. and Kleinfeld, A. M.,** Emission wavelength-dependent decay of the 9-anthroyloxy-fatty acid membrane probes, *Biophys. J.*, 35, 215, 1981.
47. **Mathews, F. S., Czerwiski, E. W., and Argos, P.,** The x-ray crystallographic structure of calf liver cytochrome b5, in *The Porphyrins*, Vol. 3, Dolphin, D., Ed., Academic Press, New York, 1979, 107.
48. **Mueckler, M., Caruso, C., Baldwin, S. A., Panico, M., Blench, I., Morris, H. R., Allard, W. J., Lienhard, G. E., and Lodish, H. F.,** Sequence and structure of a human glucose transporter, *Science*, 229, 941, 1985.
49. **Neori, H. B. and Montal, M.,** Rhodopsin in reconstituted phospholipid vesicles. I. Structural parameters and light-induced conformational changes detected by resonance energy transfer and fluorescence quenching, *Biochemistry*, 22, 197, 1983.
50. **Noda, M., Shimizu, S., Tanabe, T., Takai, T., Kayano, T., Ikeda, T., Takahashi, H., Nakayama, H., Kanoka, Y., Minamino, N., Kangawa, K., Miyata, T., and Numa, S.,** Primary structure of *Electrochorus electricus* sodium channel deduced from cDNA sequence, *Nature (London)*, 312, 121, 1984.
51. **Nabedryk, E., Bardin, A. M., and Breton, J.,** Further characterization of protein secondary structures in purple membrane by circular dichroism and polarized infrared spectroscopies, *Biophys. J.*, 48, 873, 1985.
52. **Ovchinnicov, Yu. A., Abdulaev, N. G., Feigina, M. Yu., Kiselev, A. V., and Lobanov, N. A.,** The structural basis of the functioning of bacteriorhodopsin: an overview, *FEBS Lett.*, 100, 219, 1979.
53. **Rao, A. G. and Neet, K. E.,** Tryptophan residues of the gamma subunit of 7S nerve growth factor: intrinsic fluorescence, solute quenching, and N-bromosuccinimide oxidation, *Biochemistry*, 21, 6843, 1982.
54. **Rao, A., Martin, P., Reithmeier, R. A. F., and Cantley, L. C.,** Location of the stilbenedisulfonate binding site of the human erythrocyte anion-exchange system by resonance energy transfer, *Biochemistry*, 18, 4505, 1979.
55. **Robson, R. J., Radhakrishnan, R., Ross, A. H., Takagaki, Y., and Khorana, H. G.,** Photochemical cross-linking in studies of lipid-protein interactions, in *Lipid-Protein Interactions*, Vol. 2, Jost, P. C. and Griffith, O. H., Eds., John Wiley & Sons, New York, 1982, 149.
56. **Ross, J. B. A., Rousslang, K. W., and Brand, L.,** Time resolved anisotropy decay of the tryptophan in adrenocorticotropin-(1-24), *Biochemistry*, 20, 4361, 1981.
57. **Ross, J. B. A., Schmidt, C. J., and Brand, L.,** Time-resolved fluorescence of the tryptophans in horse liver alcohol dehydrogenase, *Biochemistry*, 20, 4369, 1981.
58. **Rothschild, K. J. and Clark, N. A.,** Polarized infrared spectroscopy of oriented purple membrane, *Biophys. J.*, 25, 473, 1979.
59. **Senior, A. E.,** Secondary and tertiary structure of membrane proteins involved in proton translocation, *Biochim. Biophys. Acta*, 726, 81, 1983.
60. **Shaklai, N., Yguerabide, J., and Ranney, H. M.,** Interaction of hemoglobin with red blood cell membranes as shown by a fluorescent chromophore, *Biochemistry*, 16, 5585, 1977.
61. **Shull, G. E., Schwartz, A., and Lingrel, J. B.,** Amino-acid sequence of the catalytic subunit of the ($Na^+$ + $K^+$) ATPase deduced from a complementary DNA, *Nature (London)*, 316, 691, 1985.
62. **Sklar, L. A., Hudson, B. S., and Simoni, R. D.,** Conjugated polyene fatty acids as fluorescent probes: binding to bovine serum albumin, *Biochemistry*, 16, 5100, 1977.
63. **Spande, T. F., Green, N. M., and Witkop, B.,** The reactivity toward N-bromosuccinimide of tryptophan in enzymes, zymogens, and inhibited enzymes, *Biochemistry*, 5, 1926, 1966.
64. **Stoeckenius, W. and Bogomolni, R. A.,** Bacteriorhodopsin and related pigments of *Halobacteria*, *Annu. Rev. Biochem.*, 52, 587, 1982.
65. **Storch, J. and Kleinfeld, A. M.,** Transfer of long chain fluorescent free fatty acids between unilamellar vesicles, *Biochemistry*, 25, 1717, 1986.
66. **Stryer, L.,** Fluorescence energy transfer as a spectroscopic ruler, *Annu. Rev. Biochem.*, 47, 819, 1978.
67. **Takagaki, Y., Radhakrishnan, R., Gupta, C. M., and Khorana, G. J.,** Membrane-embedded segment of cytochrome $b_5$ as studied by cross-linking with photoactivatable phospholipids, *J. Biol. Chem.*, 258, 9128, 1983.
68. **Urry, D. W.,** The gramicidin A transmembrane channel: a proposed Pi(L,D) helix, *Proc. Natl. Acad. Sci. U.S.A.*, 68, 672, 1971.

69. **Vincent, M., Foresta, B., Gallay, J., and Alfsen, A.,** Nanosecond fluorescence decays of n-(9-anthroyloxy) fatty acids in dipalmitoylphosphatidylcholine vesicles with regard to isotropic solvents, *Biochemistry*, 21, 708, 1982.
70. **Weinstein, S., Durkin, J. T., Veatch, W. R., and Blout, E. R.,** Conformation of the gramicidin A channel in phospholipid vesicles: a fluorine-19 nuclear magnetic resonance study, *Biochemistry*, 24, 4374, 1985.
71. **Wilson, I. A., Skehel, J. J., and Wiley, D. C.,** Structure of the haemagglutin membrane glycoprotein of influenza virus at 3A resolution, *Nature (London)*, 289, 366, 1981.
72. **Wolber, P. K. and Hudson, B. S.,** Bilayer acyl chain dynamics and lipid-protein interaction. The effect of the M13 bacteriophage coat protein on the decay of the fluorescence anisotropy of parinaric acid, *Biophys. J.*, 37, 253, 1982.
73. **Wolosin, J. M., Ginsburg, H., and Cabantchik, Z. I.,** Functional characterization of anion transport system isolated from human erythrocyte membranes, *J. Biol. Chem.*, 252, 2419, 1977.

Chapter 5

# STRUCTURAL MAPPING OF MEMBRANE-ASSOCIATED PROTEINS: A CASE STUDY OF THE IgE-RECEPTOR COMPLEX

**Barbara Baird and David Holowka**

## TABLE OF CONTENTS

# I. INTRODUCTION

Elucidation of the structure and function of membrane-associated proteins has emerged as one of the major new challenges in protein chemistry during the past decade. Detailed three-dimensional structural information of these proteins is lacking for all except a few cases, in part because the amphipathic nature of most integral membrane proteins makes them difficult to purify and crystallize, and also because these proteins are often found organized into large multimeric complexes of dissimilar subunits that depend on specific protein-protein and protein-lipid interactions for functional integrity. The class of membrane proteins on cell surfaces that act as receptors has been particularly difficult to study by powerful physical methods such as magnetic resonance and X-ray crystallography for the reasons given above and also because of the paucity of these components relative to other cellular proteins. Typical cell surface receptors for hormonal or immunological stimuli range in amounts from $10^4$ to $10^5$ copies per cell and generally less than 0.01% by weight of the total cell protein. Therefore, purifications of as much as $10^5$ are required to achieve homogeneity. Technological advances such as affinity chromatography and high-performance liquid chromatography have made such feats possible with some receptors,[1,2] but even in most of these cases the submilligram quantities usually obtainable prohibit the use of standard chemical and physical approaches for probing detailed structural questions.

Our interest in the structure-function relationships of cell surface receptors has led us to investigate the three-dimensional structure of the receptor for immunoglobulin E (IgE). This system has several features that make it experimentally attractive, as discussed below, and we have capitalized on methodological tools that have allowed specific and sensitive measurements to be made. In particular, we have used fluorescence methods to map the spatial arrangement of the membrane-bound IgE-receptor complex. Fluorescence allows quantitative measurements to be made on receptor samples at typical nanomolar concentrations, and resonance energy transfer from fluorescent donors to acceptor chromophores can be measured and used as a "ruler" to determine distances between these probes located specifically on the macromolecule. As a versatile alternative to chemical modification, we have also employed monoclonal antibodies to place fluorescent probes at or near specific antigenic sites on the IgE-receptor complex. Our experience indicates that the use of labeled monoclonal antibodies and resonance energy transfer in concert should be widely applicable for structural analyses, and the studies we have carried out on the IgE-receptor system that are described in the following pages will serve to illustrate the experimental approaches that can be successfully employed for other membrane-associated proteins.

# II. EXPERIMENTAL SYSTEM

Numerous accounts of the basic theory and experimental measurements underlying application of fluorescence spectroscopy to study protein structure have been published (e.g., References 3 and 4), and the fundamental equations describing energy transfer and depolarization will not be repeated here. Before describing our fluorescence experiments, background structural information on the IgE-receptor system will be reviewed briefly. More complete reviews on IgE and its high-affinity receptor on mast cells and basophils are presented in Dorrington and Bennich[5] and Metzger,[6] respectively. Structural studies on this system are the most extensive for any immunological receptor, and this can be mainly attributed to two factors. One is the availability of a homogeneous tumor cell line, the rat basophilic leukemia (RBL) cells, that contain receptors in relatively large quantities, typically 2 to $5 \times 10^5$ copies per cell. The second factor is the extremely tight binding of IgE to this receptor, with an equilibrium binding constant of $K > 10^{10}$/mol.[7] This, together with a very

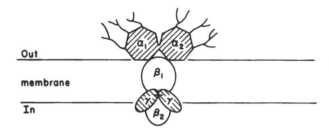

FIGURE 1. Schematic diagram showing the structure of the high-affinity receptor for IgE. Areas of the $\alpha$, $\beta$ and $\gamma$ subunits are drawn to be roughly proportional to mass. $\alpha_1$, $\alpha_2$ and $\beta_1$, $\beta_2$ designate separate domains within a single polypeptide chain in each case, and two $\gamma$ subunits are disulfide bonded to each other. Wavy lines represent carbohydrate chains on $\alpha$.

slow rate of dissociation ($k \leq 10^{-5}/\text{sec}$)[7,8] allows purification of receptors via the interaction with bound IgE using an appropriate affinity column.[1]

It is well-established that a single IgE molecule binds to each receptor on the cell surface[9,10] and, that the crosslinking of those complexes by multivalent antigens is the event that initiates the cellular response in vivo: there is exocytosis of storage granules containing histamine and serotonin, as well as synthesis and secretion of leukotrienes and prostaglandins, all of which combine to elicit the symptoms of allergy and asthma.[11] Besides antigen binding to the combining sites of receptor-bound IgE, various other reagents that act to cross-link receptors can be used to generate a triggering signal, including, oligomeric IgE,[12,13] anti-IgE antibodies,[14] and anti-receptor antibodies, in the absence of any IgE.[15,16] Aggregation of receptors as an initiating event in transmembrane signaling now appears to be a common feature in the mechanisms of immunological receptors[17], as well as receptors mediating other physiological processes.[18,19] As will be discussed in the following pages, questions regarding possible requirements for specific interactions and/or orientations between cross-linked receptors for the delivery of a triggering signal are just beginning to be approached.

Over a decade of experimental effort by several laboratories has led to a picture of the IgE receptor that is shown schematically in Figure 1. This receptor appears to have a particularly complex structure in that it contains three nonidentical subunits. The $\alpha$ subunit is a glycopeptide of about 50 kdaltons that is exposed on the outside surface of the cell and contains the high-affinity binding site for IgE. Proteolytic digestion studies[20] have suggested the presence of two similar, but nonidentical domains in this subunit, as indicated by the $\alpha_1$, $\alpha_2$ notation in Figure 1. Both the $\beta$ subunit (30 to 33 kdaltons and the two disulfide-bonded $\gamma$ subunits (7 to 10 kdaltons each) have hydrophobic regions that are embedded in the membrane bilayer,[21] as well as segments that are exposed on the cytoplasmic side of the plasma membrane.[22] Recent studies in our laboratory have shown that a $\beta$-chain segment of about 18 to 20 kdaltons (designated $\beta_1$ in Figure 1) remains associated with the $\alpha$ chain and the membrane after extensive enzymatic digestion of inside-out plasma membrane vesicles, suggesting the presence of a segment of about 10 to 15 kdaltons ($\beta_2$) that extends into the cytoplasm which may participate in the delivery of the transmembrane signal.[81] Both $\beta$ and $\gamma$ have sites of phosphorylation that are probably on the cytoplasmically exposed portions.[23,24] Neither of these subunits has been shown to be exposed on the outer surface of the cell and, their roles in the function of this receptor are as yet unknown. Although the amino acid composition of the $\alpha$ chain has been determined,[25] sequence information is still unavailable for any of the receptor subunits. Attempts to clone the gene(s) are in progress[26] and success in this area should lead to new structural insights. Strong evidence has been presented for the interaction of aggregated receptors with other cellular components during the delivery of a triggering signal[27,28] and, experiments are in progress in our laboratory to

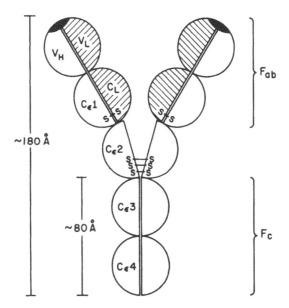

FIGURE 2. A schematic drawing of IgE. The light chains (hatched) and ε chains are arranged in domains, and the segments designated Fab and Fc are shown. The antibody-combining sites at the top of the Fab segments (solid) and approximate locations of the interchain disulfide bonds are indicated. The length of the whole IgE molecule and the length of two domains are estimated from X-ray crystallographic studies of immunoglobulin G. (Reprinted with permission from Baird, B., and Holowka, D., *Biochemistry*, 24, 6252, 1985. Copyright 1985 American Chemical Society.)

identify these components and their sites of interaction with the receptor. Possible candidates for functionally important coupling to the receptors include a $Ca^{2+}$ channel,[29] phospholipase C,[30] GTP-binding proteins,[31] and components of the cytoskeleton.[27,28,82]

Our initial fluorescence-mapping studies have focused on the nature of the interaction of IgE with the receptor. Early studies by K. Ishizaka, T. Ishizaka, and colleagues on human IgE had indicated that the Fc region of the immunoglobulin is primarily responsible for its binding to high-affinity receptors on mast cells,[32] and Dorrington and Bennich showed that the well-documented heat lability of this interaction is due, in large part, to unfolding of the $C_\epsilon 3$ and $C_\epsilon 4$ domains.[33] Figure 2 is a schematic diagram of IgE that identifies these regions of the protein molecule, as well as other domains and the two Fab segments that contain the antigen-binding sites at their N-terminal tips. The location of interchain-disulfide bonds between heavy (ε) chains and also between light (L) and heavy chains, as based on DNA sequence data,[34,35] are shown for rodent IgE; there are some differences for human IgE.[5] Perez-Montfort and Metzger[36] examined the effect of rodent IgE-binding to the α subunit of the receptor on the susceptibility of IgE to digestion by trypsin. They found that upon binding, mouse IgE was only slightly protected from digestion at several sites in the Fc region, including one at the $C_\epsilon 2$ to $C_\epsilon 3$ junction. Similarly, rat IgE was only minimally protected from digestion at sites in Fc, except the $C_\epsilon 2$ to $C_\epsilon 3$ cleavage site was substantially protected due to interaction with the receptor. These results indicate that the Fc region is probably not engulfed by the receptor upon binding, yet, it appears that a site near the $C_\epsilon 2$ to $C_\epsilon 3$ junction and quite far from the C-terminal end of rat IgE is involved in the binding interaction and thereby protected. This view of a partially exposed Fc is consistent with the ability of several monoclonal anti-Fc antibodies to bind to receptor-bound IgE.[37,38] The implications of these findings for a model of receptor-IgE interaction will be subsequently considered.

The biochemical experimentation previously discussed has provided valuable insight, but very few structural details about the interaction between IgE and its high-affinity receptor. We began investigating this structure with resonance energy transfer in a rather indirect manner: by measuring the distances from selectively labeled sites on receptor-bound IgE to the membrane surface. With a series of sites ranging over the IgE molecule we have obtained a picture of its conformation and orientation with respect to the membrane. These experimental measurements have been made possible because of several developments. One is the availability of large (milligram) quantities of purified monoclonal IgE antibodies that can be fluorescently modified and characterized as to the specificity and spectral properties of the attached probes. We can capitalize on the tight interaction of IgE with its receptor to specifically bind these fluorescent IgE derivatives and thoroughly remove any unbound IgE. Unlike many ligand-receptor interactions, the slow dissociation of IgE permits measurements over periods of several hours with no need to correct for unbound IgE. Also useful to us in designing our experimental approach was the prior development of readily applicable theoretical treatments of energy transfer between multiple donors and acceptors in a single plane and, between multiple donors and acceptors in separate planes.[39-44] The latter case provides a model which allows calculation of the distance L between a donor-labeled site on IgE and the plane of the membrane containing acceptors, but it requires knowledge of the surface density of acceptors. This density can be obtained using the model of donors and acceptors in the same plane.

A third development is our ability to prepare highly purified plasma membranes from RBL cells.[45] These are primarily right-side-out vesicles that are virtually free of intracellular membranes and organelles. Our procedure for preparing these vesicles was adapted from the method of Scott[46] for fibroblasts, and it is based on inducing vesiculation or blebbing of adherent cells using sulfhydryl-reactive reagents followed by two centrifugations of the supernatant to isolate the detached vesicles. The absence of intracellular membranes, which normally comprise at least 90% of the total cellular membranes, is important in our studies. This is because intracellular membranes take up most of the amphipathic probes employed, and accurate determination of the distance from a site above the plasma membrane requires measurement of the density of probes exclusively in this membrane. Another feature of these vesicles that makes them different from cells is that the membrane proteins (including the IgE receptor), as well as the lipids, have rapid lateral mobility.[47] This may fortuitously help to satisfy the assumption of the models that the membrane probes are uniformly distributed over the entire membrane surface, including in the vicinity of the receptors.

A fourth development is the availability of two monoclonal anti-IgE antibodies that have allowed us to measure intramolecular distances within the IgE-receptor complex. One monoclonal antibody is specific for the Fc region of IgE (A2), and the other is specific for the Fab region (B5).[37] These antibodies have the useful feature that both bind equally well to IgE free in solution, as well as to receptor-bound IgE. Placement of donor or acceptor probes at either the N-terminal or the C-terminal end of these antibodies or their Fab' fragments, as described in the following pages, has provided sufficiently localized labeling that subtle changes have been detected in IgE conformation upon binding to its receptor.

## III. ENERGY TRANSFER EXPERIMENTS

Calculation of distances from donor probes at sites on receptor-bound IgE to amphipathic acceptor probes at the membrane surface assumes that the acceptors are uniformly distributed and noninteracting and that the surface density of the acceptors is known. Our characterization of the acceptors in these structural mapping studies will be discussed first. The amphipathic

Amphipathic Fluorescent Probes

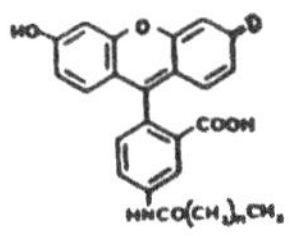

5-(N-hexadecanoyl)-
aminofluorescein (HAF)

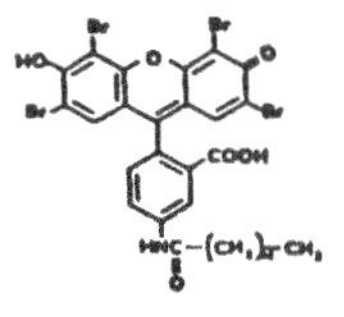

5-(N-hexadecanoyl)-
aminoeosin (HAE)

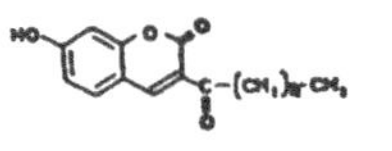

3-hexadecanoyl-7-
hydroxycoumarin (HHC)

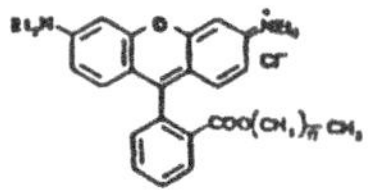

octadecyl rhodamine B
(ORB)

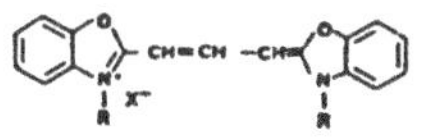

3,3'-dialkyloxacarbocyanine
(DiOC$_n$)

FIGURE 3.    Chemical structures of amphipathic fluorescent probes used as donors or acceptors at the membrane surface for resonance energy transfer studies.

probes we have employed are shown in Figure 3. These molecules all have one or two long hydrocarbon "tails" that spontaneously insert into the membrane bilayer, as well as a chromophoric "head group" that is charged at physiological pH and, therefore, confined to the membrane-aqueous solution interface. The amphipathic probes have spectral properties such that some can act as resonance energy transfer donors to others acting as acceptors, and, therefore, energy transfer measurements can be used to monitor the presence of these probes in the membrane and their spatial arrangement. For example (Figure 4), when 5-(N-hexadecanoyl)aminoeosin (HAE; acting as acceptor in this case) is added to a suspension of RBL cell plasma membrane vesicles previously equilibrated with 5-(N-hexadecanoyl)aminofluorescein (HAF; acting as donor), the HAF fluorescence is quenched. This quenching reflects the partitioning of HAE into the membrane bilayer as also revealed by the fluorescence enhancement of HAE due to relief of the self-quenching of its aqueous micellar state: HAF quenching and HAE enhancement occur at the same rate. The rate of insertion of the amphipathic probes and the extent of fluorescence enhancement are characteristically different for each of the probes used (Figure 3) and depend, both on the length of the alkyl chains(s) and the nature of the head group. The observed quenching of donor fluorescence in titrations such as that of Figure 4 can be attributed to resonance energy transfer by two criteria:[45] 1) sensitized emission of HAE was found to correspond exactly to HAF quenching; and 2) the magnitude of quenching was found to be independent of donor density over a fivefold range, as predicted by theory.[39]

Evidence that the acceptor probes are partitioning into the membrane in a noninteracting manner is shown in Figure 5A where it can be seen that the enhanced fluorescence of inserted HAE is linear with concentration over a range where substantial energy transfer from HAF

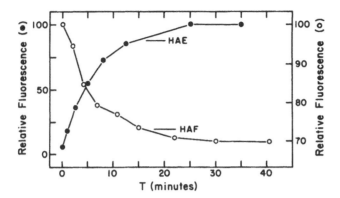

FIGURE 4.   Time course of the binding of HAE (67 μ*M*) to membrane vesicles (0.4 pmol of receptor per milliliter) containing bound HAF (53 μ*M*). ●, enhancement of HAE fluorescence (ex 525 nm, em 550 nm) accompanying insertion into bilayer; ○, quenching of HAF fluorescence (ex 480 nm, em 515 nm) due to energy transfer to HAE. (Reprinted with permission from Holowka, D. and Baird, B., *Biochemistry*, 22, 3466, 1983. Copyright 1983 American Chemical Society.)

occurs. Figure 5B shows that energy transfer, as measured by donor quenching, corresponds very well with the theoretical curve derived by Wolber and Hudson[40] for the case of donors and acceptors that are mobile and randomly distributed in a single plane. Use of this theoretical curve provides direct conversion of energy transfer efficiency at a given bulk concentration of acceptor to the surface density of that acceptor in the plane of the membrane, as shown by the upper abscissa in Figure 5B. Thus, the density of these amphipathic acceptor probes can be assessed directly without the need to make assumptions about membrane surface area or the uniformity of their distribution. Donor-acceptor pairs, other than HAF and HAE, were found to behave in a similar manner, and some of their spectral characteristics are listed in Table 1. It is likely that most of the probes we have employed are present on both halves of the lipid bilayer, but neglecting transbilayer transfer results in an overestimation of the acceptor density by only about 20% for donor-acceptor pairs with an $R_0$ (distance corresponding to 50% efficiency of energy transfer) of about 50 Å.[39]

In the following we will describe energy transfer measurements between donors on receptor-bound IgE and amphipathic acceptors at the membrane surface, giving several examples for illustrative purposes, and then we will summarize all of our experiments. Figure 6 shows an experiment for which IgE was modified at its interheavy chain disulfide bonds in $C_e2$ (see Figure 2) by reduction and alkylation with coumarin phenylmaleimide (CPM(+)IgE), and this derivative was bound to receptors on the vesicles. Then donor CPM fluorescence was monitored as the acceptor probe 3,3'-didecanoyloxacarbocyanine ($DiOC_{10}$) was titrated into this sample (Sample a; ●), as well as into two control samples. One control sample (Sample b; ○) contained the same amount of CPM(+)IgE free in solution and vesicles that had receptors blocked with unmodified IgE. The other (Sample c; ■) contained vesicles bound with IgE that had been modified with CPM in the absence of reduction (CPM(−)IgE) in order to correct for a small amount of nonspecific donor label, as well as the background signal. Several points can be made by this figure. (1) Significant quenching of fluorescence occurs in Sample b, which contains free CPM(+)IgE, suggesting some direct interaction between donor and acceptor. The magnitude of this type of quenching varies unpredictably with both donor and acceptor probes, so this control is essential in order to correct for acceptors not at the membrane surface. (2) The background signal due to light scattering, endogenous membrane fluorescence, and nonspecific CPM fluorescence (Sample c) is a

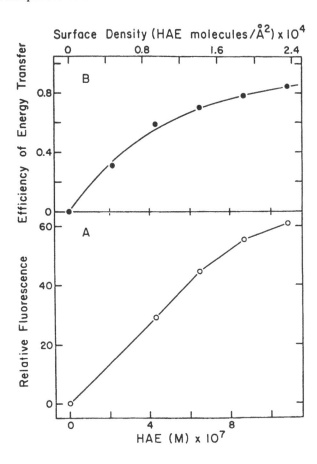

FIGURE 5. Titration of amphipathic acceptor (HAE) into membrane vesicles (1.6 pmol receptor per milliliter) containing donor (320 n*M* HAF). A) The fluorescence intensity of bound HAE (ex 520 nm, em 555 nm) as a function of HAE concentration. B) Efficiency of energy transfer at each addition (●) as determined from the quenching of vesicle-bound HAF fluorescence (ex 480 nm, em 520) by bound HAE. In (B) the solid line represents the two-exponential solution to energy transfer in two dimensions derived by Wolber and Hudson.[40] The upper abscissa is calibrated in units of surface density according to this fit and $R_0 = 50.1$ Å (Table 1). (Reprinted with permission from Holowka, D. and Baird, B., *Biochemistry*, 22, 3466, 1983. Copyright 1983 American Chemical Society.)

significant fraction of the total signal, and this must be carefully subtracted in order to measure accurately the quenching due to energy transfer. Good fits of the corrected experimental data by theoretical curves indicate that proper correction has been made (see material that follows). A fourth sample (Sample d) containing vesicles preequilibrated with an appropriate amphipathic donor probe is titrated with the amphipathic acceptor in parallel with the other three samples. This sample is used to obtain the density of acceptors at each point.

The efficiency of energy transfer is measured in terms of the ratio of the quantum yield of the donor in the absence and presence of the acceptor, $Q_D/Q_{DA}$. For experiments like that shown in Figure 6, $Q_D = I^b - I^c$ and $Q_{DA} = I^a - I^c$, where I is the measured donor fluorescence of the designated sample for each point in the titration curve corresponding to a different surface density of acceptors. To relate the experimental measurements to a distance L between

**Table 1**
## SPECTRAL PROPERTIES OF THE MEMBRANE-BOUND AMPHIPATHIC DONOR-ACCEPTOR PAIRS

| Donor | $Q_D$[b] | $\bar{A}_D$[c] | Acceptor | $\epsilon max$ ($\lambda max$) | $\bar{A}_A$[c] | $R_0(\text{Å})$[d] |
|-------|------|------|----------|--------------------|------|------------|
| HAF | 0.34 | 0.25 | HAE | $8.5 \times 10^4$; 528 nm | 0.16 | 50.1 |
| | 0.34 | 0.25 | ORB | $1.07 \times 10^5$; 538 nm | ND[e] | 54.3 |
| HHC | 0.065 | 0.28 | DiOC$_6$ | $1.40 \times 10^5$; 490 nm | ND | 38.1 |
| | 0.065 | 0.28 | DiOC$_{10}$ | $1.40 \times 10^5$; 490 nm | ND | 38.1 |

[a] Values were determined as described in Reference 45.
[b] Quantum yield of membrane-bound donors in the absence of acceptor.
[c] Steady-state anisotropy at wavelengths of maximum excitation and emission and 20°C.
[d] Distance between donor and acceptor corresponding to efficiency of energy transfer (E) = 0.50.
[e] Not determined.

donor and plane of acceptors, the titration curve is replotted as $Q_D/Q_{DA}$ vs. acceptor density and compared to theoretical curves. Useful formulations of this type of curve are provided by Wolber and Hudson[40] and Dewey and Hammes.[41] An alternative is carrying out Monte Carlo calculations of energy transfer for donors and acceptors in separate planes set apart by fixed distances.[43] As L increases, theoretical plots of $Q_D/Q_{DA}$ vs. acceptor density show decreasing curvature. If there is little energy transfer, which occurs when L is large compared to $R_0$, then a linear approximation can be made,[42] and L is easily determined from the slope of the line: when density is given as acceptors/Å,$^2$

$$\text{slope} = \frac{\pi R_0^6}{2L^4} \qquad (1)$$

The fluorescence-quenching data from the experiment of Figure 6 (●) along with that from a similar experiment using the same donor but with HAF as the amphipathic acceptor (○) are plotted in Figure 7 as $Q_D/Q_{DA}$ vs. normalized acceptor density (acceptors/$R_0^2$). These data are compared to curves generated by Monte Carlo calculations for different values of $L/R_0$.[43] For each experiment there is good correspondence with the model of energy transfer between separated planes, except at the highest density tested in the case of CPM(+)IgE and DiOC$_{10}$ (●), and this discrepancy is probably due to a nonrandom distribution of acceptor probes at this point in the titration. Such deviation is seen to occur at very high concentrations of several of the amphipathic acceptors used (e.g., see Figure 5A), and these data points are omitted in the determination of L if they are clearly aberrant from the theoretical curves. Confidence in the use of energy transfer measurements to determine accurate distances comes from good agreement between data sets for different donor-acceptor pairs. Such agreement is seen for the two experiments presented in Figure 7 (For the donor-acceptor pair CPM(+) and DiOC$_{10}$, $R_0 = 51.7$ Å; for CPM(+) and HAF, $R_0 = 47.4$ Å). This provides the best test that results obtained from a particular acceptor are not highly dependent on the dipole-dipole orientation factor ($\kappa^2$) nor influenced by possible binding of an acceptor at some site in the

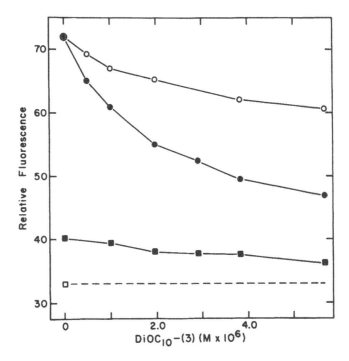

FIGURE 6.   Fluorescence data (ex 395 nm, em 470 nm) from a representative experiment in which energy transfer was measured between CPM (donor) at the interheavy chain disulfides of receptor-bound IgE (CPM(+)IgE) and DiOC$_{10}$ (acceptor) in the plane of the membrane bilayer as a function of acceptor concentration. ●, Sample a in which CPM(+)IgE is bound to the receptor; ○, Sample b in which receptor is blocked with unmodified IgE and CPM(+)IgE is present in solution; ■, Sample c in which CPM(−)IgE is bound to the receptor to correct for membrane fluorescence, light scattering, and energy transfer from CPM at noninterheavy chain sites. The level of background membrane fluorescence and light scattering from these samples is indicated (□). (Reprinted with permission from Holowka, D. and Baird, B., *Biochemistry*, 22, 3475, 1983. Copyright 1983 American Chemical Society.)

IgE-receptor complex. Unusual behavior by a particular acceptor may be noteworthy. For example, one acceptor tested appeared to interact with receptor-bound, but not free, CPM(+)IgE, suggesting a conformationally sensitive site on this IgE that became exposed upon binding to receptor.[48]

Some of the acceptor probes employed in these studies partition into the vesicle bilayer at such a slow rate that it is not practical to carry out titrations on the four samples by a series of additions as described for CPM(+)IgE and DiOC$_{10}$. Alternatively, an acceptor can be added as a single addition to each sample, and energy transfer can be monitored in each as a function of time where increasing time corresponds to increasing surface density. This approach is illustrated in Figure 8. This experiment was performed to measure energy transfer between 5-dimethylaminonaphthalene-1-sulfonyl-L-lysine (DNS-lys) in the combining site of monoclonal anti-DNS IgE bound to receptor and the amphipathic acceptor octadecyl rhodamine B (ORB) in the membrane. Figure 8B shows the enhancement of ORB fluorescence as it partitions into the bilayer (○) and the simultaneous quenching of the amphipathic donor HAF present in the vesicles in Sample d (□). Figure 8A shows that very little quenching of DNS-lys fluorescence accompanies the insertion of ORB (Sample a, ■). Correction of Sample a and Sample b (▲) fluorescence at each point with control Sample c (●) together

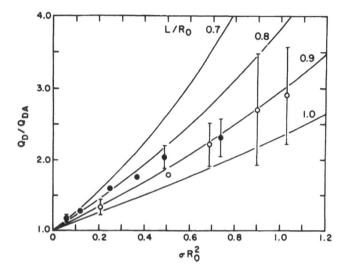

FIGURE 7. Energy transfer between CPM at the interheavy chain disulfides of receptor-bound IgE and acceptors at the membrane surface. The ratio of donor quantum yield in the absence ($Q_D$) and presence ($Q_{DA}$) of acceptors (●, $DiOC_{10}$; O, HAF) is plotted as a function of acceptor density (acceptors/$R_0^2$). The representative error bars were calculated from a propagation of error analysis, and the curves correspond to different values of $L/R_0$ based on Monte Carlo calculations of energy transfer between donors and acceptors in separate planes. (Reprinted with permission from Holowka, D. and Baird, B., *Biochemistry*, 22, 3475, 1983. Copyright 1983 American Chemical Society.)

with conversion of time to acceptor density using Sample d allows construction of the standard $Q_D/Q_{DA}$ plot. Plots for this and a duplicate experiment are shown in Figure 9. In this case, very little energy transfer from DNS-lys to ORB occurs, and the linear approximation is valid. The slope of the best straight line through the data points indicates the distance between the combining site of receptor-bound IgE and the membrane surface is greater than 100 Å. This experiment demonstrates that small amounts of energy transfer can be measured in these complex membrane samples, and, therefore, distances of 100 Å and more can be assessed if donor-acceptor pairs with large values of $R_0$ are used.

It is obvious that calculation of meaningful distances from energy transfer measurements requires knowledge of the location of donor and acceptor probes. However, the sites of modification need not be absolutely specific in order for useful structural information to be obtained if selectivity for a particular region within the macromolecule can be determined. In our initial exploratory studies IgE was labeled with fluorescein-5-isothiocyanate (FITC) at pH 9 with the expectation that there would be random labeling of amino groups.[49]

The stoichiometry of labeling was 2 FITC per IgE and analysis of proteolytic fragments of this derivative revealed that >90% of the fluorescence was located in the F(ab')$_2$ portion (see Figure 2). This localization is consistent with our observation that FITC fluorescence is significantly quenched when 2,4-dinitrophenyl (DNP) ligands bind to the combining sites of anti-DNP FITC-IgE[49] (for FITC and DNP, $R_0$ = 17.3 Å). It is likely that a significant fraction of the FITC labeling occurs at the N-terminal amino groups, which are near the antibody-combining sites and are preferentially reactive, especially at lower pH.[50] As discussed below, we have utilized this apparent selectivity in subsequent strategies for labeling the monoclonal anti-IgE antibodies in order to incorporate preferentially donor or acceptor probes at N-terminal residues and thereby allow placement in proximity to the antigenic determinant on IgE.

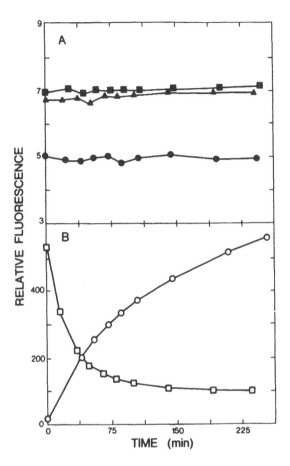

FIGURE 8.   A and B) Data from experiment to measure energy transfer between DNS-lys in the IgE-combining sites and ORB at the membrane surface. Shown are fluorescence changes of probes accompanying the time-dependent partitioning of ORB into vesicle membranes of different samples measured in parallel. A single addition of ORB (20 μ*M*) was made to each sample at time = 0, and its fluorescence became enhanced as insertion into the membrane bilayer occurred (○; ex 340 nm, em 580 nm). Sample a (■; ex 340 nm, em 505 nm) contained vesicles with receptor-bound anti-DNS IgE that had DNS-lys in the antibody-combining sites. Sample b (▲; ex 340 nm, em 505 nm) contained vesicles with receptors saturated by anti-DNP IgE and DNS-lys bound to anti-DNS IgE was present in solution. Sample c (●; ex 340 nm, em 505 nm) contained vesicles with receptors saturated by anti-DNP IgE, and DNS-lys was present in solution, but no anti-DNS IgE, was present. Sample d (□, ex 470 nm, em 505 nm) contained vesicles that had the amphipathic donor HAF (4 μ*M*) inserted in the bilayer. (Reprinted with permission from Baird, B. and Holowka, D., *Biochemistry*, 24, 6252, 1985. Copyright 1985 American Chemical Society.)

Results from a number of experiments measuring distances from sites on IgE to the membrane surface are summarized in Table 2. The distance from DNS-lys in the anti-DNS-combining sites of receptor-bound IgE is greater than 100 Å as determined using both ORB and HAE as acceptors (Table 2, Line 1). The average distance from FITC probes located in the Fab segments of IgE is only slightly less than that for DNS-lys, but the precise

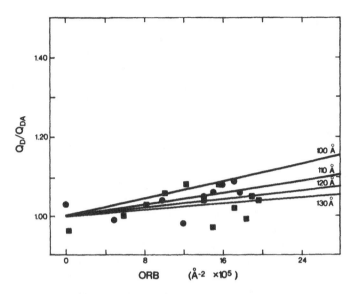

FIGURE 9. Energy transfer between DNS-lys in the antibody-combining site of receptor-bound anti-DNS IgE and ORB acceptors at the membrane surface. The ratio of donor quantum yield in the absence of acceptors ($Q_D$) to that in their presence ($Q_{DA}$) is plotted as a function of acceptor density (acceptor molecules per $Å^2$). The ● points were calculated from the experiment shown in Figure 8, and the ■ points were calculated from a separate experiment carried out in the same manner. The solid lines are drawn from Equation 1 using $R_0 = 57.2$ Å (Table 2) and different possible values of L. (Reprinted with permission from Baird, B. and Holowka, D., *Biochemistry*, 24, 6252, 1985. Copyright 1985 American Chemical Society.)

## Table 2
### SUMMARY OF DISTANCES TO THE MEMBRANE SURFACE (L) MEASURE BY RESONANCE ENERGY TRANSFER

| | Donor | Location[a] | Acceptor[b] | $R_0$(Å) | Distance (Å) | Ref. |
|---|---|---|---|---|---|---|
| 1. | DNS-lysine | anti-DNS IgE combining sites | HAE, ORB | 53.7, 57.2 | 100 to 120 | 48 |
| 2. | FITC | Fab$_\epsilon$ | HAE, ORB | 53.0, 53.2 | 86 to 91 | 49 |
| 3. | CPM(−) | Fab (C$\epsilon$l domain) | HAF, DiOC$_6$ | 50.5, 55.7 | 75 to 87 | 48 |
| 4. | FITC | B5 or B5 F(ab')$_2$ bound to Fab$_\epsilon$ | HAE | 55.0, 58.2 | 78 to 87 | 52 |
| 5. | | A2 Fab' bound to FC$_\epsilon$ | HAE | 46.0 | 54 | 52 |
| 6. | CPM(+) | C$_\epsilon$ domain | HAF, DiOC$_{10}$ | 47.4, 51.7 | 43 to 44 | 49 |

[a]  Sites on receptor-bound IgE on membrane vesicles.
[b]  Acceptors located at the bilayer surface.

Reprinted with permission from Holowka, D., Conrad, D. H., and Baird, B., *Biochemistry*, 24, 6260, 1985. Copyright 1985 American Chemical Society.

locations of FITC molecules are unknown. Calculation of an average distance from multiple donors weights more heavily those that are closer to the plane of acceptors in these experiments.[49] Therefore, the magnitude of the distance calculated for FITC-IgE (86 to 91 Å, Line 2) indicates that most of the FITC molecules are closer to the IgE-combining sites than to the interheavy chain disulfide bonds which are labeled by CPM(+) and are rather close to the membrane surface (43 to 44 Å, Line 6).

An additional site in the Fab segments of IgE was labeled by CPM in the absence of reduction (CPM($-$)), and this modification cannot be blocked by pretreatment with $N$-ethylmaleimide, nor, with other alkylating reagents. These results indicate that the residue labeled is not an easily accessible sulfhydryl, but, that the covalent reaction requires initial noncovalent binding via the ring structure of the chromophoric end of the CPM molecule. Several other fluorescent alkylating reagents show similar reactivity patterns, and the site of labeling seems to be confined to the $C_\epsilon 1$ domain in the Fab segment based on its distance from the antibody-combining sites and analysis of proteolytic fragments.[48] The $C_\epsilon 1$ domain is also the site of binding by the B5 monoclonal anti-IgE antibody. The F(ab')$_2$ fragment of B5 was labeled with FITC at pH 9 and bound to IgE on the vesicles, and the distance from the membrane of donors on this derivative was found to be very similar to the distance of the CPM($-$) site (Table 2, Lines 3 and 4). These distances are intermediate between those for the probes near the tips of the Fb segments (DNS-lys and FITC) and CPM($+$) at the interheavy chain disulfides.

We prepared a second FITC derivative of B5, carrying out the modification reaction at pH 7 in order to label preferentially the N-terminal $\alpha$-amino groups.[50] Edman degradation and proteolytic digestion analyses of this derivative showed that 40% of the FITC label is confined to the N-terminal residues, and less than 10% is located in the Fc region.[83] The measured distance to the membrane surface of this derivative bound to IgE is similar to that for the FITC-B5 F(ab')$_2$ derivative for which the sites of modification are not known (Table 2, Line 4). Since there is probably less selectivity of amino groups with the pH 9 modification, this agreement indicates that the distances measured are not sensitive to the distribution of label in the Fab portions of IgE-bound B5. These observations sugest that measurements made with a donor-labeled monoclonal antibody fragment that binds to a cell surface antigen can be interpreted meaningfully in mapping experiments, particularly when several different derivatives of that antibody can be employed. An informative comparison can be made with Fab' fragments preferentially labeled at the N-terminal residues near the combining sites and Fab' fragments labeled near the C-terminal end by reduction and alkylation of interchain disulfide bonds. The distances measured from the membrane surface to these two donor locations will indicate the orientation of the Fab upon binding, as well as providing a distance to the antigenic determinant.

A second monoclonal antibody, A2, employed in these energy transfer experiments is specific for the Fc region of IgE. The Fab' fragment of A2 was labeled preferentially at its N-terminal $\alpha$-amino groups, and the distance from these sites to the membrane surface when this fragment is bound to the IgE-receptor complex was found to be slightly greater than the distance from the membrane surface to the interheavy chain disulfide bonds in $C_\epsilon 2$ (Table 2, Lines 5 and 6). This result is somewhat surprising since it suggests that when IgE binds to receptor the A2 antigenic site in the Fc region might be as far or farther away from the membrane surface as the $C_\epsilon 2$ domain (see Figure 2). Only one derivative of A2 has been used to place donor probes thus far, and, we cannot rule out the possibility that the distance calculated reflects the orientation of the bound Fab and some distribution of the FITC label along its length; sites nearer the membrane would be weighted more heavily. Although more experiments are required, a cautious interpretation is that donors near the antigenic site of A2 are not a lot closer to the membrane surface than the interheavy disulfides in $C_\epsilon 2$.

The set of distances together, as listed in Table 2, provide confidence in the individual measurements. The distances from the membrane decrease with the progression of donor-labeled sites on IgE from the most distal at the tips of the Fab segments (Line 1) to the most proximal at the interheavy chain disulfides in the $C_\epsilon 2$ domains (Line 6). Sources of error in these experiments have been discussed previously,[49] and the accuracy of any measurement is generally considered to be $\pm 20\%$. The range of distances obtained for any one donor-labeled site in Table 2 reflects this error, as well as, variation arising from use of

A

IgE

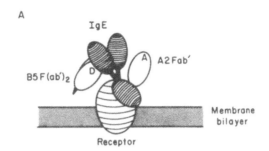

B5 F(ab')₂

A2 Fab'

Membrane
bilayer

Receptor

B

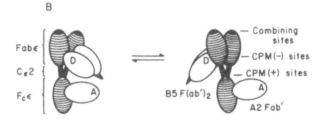

Fabε

Cε2

Fcε

— Combining
sites
— CPM(–) sites
— CPM(+) sites

B5 F(ab')₂

A2 Fab'

FIGURE 10.   Models for the structures of A2 Fab' and B5 F(ab')₂
bound to A) IgE-receptor complex on the plasma membrane and B)
IgE in solution. D and A represent the approximate locations of the
donor and acceptor probes, respectively, for the experiment shown in
Figure 12. See text for explanation. (Reprinted with permission from
Holowka, D., Conrad, D. H., and Baird, B., *Biochemistry,* 24, 6260,
1985. Copyright 1985 American Chemical Society.)

more than one acceptor. The point to be emphasized is that the reliability of donor quenching
as a monitor of resonance energy transfer is founded on the carefully controlled experimental
design as described above. Use of more than one type of acceptor probe (preferably of
opposite charges) is also very important in order to rule out complications due to selective
binding, nonrandom dipole orientations, or nonuniform distribution in the vicinity of the
receptors. A further point should be made that any distance measured is one of "closest
approach" between the donor-labeled site on IgE and acceptors in the lipid bilayer, and this
may be greater than the vertical distance, depending on the unknown width of the receptor
at the membrane surface.

## IV. STRUCTURAL MODELS AND FURTHER TESTS

Based on some of the results described above, we suggested two possible models for the
conformation of receptor-bound IgE.[49] In one, IgE is fairly rigid, and its long twofold axis
of symmetry is extended at some angle with respect to the plasma membrane such that, the
combining sites are held at a distance of greater than 100 Å from the membrane surface,
while the $C_ε4$ domain is held close to that surface. In a second model, IgE bends in order
to bind to its receptor such that, it no longer has a twofold axis of symmetry. In this
arrangement the Fab segments point away from the membrane surface while the Fc segment
is closer, but its long axis is oriented at some nonperpendicular angle to the plane to the
membrane (see Figure 10A). This second model predicts that the IgE must undergo a change
in average conformation upon binding to its receptor but does not restrict potential segmental
flexibility of the Fab' segments.

We have recently tested these alternative models with two different types of experiments.
In one set of experiments, the segmental flexibility of anti-DNS IgE in solution was compared
to that for IgE bound to receptor on vesicles.[51] DNS-lys in the antibody-combining sites

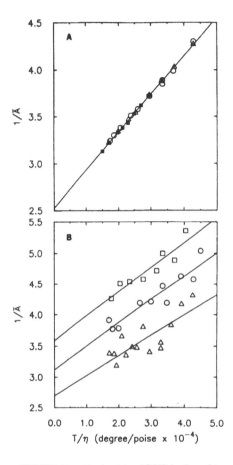

FIGURE 11.   Perrin plots of DNS-lys bound to
anti-DNS IgE obtained by varying the temperature except where noted. A) Anti-DNS IgE in
solution: Exp 1 (○), Exp 2 (△), and Exp 3 (■);
carried out isothermally at 20°C by addition of
glycerol) are fit simultaneously by a single best
straight line. B) Anti-DNS IgE bound to vesicles:
Exp 4 (○) Exp 5 (△), and Exp 6 (□) are fit
separately with best straight lines. Ā is the steady-
state anisotropy (ex 340 nm, em 510 nm), T is
the temperature in K and η is the viscosity of
the solution in poise. (Reprinted with permission
from Slattery, J., Holowka, D., and Baird, B.,
*Biochemistry*, 24, 7810, 1985. Copyright 1985
American Chemical Society.)

served as the fluorescent probe, and Perrin plot analyses of the steady-state depolarization
data are presented in Figure 11. Figure 11A shows data from several experiments in solution
for which either temperature or viscosity of the solution was varied. All points lie along a
single straight line that yields an average rotational correlation time ($\phi$) of $54 \pm 1$ nsec.
Values of $\phi$ calculated for an equivalent sphere with the size and hydration of IgE are
significantly larger (Table 3) indicating that IgE in solution exhibits some segmental flexibility that is probably due in large part to independent rotational motion of the Fab arms.
Data from separate experiments with vesicle-bound IgE form a set of roughly parallel lines
as seen in Figure 11B. The lack of co-linearity of all the points was shown to result from
differential amounts of depolarization caused by light scattering of the vesicles; this contri-

**Table 3**
**SUMMARY OF ROTATIONAL CORRELATION**
**TIMES**

| Sample | Average φ (nsec)[a] | |
|---|---|---|
| | In solution | On membranes |
| DNS-lys/anti-DNS IgE | 54 ± 1 (3)[b] | 74 ± 3 (3) to 89[c] |
| PM-IgE[d] | 65 ± 6 (5) to 75[e] | 75 ± 20 (5) to 86[e] |
| IgE, equivalent sphere | 78 to 155[f] | >1000[g] |

[a] At 25°C in 135 m$M$ NaCl, 5 m$M$ KCl, 10 m$M$ HEPES, 0.0170 NaN$_3$, pH 7.4.

[b] Error represents SD and the number of experiments is indicated in parentheses.

[c] Upper limit of range includes possible error due to improper background correction (see Reference 51).

[d] IgE modified in its Fab segments by $N$-(1-pyrene) maleimide (see Reference 51).

[e] Upper limit of range includes possible error in lifetime determinations (see Ref. 51).

[f] Range of values calculated as described in Reference 51.

[g] Assumes rotational motion limited to in-plane rotation of the membrane-bound IgE-receptor complex.

bution can vary with experiment, but does not affect the measured φ. The values of φ obtained for IgE bound to receptor on the vesicles are only slightly larger than those for IgE in solution and much less than >10³ nsec expected for an IgE molecule that rotates in the plane of the membrane as a rigid extention of the complex formed with receptor (Table 3). Similar results were obtained in depolarization experiments using IgE modified in its Fab segments by $N$(1-pyrene)maleimide (Table 3). From a comparison of these results we conclude that IgE segmental flexibility is not greatly altered upon binding to high-affinity receptors. This observation is not consistent with the first structural model described above that has IgE extended at full length from its receptor-binding site, since any rotation of the Fab segments toward the plane of the membrane would result in a combining site to membrane surface distance that is significantly less than 100 Å (see Table 2).

A second approach toward distinguishing structural models for IgE-receptor interaction employed the monoclonal anti-IgE antibodies, A2 and B5. In energy transfer experiments we attached donor and acceptor probes to separate fragments of these antibodies and compared the average distance between these probes for receptor-bound IgE and IgE in solution; this was possible because A2 and B5 bind simultaneously in both cases. In the most unambiguous of these experiments FITC donor probes were reacted at pH 9 with B5 F(ab')$_2$ (anti-Fab$_c$), and iodoacetamido rhodamine acceptor probes were introduced at the C-terminal sulfhydryl groups of A2 Fab' (anti-Fc$_c$) by selective reduction and alkylation of the F(ab')$_2$ interchain disulfide bonds. Figure 12 has the fluorescence-quenching data for these experiments which show significant energy transfer occurred for IgE in solution (△,▲), but not for receptor-bound IgE (□,■), although both donor- and acceptor-labeled antibodies were shown to be binding under these conditions.[52] The data were interpreted as indicating that some change in the average conformation of IgE accompanies its binding to receptor such that, the average

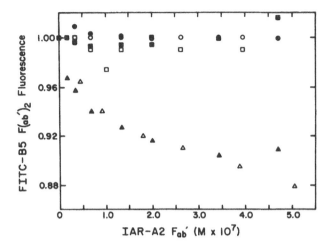

FIGURE 12. Quenching of donor fluorescence of FITC-B5 F(ab')$_2$ due to binding of the acceptor probe IAR-A2 Fab' to IgE. Donors bound to IgE in solution in the presence of membrane vesicles, ($\triangle$,$\blacktriangle$); donors bound to IgE on vesicle-associated receptors, ($\square$,$\blacksquare$); donors and acceptors in solution in the absence of IgE, but in the presence of membrane vesicles, ($\circ$,$\bullet$). Open and closed symbols represent data from separate experiments with different preparations of vesicles; the concentration of receptor-bound IgE was 14 to 19 n$M$, and IgE in solution was 63 to 71. (Reprinted with permission from Holowka, D., Conrad, D. H., and Baird, B., *Biochemistry*, 24, 6260, 1985. Copyright 1985 American Chemical Society.)

distance between probes on B5 bound to the Fab segments of IgE and A2 bound to the Fc segment is altered.[52] This result then supports the model of a "bent" IgE binding to receptor, and this structure, together with bound A2 and B5 fragments, is depicted in Figure 10A. The model is consistent with all of our energy transfer measurements, as well as the biochemical evidence from several laboratories previously discussed (Section II). Features of the structure and its functional implications deserve further comment:

1. The intact A2 antibody cross-links IgE-receptor complexes on the same membrane[37] and also agglutinates vesicles and cells bearing IgE (unpublished observations). Therefore, the orientation of the A2 Fab' segment is shown with its C-terminal end pointed away from the membrane surface such that, an adjoining Fab' would be in position to bind to IgE on the same or on an apposed membrane.
2. Bivalent B5 does not readily cross-link adjacent IgE-receptor complexes nor, does it cause membrane vesicle agglutination, so it is shown to be cross-linking the IgE Fab segments intramolecularly.
3. The model provides a possible explanation for the extremely slow rate of IgE binding to receptors on cells and vesicles.[7,53] The measured forward rate constant for binding, $k \simeq 10^5$ mol$^{-1}$sec$^{-1}$ is several orders of magnitude less than that expected for a diffusion-controlled reaction and strongly suggests that IgE or receptor (or both) must undergo a conformational transition before a stable association occurs.[54] The forward rate constant for IgE-receptor binding in solution is considerably faster than on membranes,[53] and it will be interesting to see whether a different conformation of bound IgE is detected by energy transfer measurements in this case.
4. The model suggests that the bound IgE molecules might direct the orientation of receptor-receptor interactions that lead to the delivery of a cell-triggering signal. A similar suggestion is made by other studies in our laboratory in which we have tested

the A2 and B5 antibodies and compared the amount of cross-linking required for triggering degranulation.[28] A very small amount of cross-linking by B5 is as effective as a large amount of cross-linking by A2 and, this may reflect the importance of proper orientation of the aggregated receptors.

The cartoon in Figure 10B attempts to account for the result shown in Figure 12 that energy transfer from FITC-B5 F(ab')₂ to IAR-A2 Fab' is greater for IgE free in solution than for IgE that is bound to receptor in a bent conformation. We suggest that in solution A2 and B5 can bind simultaneously, either on the same side, or, on the opposite sides of IgE, while binding can occur only on opposite sides of receptor-bound IgE (Figure 10A). Consequently, the *average* distance between the probes on IgE in solution is less, and therefore, the energy transfer would be greater. The models of Figure 10 will be subject to further testing in future energy transfer experiments, including measurements using less bulky probes which will reduce some complications in the interpretation of results.

## V. STUDIES ON OTHER SYSTEMS

There have been a number of other studies carried out to map the structure of membrane proteins with resonance energy transfer measurements. Work prior to 1978 has been reviewed by Stryer,[55] and studies utilizing endogenous tryptophans with lipophilic probes to investigate the spatial properties of purified membrane proteins in lipid bilayers are discussed in another chapter of this volume.[56] Several studies have exploited the uniquely located retinal chromophores in bacteriorhodopsin to study protein-membrane[57,58] and protein-protein[59,60] interactions in this system. Structural mapping of the reconstituted chloroplast coupling factor has been carried out using both fluorescently labeled antibodies[61] and the chemically modified enzyme.[62] Cantley and co-workers have measured an extensive set of distances between specific sites on Band 3 of human erythrocytes[63,64] and also on the purified Na$^+$/K$^+$ ATPase from canine kidney cells,[65] while Angelides and colleagues have mapped spatial relationships between various toxin-binding sites on the Na$^+$ channel from rat brain.[66,67] In one of the small number of energy transfer studies to be carried out on cell surface receptors to date, Johnson et al. have measured separation distances between α-bungarotoxin binding sites on adjacent acetylcholine receptors from *Torpedo* membranes.[68] Detailed descriptions of these studies is beyond the scope of this chapter, but it is clear that successful investigations of structure of membrane proteins by energy transfer measurements are just beginning to be reported and, they are still quite limited compared to the extensive studies carried out on a variety of soluble proteins.[55] The additional technical limitations of studying membrane proteins, including difficulties in isolating sufficient amounts of protein in a functionally intact state and in the placement of specific probes for reasonably precise spectroscopic measurements on highly scattering membrane suspensions are formidable, but are gradually being overcome.

## VI. FUTURE DIRECTIONS

The structural model proposed for receptor-bound IgE is subject to continued testing and refinement by additional energy transfer measurements, as well as by other techniques. The set of distances listed in Table 2 can be extended, and a particularly useful site to label for this purpose is the C-terminal end of IgE. Distances from a probe at this location to the membrane surface, as well as to other sites on receptor-bound IgE would help to confirm the position of the Fc segment. A good possibility for labeling the C-terminal amino acids is by forcing the back reaction of carboxypeptidase Y in the presence of a fluorescent peptide.[69] Distances between the same pairs of sites measured for IgE bound to receptor on the vesicles can also be measured for solubilized IgE-receptor complexes. A comparison of these results will reveal whether the apparent conformational change in IgE that was characterized is unique to the membrane-bound form. The sites of binding and the orientations of the monoclonal anti-IgE antibodies used for placing probes on IgE in our energy transfer measurements are major factors in the interpretation of results in terms of apparent distances. In addition to assessment by spectroscopic measurements and fluorescence microscopy as described in Section IV, a direct visualization of A2 and B5 binding to IgE and IgE-receptor complexes may be obtainable using electron microscopy and negative staining techniques.[70]

Extending our structural map will require placement of probes on the receptor subunits (Figure 1) and investigating the spatial relationships between these sites and sites on IgE. Specific and efficient modification of integral membrane proteins by chemical means is usually very difficult to achieve, and our own limited attempts to label the $\beta$ subunit with fluorescent hydrophobic probes have been only marginly successful.[83] However, fluorescently labeled monoclonal antibodies can be exploited as powerful reagents for this purpose. Several monoclonal antibodies have been reported to bind to the receptor at or near the site of IgE binding,[71] and we have recently succeeded in producing several monoclonal antibodies that may be specific for the cytoplasmically exposed regions[22] of the receptor.[83] The use of antireceptor Fab fragments that are labeled at either their N-terminal or C-terminal residues, as described in Section IV for the monoclonal anti-IgE antibodies (A2 and B5) should allow reasonably accurate estimation of distances between pairs of sites on the IgE-receptor complex of 50 Å and greater.

Use of monoclonal antibodies for the placement of specific probes should greatly facilitate similar structural mapping studies on other complex receptor systems, especially in those cases where purification to homogeneity is not possible or not desirable because of accompanying loss of function. For the measurement of distances between receptor sites and the membrane surface it is highly advantageous to be able to prepare homogeneous, right-side-out plasma membrane vesicles, as we have done for the IgE receptor complex on RBL cells (Section II). This may not be possible for all types of cells, but the use of intact cells for these measurements is feasible if appropriate amphipathic probes can be found that partition into the plasma membrane, but do not readily pass into internal membranes. Several possible candidates exist: Sleight and Pagano have reported that 1-acyl-2-(N-4-nitrobenzo-2-oxa-1,3-diazole-aminocaproyl phosphatidylcholine remains in the outer plasma membrane below 4°C,[72] and Wolf has reported similar localization for certain carbocyanine dyes in several cell types.[73]

Fluorescently labeled monoclonal antibodies should also provide a versatile alternative to chemical modification for placement of probes and measuring distances between antigenic sites within a membrane-associated complex. In this regard, the procedure we have used with A2 and B5 for the IgE-receptor complex on both vesicles and whole cells should be generally applicable. Measurement of distances between separate membrane proteins is of interest in some systems, and some pioneering efforts have been reported, using as probes fluorescent concanavalin A molecules[74,75] or concanavalin A molecules together with monoclonal antibodies against histocompatibility antigens[76] that were nonspecifically labeled with donors or acceptors and bound to cells. Distances between cell surface receptors are of particular interest in systems such as the IgE receptor system where receptor aggregation leads to signal transduction. We have used IgE labeled at the interheavy disulfide bonds with donors or acceptors and compared energy transfer before and after cross-linking IgE-receptor complexes with A2 (Menon et al., unpublished results). Preliminary results from these studies indicate that the average distances between aggregated receptors is too large to be measured accurately by conventional donor-acceptor pairs with $R_0 < 60$ Å. For situations where a precise site-site distance measurement is not as important as observing a change in the average distance between receptors, monoclonal antibodies or their Fab' fragments labeled with selected phycobiliproteins hold great promise. This set of highly fluorescent phycobiliproteins that are purified from photosynthetic light harvesting complexes have spectral properties that yield $R_0$ values upward of 80 Å, and surface-to-surface separation distances as great as 150 Å between these probes can be measured.[77]

The IgE receptor system on RBL cells has proven to be amenable to detailed structural mapping by energy transfer measurements using a variety of probes and experimental approaches. A large part of our interest in extending this work is to develop methodologies that have general applicability. While continuing to use the RBL cells and IgE receptor system as a kind of "proving ground" we are beginning to test some of the experimental approaches on more technically difficult receptors, such as the receptor for antigen on human T cells[78] and the low-affinity receptor for IgE on murine B cells.[79] We anticipate that these techniques will be valuable for structural studies on a wide variety of membrane proteins.

## ACKNOWLEDGMENTS

Research in our laboratory has been supported by grants AI18306 and AI18610 from the National Institutes of Health and in part by a grant from the Cornell Biotechnology Program, which is sponsored by the New York State Science and Technology Foundation and a consortium of industries.

## REFERENCES

1. **Kanellopoulos, J., Rossi, G., and Metzger, H.,** Preparative isolation of the cell receptor for immunoglobulin E, *J. Biol. Chem.*, 254, 7691, 1979.
2. **Shorr, R. G. L., Heald, S. L., Jeffs, P. W., Lavin, T. H., Strohsacher, M. W., Lefkowitz, R. J., and Caron, M. C.,** The β-adrenergic receptor: rapid purification and covalent labeling by photoaffinity crosslinking, *Proc. Natl. Acad. Sci. U.S.A.*, 79, 2778, 1982.
3. **Hammes, G. G.,** Fluorescence methods, in *Protein-Protein Interactions*, Frieden, C. and Nichol, L. W., Eds., Wiley-Interscience, New York, 1981, 257.
4. **Lakowicz, J. R.,** *Principles of Fluorescence Spectroscopy*, Plenum Press, New York, 1983.
5. **Dorrington, K. J. and Bennich, H. H.,** Structure-function relationships in human immunoglobulin E, *Immunol. Rev.*, 41, 3, 1978.
6. **Metzger, H.,** The receptor on mast cells and related cells with high affinity for IgE, *Contemp. Top. Mol. Immunol.*, 9, 115, 1983.
7. **Kulczycki, A., Jr., and Metzger, H.,** The interaction of IgE with rat basophilic leukemia cells. II. Quantitative aspects of the binding reaction, *J. Exp. Med.*, 140, 1676, 1974.
8. **Isersky, C., Rivera, J., Mims, S., and Triche, T. J.,** The fate of IgE bound to rat basophilic leukemia cells, *J. Immunol.*, 122, 1926, 1979.
9. **Mendoza, G. and Metzger, H.,** Distribution and valency of receptor for IgE on rodent mast cells and related tumor cells, *Nature (London)*, 264, 548, 1976.
10. **Schlessinger, J., Webb, W. W., Elson, E. L., and Metzger, H.,** Lateral motion and valence of $F_c$ receptors on rat peritoneal mast cells, *Nature (London)*, 264, 550, 1976.
11. **Becker, E. L., Simon, A. S., and Austen, K. F., Eds.,** *Biochemistry of the Acute Allergic Reactions*, Alan R. Liss, New York, 1981.
12. **Segal, D. M., Taurog, J. D., and Metzger, H.,** Dimeric immunoglobulin E serves as a unit signal for mast cell degranulation, *Proc. Natl. Acad. Sci. U.S.A.*, 74, 2993, 1977.
13. **Fewtrell, C. and Metzger, H.,** Larger oligomers of IgE are more effective than dimers in stimulating rat basophilic leukemia cells, *J. Immunol.*, 125, 701, 1980.
14. **Ishizaka, T. and Ishizaka, K.,** Biology of immunoglobulin E: molecular basis of reaginic hypersensitivity, *Prog. Allergy*, 19, 60, 1975.
15. **Ishizaka, T., Chang, T. H., Taggert, M., and Ishizaka, K.,** Histamine release from rat mast cells by antibodies against rat basophilic leukemia cell membrane, *J. Immunol.*, 119, 1589, 1977.
16. **Isersky, C., Taurog, J. D., Poy, G., and Metzger, H.,** Triggering of cultured neoplastic mast cells by antibodies to the receptor for IgE, *J. Immunol.*, 121, 549, 1978.

17. **Metzger, H.,** The effect of antigen on antibodies: recent studies, *Contemp. Top. Mol. Immunol.,* 7, 119, 1978.
18. **Kahn, C. R., Baird, K. K., Jarrett, D. B., and Flier, J. S.,** Direct demonstration that receptor crosslinking or aggregation is important in insulin action, *Proc. Natl. Acad. Sci. U.S.A.,* 75, 4209, 1978.
19. **Schreiber, A. B., Libermann, T. A., Lax, I., Yarden, Y., and Schlessinger, J.,** Biological role of epidermal growth factor-receptor clustering. Investigation with monoclonal anti-receptor antibodies, *J. Biol. Chem.,* 258, 846, 1983.
20. **Goetze, A., Kanellopoulos, J., Rice, D., and Metzger, H.,** Enzymatic cleavage products of the α-subunit of the receptor for IgE, *Biochemistry,* 20, 6341, 1981.
21. **Holowka, D., Gitler, C., Bercovici, T., and Metzger, H.,** Reaction of 5-iodonaphtlyl-1-nitrene with IgE receptor on normal and tumor mast cells, *Nature (London),* 289, 806, 1981.
22. **Holowka, D. and Baird, B.,** Lactoperoxidase-catalyzed iodination of the receptor for immunoglobulin E at the cytoplasmic side of the plasma membrane, *J. Biol. Chem.,* 259, 3720, 1984.
23. **Fewtrell, C., Goetze, A., and Metzger, H.,** Phosphorylation of the receptor for immunoglobulin E, *Biochemistry,* 21, 2004, 1982.
24. **Perez-Monfort, R., Kinet, J.-P., and Metzger, H.** A previously unrecognized subunit of the receptor for immunoglobulin E, *Biochemistry,* 22, 5722, 1983.
25. **Kannellopoulos, J. M., Liu, T. Y., Poy, G., and Metzger, H.,** Composition and structure of the cell receptor for immunoglobulin E, *J. Biol. Chem.,* 255, 9060, 1980.
26. **Liu, F.-T. and Orida, N.,** Synthesis of surface immunoglobulin E receptor in xenopus oocytes by translation of mRNA from rat basophilic leukemia cells, *J. Biol. Chem.,* 259, 10649, 1984.
27. **Menon, A. K., Holowka, D., Webb, W. W., and Baird, B.,** Clustering, mobility, and triggering activity of small oligomers of immunoglobulin E on rat basophilic leukemia cells, *J. Cell Biol.,* 102, 534, 1986.
28. **Menon, A. K., Holowka, D., Webb, W. W., and Baird, B.,** Crosslinking of receptor-bound immunoglobulin E to aggregates larger than dimers leads to rapid immobilization, *J. Cell Biol.,* 102, 541, 1986.
29. **Mazurek, N., Schindler, H., Schürholz, Th., and Pecht, I.,** The cromolyn binding protein constitutes the $Ca^{2+}$ channel of basophils opening upon immunological stimulation, *Proc. Natl. Acad. Sci. U.S.A.,* 21, 6841, 1984.
30. **Beaven, M. A., Moore, J. P., Smith, G. A., Hesketh, T. R., and Metcalfe, J. C.,** The calcium signal and phosphatidylinositol breakdown in 2H3 cells, *J. Biol. Chem.,* 259, 7137, 1984.
31. **Gomperts, B. D.,** Involvement of guanine nucleotide binding protein in the gating of $Ca^{2+}$ by receptors, *Nature (London),* 306, 64, 1983.
32. **Ishizaka, K., Ishizaka, T., and Lee, E.,** Biologic function of the Fc fragments of E myeloma protein, *Immunochemistry,* 7, 687, 1970.
33. **Dorrington, K. J. and Bennich, H.,** Thermally induced strutural changes in immunoglobulin E, *J. Biol. Chem.,* 248, 8378, 1973.
34. **Ishida, N., Veda S., Hayashida, H., Miyata, T., and Honjo, T.,** The nucleotide sequence of the mouse immunoglobulin epsilon gene: comparison with the human epsilon gene sequence, *EMBO J.,* 1, 1117, 1982.
35. **Liu, F.-T., Albrandt, K., Sutcliffe, J. G., and Katz, D. H.,** Cloning and nucleotide sequene of mouse immunoglobulin ε chain c DNA, *Proc. Natl. Acad. Sci. U.S.A.,* 79, 7852, 1982.
36. **Perez-Montfort, R. and Metzger, H.,** Proteolysis of soluble IgE-receptor complexes: localization of sites on IgE which interact with the Fc receptor, *Mol. Immunol.,* 19, 1113, 1982.
37. **Conrad, D. H., Studer, E., Gervasoni, J., and Mohanakumar, T.,** Properties of two monoclonal antibodies directed against the Fc and Fab' Regions of Rat IgE, *Int. Arch. Allergy Appl. Immunol.,* 70, 352, 1983.
38. **Baniyash, M. and Eshhar, Z.,** Inhibition of IgE binding to mast cells and basophils by monoclonal antibodies to murine IgE, *Eur. J. Immunol.,* 14, 799, 1984.
39. **Fung, B. K-K. and Stryer, L.,** Surface density determination in membranes by fluorescence energy transfer, *Biochemistry,* 17, 5241, 1978.
40. **Wolber, P. K. and Hudson, B. S.,** An analytical solution to the Förster energy transfer problem in two dimensions, *Biophys. J.,* 28, 197, 1979.
41. **Dewey, T. G. and Hammes, G. G.,** Calculation of fluorescence resonance energy transfer on surfaces, *Biophys. J.,* 32, 1023, 1980.
42. **Shaklai, N., Yguerabide, J., and Ranney, H. M.,** Interaction of hemoglobulin with red blood cell membranes as shown by a fluorescent chromophore, *Biochemistry,* 16, 5585, 1977.
43. **Snyder B. and Freire, E.,** Fluorescence energy transfer in two dimensions. A numeric solution for random and nonrandom distributions, *Biophys. J.,* 40, 137, 1982.
44. **Koppel, D. E., Fleming, P. J., and Strittmatter, P.,** Intramembrane positions of membrane-bound chromophores determined by èxcitation energy transfer, *Biochemistry,* 18, 5450, 1979.

45. **Holowka, D. and Baird, B.**, Structural studies on the membrane-bound immunoglobulin E-receptor complex. I. Characterization of large plasma membrane vesicles from rat basophilic leukemia cells and insertion of amphipathic fluorescent probes, *Biochemistry*, 22, 3466, 1983.

46. **Scott, R. E.**, Plasma membrane vesiculation: a new technique for isolation of plasma membranes, *Science*, 194, 743, 1976.

47. **Tank, D. W., Wu, E.-S., and Webb, W. W.**, Enhanced molecular diffusibility in muscle membrane blebs: release of lateral constraints, *J. Cell Biol.*, 92, 207, 1982.

48. **Baird, B. and Holowka, D.**, Structural mapping of Fc receptor-bound immunoglobulin E: proximity to the membrane surface of the antibody combining site and another site in the Fab segments, *Biochemistry*, 24, 6252, 1985.

49. **Holowka, D. and Baird, B.**, Structural studies on the membrane-bound immunoglobulin E-receptor complex. II. Mapping of distances between sites on IgE and the membrane surface, *Biochemistry*, 22, 3475, 1983.

50. **Kaplan, H., Long, B. G., and Young, N. M.**, Chemical properties of functional groups of mouse immunoglobulin of the IgA, IgG2a, and IgM classes, *Biochemistry*, 19, 2821, 1980.

51. **Slattery, J., Holowka, D., and Baird, B.**, Segmental flexibility of receptor-bound immunoglobulin E, *Biochemistry*, 24, 7810, 1985.

52. **Holowka, D., Conrad, D. H., and Baird, B.**, Structural mapping of membrane-bound immunoglobulin E-receptor complexes: use of monoclonal anti-IgE antibodies to probe the conformation of receptor-bound IgE, *Biochemistry*, 24, 6260, 1985.

53. **Wank, S. A., De Lisi, C., and Metzger, H.**, Analysis of the rate-limiting step in a ligand-cell receptor interaction: the immunoglobulin E system, *Biochemistry*, 22, 954, 1983.

54. **Koren, R. and Hammes, G. G.**, A kinetic study of protein-protein interactions, *Biochemistry*, 15, 1165, 1976.

55. **Stryer, L.**, Fluorescence energy transfer as a spectroscopic ruler, *Annu. Rev. Biochem.*, 47, 819, 1978.

56. **Kleinfeld, A.**, Tertiary structure of membrane proteins determined by fluorescence resonance energy transfer, in *Spectroscopic Membrane Probes*, Loew, L., Ed., CRC Press, Boca Raton, Fla., in press.

57. **Rehorek, M., Dencher, N. A., and Heyn, M. P.**, Fluorescence energy transfer from diphenylhexatriene to bacteriorhodopsin in lipid vesicles, *Biophys. J.*, 43, 39, 1983.

58. **Thomas, D. D. and Stryer, L.**, Transverse location of the retinal chromophore of rhodopsin in rod outer segment disc membranes, *J. Mol. Biol.*, 154, 145, 1982.

59. **London, E. and Khorana, H. G.**, Denaturation and renaturation of bacteriorhodopsin in detergents and lipid-detergent mixtures, *J. Biol. Chem.*, 257, 7003, 1982.

60. **Hasselbacher, C. A., Street, T. L., and Dewey, T. G.**, Resonance energy transfer as a monitor of membrane protein domain segregation. Application to the aggregation of bacteriorhodopsin reconstituted into phospholipid vesicles, *Biochemistry*, 23, 6445, 1984.

61. **Baird, B. A., Pick, U., and Hammes, G. G.**, Structural investigation of reconstituted chloroplast ATPase with fluorescence measurements, *J. Biol. Chem.*, 254, 3818, 1979.

62. **Cerione, R. A., McCarty, R. E., and Hammes, G. G.**, Spatial relationships between specific sites on reconstituted chloroplast $H^+$-ATPase and the phospholipid vesicle surface, *Biochemistry*, 22, 769, 1983.

63. **Rao, A., Martin, P., Reithmeier, A. F., and Cantley, L. C.**, Location of the stilbene disulfonate binding site of the human erythiocyte anion-exchange system by resonance energy transfer, *Biochemistry*, 18, 4505, 1979.

64. **Macara, I. G. and Cantley, L. C.**, Interactions between transport inhibitors at the anion binding sites of the band 3 dimer, *Biochemistry*, 20, 5095, 1981.

65. **Carilli, C. T., Farley, R. A., Perlman, D. M., and Cantley, L. C.**, The active site structure of $Na^+$- and $K^+$-stimulated ATPase. Location of a specific fluorescein isothiocyanate reactive site, *J. Biol. Chem.*, 257, 5601, 1982.

66. **Angelides, K. J. and Nutter, T. J.**, Molecular and cellular mapping of the voltage-dependent $Na^+$ channel, *Biophys. J.*, 45, 31, 1984.

67. **Angelides, K. J. and Brown, G. B.**, Fluorescence resonance energy transfer on the voltage-dependent sodium channel, *J. Biol. Chem.*, 259, 6117, 1984.

68. **Johnson, D. A., Voet, J. G., and Taylor, P.**, Fluorescence energy transfer between cobra α-toxin molecules bound to the acetylcholine receptor, *J. Biol. Chem.*, 259, 5717, 1984.

69. **Widmer, F. and Johansen, J. T.**, Carboxypeptidase Y as a catalyst for peptide synthesis, *Carlsberg Res. Commun.*, 44, 37, 1979.

70. **Roux, K. H. and Metzger, D. W.**, Immunoelectron microscopic localization of idiotypes and allotypes on immunoglobulin molecules, *J. Immunol.*, 129, 2548, 1982.

71. **Basciano, L. K., Berenstein, E. H., Kmak, L., and Siraganian, R. P.**, Monoclonal antibodies to the high affinity Fc receptor on rat basophilic leukemia cells, *Fed. Proc. Fed. Am. Soc. Exp. Biol.*, 43 (Abstr. 1221), 1625, 1984.

72. **Sleight, R. G. and Pagano, R. E.**, Transport of a fluorescent phosphatidylcholine analog from the plasma membrane to the golgi apparatus, *J. Cell Biol.*, 99, 742, 1984.

73. **Wolf, D. E.**, Determination of the siddeness of carbocyanine dye labeling of membranes, *Biochemistry*, 24, 582, 1985.

74. **Fernandez, S. M. and Berlin, R. D.**, Cell surface distribution of lectin receptors determined by resonance energy transfer, *Nature (London)*, 264, 411, 1976.

75. **Dale, R. E., Norvos, J., Roth, S., Edidin, M., and Brand, L.**, Application of Förster long-range excitation energy transfer to the determination of distributions of fluorescently labeled concavavalin A-receptor complexes at the surfaces of yeast and of normal and malignant fibroblasts, in *Fluorescent Probes*, Beddard, G. S. and West, M. A., Eds., Academic Press, New York, 1980, 159.

76. **Damjanovich, S., Tron, L., Szollosi, J., Zidovetski, R., Vaz, W. L. C., Regateiro, F., Arndt-Jovin, D. J., and Jovin, T. M.**, Distribution and mobility of murine histocompatibility H-2K$^k$ antigen in the cytoplasmic membrane, *Proc. Natl. Acad. Sci. U.S.A.*, 80, 5985, 1983.

77. **Glaser, A. N. and Stryer, L.**, Phycofluor probes, *Trends Biochem. Sci.*, 9, 423, 1984.

78. **Meuer, S. C., Acuto, O., Hercend, T., Schlossman, S., and Reinherz, E.**, The human T-cell receptor, *Annu. Rev. Immunol.*, 2, 23, 1984.

79. **Conrad, D. H. and Peterson, L. H.**, The murine lymphocyte receptor for IgE. I. Isolation and characterization of the murine B cell Fc receptor and comparison with Fc receptors from rat and human, *J. Immunol.*, 132, 796, 1984.

80. **Amzel, L. M. and Poljak, R. J.**, Three-dimensional structure of immunoglobulins, *Annu. Rev. Biochem.*, 48, 961, 1979.

81. **Hammes, S., Holowka, D., and Baird, B.**, submitted for publication.

82. **Robertson, D., Holowka, D., and Baird, B.**, Cross-linking of immunoglobulin E-receptor complexes induces their interaction with the cytoskeleton of rat basophilic leukemia cells, *J. Immunol.*, 136, 4565, 1986.

83. **Baird, B. and Holowka, D.**, unpublished results.

Chapter 6

FLUORESCENCE ASSAYS FOR MEMBRANE FUSION

Nejat Düzgüneş and Joe Bentz

TABLE OF CONTENTS

"Every judgement in science stands on the edge of error, and is personal."

J. Bronowski
*The Ascent of Man*

## I. INTRODUCTION

Many fundamental biological processes involve membrane fusion. In higher organisms life starts with sperm-egg fusion, which is preceded by the acrosome reaction in the sperm and followed by the cortical reaction in the egg, both of which are membrane-fusion reactions.[1-3] The extracellular release of neurotransmitters, hormones, and enzymes is accomplished by the fusion of secretory vesicles with the plasma membrane.[4-7] Release of inflammatory mediators by mast cells, platelets, and neutrophils also takes place by exocytosis.[8,9] Receptor-mediated and fluid-phase endocytosis as well as phagocytosis proceed via the fusion of the outer leaflet of the plasma membrane with itself following invagination.[10,11] The endocytotic vesicles thus formed fuse with each other to form endosomes, which may then fuse with primary lysosomes to form secondary lysosomes.[12-15] Receptor recycling to the plasma membrane is thought to occur by the budding off of vesicles from the endosome which involves the fusion of the lumenal surfaces of the endosome.[16] Vesicles which transport secretory proteins and newly synthesized membrane to the plasma membrane bud off from the Golgi apparatus via membrane fusion.[17,18]

Membrane fusion is also involved in a variety of pathological processes. For example, lipid-enveloped viruses infect their target cells by fusing with either the plasma membrane (Sendai virus[19,20]) or with the endosome membrane after internalization by the endocytotic pathway (influenza virus, Semliki Forest virus[21,22]). Multinucleate giant cells are formed in inflamed tissue by fusion of macrophages.[23] Irreversibly sickled cells are thought to form as a consequence of membrane loss via vesiculation.[24] Neutrophils release lytic enzymes which cause tissue damage by the fusion of secretory vesicles with phagosomes that are not completely sealed off from the extracellular milieu.[25,26] Human immunodeficiency virus (LAV, HTLV-III) causes extensive fusion of infected T-lymphocytes.[27]

Many of these fusion processes have been discovered and/or visualized by electron microscopy. Some insights into the molecular mechanisms of fusion, such as the relative roles of lipids and proteins, have been obtained by this technique.[28-34] However, observations by quick-freezing freeze-fracture electron microscopy have pointed to the possible artifacts of fixation and cryoprotectants.[35,36] The rapidity of fusion events also presents difficulties in being able to detect them with electron microscopy. Thus, stimulation and arrest by quick-freezing of exocytotic fusion events in the neuromuscular junction has to be performed in the msec range;[37,38] this is also true for the fusion of certain types of phospholipid vesicles.[39,40]

Several stages can be identified within the overall process of membrane fusion:

1. Adhesion of the membranes, or, in the case of membrane vesicles, aggregation
2. Close approach of the membranes to establish molecular contact
3. Destabilization of the membrane in the zone of contact
4. Intermixing of membrane components
5. Communication between the internal aqueous contents of the compartments bounded by the membranes

The first and second steps correspond to the formation of a pentalaminar structure as seen in transmission electron microscopy, in Palade's[4] description of exocytotic membrane fusion. The third step has been observed by Ornberg and Reese[41] during exocytosis in Limulus amebocytes as a defect in molecular packing in the region of adhesion of the secretory

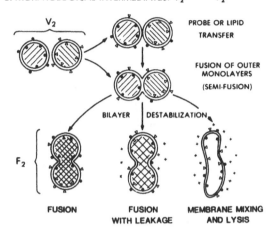

A. KINETIC MODEL: $V_1 + V_1 \rightleftharpoons V_2 \longrightarrow F_2$

B. MORPHOLOGICAL INTERMEDIATES: $V_2 \longrightarrow F_2$

$V_2$

PROBE OR LIPID TRANSFER

FUSION OF OUTER MONOLAYERS (SEMI-FUSION)

BILAYER | DESTABILIZATION

$F_2$

FUSION

FUSION WITH LEAKAGE

MEMBRANE MIXING AND LYSIS

C. CONTINUED AGGREGATION AND FUSION:
MULTIVESICULAR STRUCTURES

FIGURE 1. The mass-action kinetic model of membrane fusion (A), and the possible morphological changes during fusion (B). Vesicle aggregation may be accompanied by lipid (or probe) transfer between the membranes. The outer monolayers of the vesicles may fuse, forming a trilaminar diaphragm (semifusion). Destabilization of bilayer structure in the region of contact leads to communication between the aqueous contents (fusion). Fusion may be accompanied by leakage or it may be nonleaky. In some cases, only mixing of membrane components and lysis are observed. This case also represents the interaction of PE-containing vesicles at temperatures above that required for the phase transition to the hexagonal phase. Intermembrane contact results in lysis and the transformation of the lipid to inverted tubes. At Step (C), the continued aggregation and fusion of the vesicles often leads to leakage to contents due to the collapse of the larger multivesicular structures.

vesicle to the plasma membrane. The pentalaminar structure may transform into a trilaminar diaphragm (i.e., a single bilayer) in a process called "fusion" by Palade,[4] and "semifusion" in our description.[42] This step would be accompanied with the intermixing of lipids and proteins from the outer leaflets of the two membranes, without a communication between the aqueous compartments, i.e., the lumen of the secretory vesicle and the extracellular milieu. Exocytosis would occur when the contents of the secretory vesicle are exposed to the outside of the cell via fusion of the two membranes; this stage is termed "fission" by Palade.[4] These stages of fusion are depicted schematically in Figure 1.

The molecular mechanisms of membrane fusion are largely unknown. Considerable progress has been made by the use of phospholipid vesicles to identify the role of individual phospholipid species,[42-47] cytoplasmic proteins,[42,48] and by the study of the fusion proteins of lipid-enveloped viruses.[49] Many biochemical and biophysical techniques have been used to show if membrane fusion has occurred and to ascertain the extent of the process.[44,45,50,124] The development of fluorescence assays for membrane fusion has facilitated greatly the understanding of the mechanisms of fusion through the study of the kinetics of aggregation and fusion. Fluorescence seems to be a uniquely favored technique for studying this problem because of its sensitivity and its ability to reveal where molecules move during fusion.

In this communication, we review various assays for the intermixing of aqueous contents and membrane components of phospholipid vesicles, and address ourselves to the reliability of the assays. In Section II, we describe the experimental methodology for three well-known fusion assays. In Section III, we show how these assays can be used to measure the primary kinetic rate constants of vesicle aggregation and fusion. We then describe in Section IV the elements found to be necessary to construct properly operating vesicle fusion assays. In Section V, we critique some of the fusion assays developed to date, show comparative fusion kinetics obtained by several assays, and summarize results obtained with some phospholipid vesicle systems. We outline the application of fluorescence assays to measure the kinetics of fusion of biological membranes in Section VI.

## II. ELEMENTS OF FLUORESCENCE ASSAYS FOR VESICLE FUSION

No fusion assay can be presumed correct, *a priori*. Only after testing its responses to a wide variety of fusogenic events, can we come to trust our interpretation of its signals. Prior to the fluorescence assays, vesicle fusion was monitored either by an "irreversible" increase in average particle size, using techniques such as gel filtration, light absorbance, dynamic light scattering, electron microscopy and NMR, or by the leakage of contents, using radioactive tracers (see reviews,[42,44,45]). These techniques required large lipid concentrations, by current fluorometric standards, and could not monitor the initial events of membrane fusion. They could only describe what happened after many rounds of fusion. These techniques found that small unilamellar vesicles (SUV) composed of phosphatidylserine (PS) would collapse to anhydrous structures when treated with $Ca^{2+}$.[51,53] Perforce, their aqueous contents were lost by this time.[53-57]

The continuous monitoring of leakage of contents from SUV (PS) in the presence of $Ca^{2+}$ (using the relief of self-quenching of carboxyfluorescein (CF) encapsulated in the liposomes) showed that at low lipid concentrations the release followed second order kinetics, i.e., leakage required interbilayer contact.[53] With the advent of the Terbium (Tb)/dipicolinic acid (DPA) assay, it was shown that PS SUV underwent mixing of contents before the complete release of contents.[58-64] It was also demonstrated that under certain ionic conditions SUV (PS) could fuse and retain their contents over hours (see Figure 19[65]).

In this section we describe in detail the most rigorously examined assays for membrane fusion: two assays for the intermixing of aqueous contents of phospholipid vesicles, and one for the mixing of membrane components.

### A. The Tb/DPA Assay

The assay is based on the interaction of Tb and DPA, initially encapsulated in two separate vesicle populations. Tb is encapsulated as the citrate or nitrilotriacetate complex, since it would otherwise interact with negatively charged lipids.[59] Communication between the aqueous compartments of fusing vesicles (Figure 1) results in the formation of the fluorescent $Tb(DPA)_3^{3-}$ chelation complex.[66-68] The complex is excited at a wavelength near the absorption maximum of DPA, since the fluorescence is generated via internal energy transfer from the ligand to Tb.[67] The fluorescence intensity of Tb increases by four orders of magnitude when it forms a complex with DPA. The complex has emission maxima at 491 and 545 nm and is excited at 276 nm. To perform a fusion assay with large unilamellar vesicles (LUV), the 545 nm peak is detected by simply inserting a cut-off filter (transmitting light above 530 nm) in one emission channel of a "T-format" fluorometer. Light scattering from the sample is monitored in the other channel through a monochromator set at 276 nm. For SUV, the emission monochromator is usually set at 545 nm, with a cut-off filter (>530 nm). Light scattering in this case can be measured simultaneously in the second emission channel by the use of a UV band-pass filter.

For experiments utilizing LUV, the following solutions are encapsulated:

1.  2.5 m$M$ TbCl$_3$, 50 m$M$ Na citrate ("Tb-vesicles")
2.  50 m$M$ Na dipicolinate, 20 m$M$ NaCl ("DPA-vesicles")

In addition, the media contain appropriate buffers (usually at pH 7.4) that do not interfere with the formation of the Tb/DPA complex.[67] It is important to have a 10 to 20-fold excess of citrate (to chelate the Tb efficiently) and DPA (to ensure that all the Tb is chelated by DPA during the fusion reaction). These concentrations of material were chosen to balance the osmolality of the internal and external solutions. In our laboratory LUV are usually prepared by reverse phase evaporation followed by extrusion through polycarbonate membranes of defined pore diameter.[59,69-71]

Higher concentrations of Tb citrate and DPA can be encapsulated in SUV (approx. 30 nm diameter), since these vesicles are not osmotically active. Moreover, the small internal compartment of SUV necessitate the use of higher concentrations of the reactants to be able to attain a sufficient fluorescence signal:

1.  15 m$M$ TbCl$_3$, 150 m$M$ Na citrate
2.  150 m$M$ Na dipicolinate

SUV are prepared by sonication in a bath-type sonicator under an argon atmosphere, followed by ultracentrifugation to remove any larger vesicles.[59,72-74]

The vesicles are separated from unencapsulated material by gel filtration on Sephadex® G-75 or other appropriate filtration medium, using 100 m$M$ NaCl, 1 m$M$ EDTA (buffered to pH 7.4) as the elution buffer. EDTA is included in the buffer to prevent the binding of Tb$^{3+}$ to the membrane as the citrate is diluted during column chromatography. A portion of the Tb-vesicles freed of unencapsulated material is then chromatographed a second time on Sephadex® G-75 equilibrated with NaCl buffer to eliminate the external EDTA, since EDTA strongly interferes with the Tb-DPA reaction. These vesicles are used for calibrating the Tb fluorescence scale. This is accomplished by lysing an aliquot of these vesicles (equivalent to the amount used in the fusion assay) by cholate (0.5%, w/v) or octaethyleneglycol-dodecyl ether (C$_{12}$ E$_8$; 0.8 m$M$) in the presence of 20 μ$M$ free DPA to chelate all the Tb released from the vesicles. The fluorescence intensity observed during the assay gives the percentage of the total amount of Tb present that is associated with DPA.

Tb-vesicles and DPA-vesicles are usually mixed at a 1:1 ratio in an aqueous medium consisting of 100 m$M$ NaCl, 0.1 m$M$ EDTA, pH 7.4. EDTA and divalent cations such as Ca$^{2+}$, Ba$^{2+}$, and Sr$^{2+}$ interfere strongly with the formation of the Tb/DPA complex. Thus, any leakage of contents into the external medium during fusion does not result in the generation of fluorescence when using the Tb/DPA assay.

The kinetics of the release of aqueous contents during membrane fusion can be measured by following the increase in fluorescence of CF encapsulated in liposomes at a self-quenching concentration.[53,59,60,70,75-77] Early studies with PS vesicles indicated that fusion of SUV (PS) in the presence of Ca$^{2+}$ and 100 m$M$ NaCl is accompanied by the leakage of approximately 10% of the contents per fusion event (Figure 2A), whereas the fusion of LUV (PS) is initially nonleaky (Figure 2B).[59,61,62,64] Figure 2 shows that leakage can also be measured by the

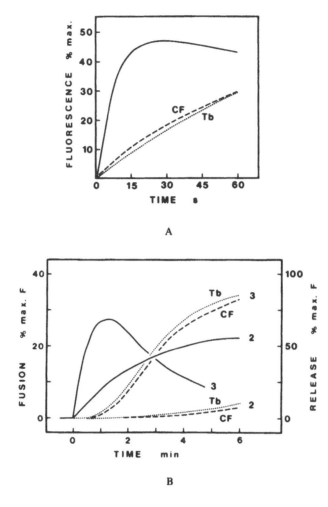

A

B

FIGURE 2.   Time-course of fusion of PS SUV (A) and LUV (B) in the presence of Ca²⁺, followed by the Tb/DPA assay (solid lines). In A, 1.5 m*M* Ca²⁺ was added at t = 0. In B, the indicated concentrations of Ca²⁺ were added at t = 0. Release of contents was followed by the CF-release (dashed line) and Tb-release (dotted line) assays. For the Tb-release assay, 100% fluorescence was determined by lysing the vesicles in the presence of 10 μ*M* DPA and the indicated Ca²⁺ concentration, without EDTA in the medium. Lipid concentrations were 50 μ*M* (From Wilschut J., Düzgüneş, N., Hoekstra, D., and Papahadjopoulos, D., *Biochim. Biophys. Acta*, 734, 309, 1983. With permission.)

release of Tb into a medium containing free DPA.[61] In this case it is also possible that DPA is entering the vesicle interior and causing an increase in the Tb fluorescence.

An alternative method for measuring the release of contents and the entry of medium into the vesicle interior is to follow the dissociation of Tb/DPA complex preencapsulated in the liposomes.[65,78-81] In this case the medium has to contain at least 0.1 m*M* EDTA to achieve the dissociation. During Ca²⁺-induced fusion of PS LUV the dissociation appears to be faster than the release of CF, indicating that the medium is entering the interior of the vesicles in addition to the leakage of internal contents.[82]

## B. The ANTS/DPX Assay

This assay is based on the quenching of 1-aminonaphthalene-3,6,8-trisulfonic acid (ANTS) fluorescence by the collisional quencher $N,N'$-$p$-xylylenebis-(pyridinium bromide) (DPX).[83] ANTS is encapsulated in one population of vesicles and DPX in another. Membrane fusion results in the intermixing of ANTS and DPX, and the quenching of fluorescence.[84-87] This assay can be used at pH down to at least 4.0, whereas the protonation of DPA around pH 5.0 prevents the formation of the Tb/DPA complex. Leakage of contents into the medium does not contribute to the decrease in fluorescence, since the collisional quenching property of DPX is highly concentration dependent.[88] Likewise, cations such as $Ca^{2+}$, $Co^{2+}$, and $La^{3+}$ do not interfere with ANTS fluorescence.[193] In a typical assay, the following solutions are encapsulated in separate populations of LUV, together with low concentrations (5 to 10 m$M$) of buffer:

1. 25 m$M$ ANTS, 40 m$M$ NaCl ("ANTS-vesicles")
2. 90 m$M$ DPX ("DPX-vesicles")

The vesicles are separated from unencapsulated material by chromatography on Sephadex® G-75 using 100 m$M$ NaCl, 1 m$M$ EDTA, and buffer as the elution medium.

In a typical experiment, the vesicles are mixed at a 1:1 molar ratio (usually 0.05 μmol lipid per m$\ell$), and the fluorescence arising from the ANTS-vesicles is set to 100%. The fluorescence of vesicles containing 12.5 m$M$ ANTS/45 m$M$ DPX/20 m$M$ NaCl (i.e., a 1:1 mixture of the media encapsulated in the individual populations) is set to 0%. This fluorescence should be very close to the residual fluorescence from the buffer. The latter type of vesicle is also used to measure the leakage of contents, since the dilution of ANTS and DPX into the medium relieves the quenching of fluorescence. The 100% value of fluorescence is set by lysing the vesicles with 0.1% (w/v) Triton® X-100. Fluorescence is excited at 384 nm[84] or 360 nm.[85-87] Typical fusion and leakage curves for LUV (PS) in the presence of 5 m$M$ $Ca^{2+}$ are shown in Figure 3. For SUV, systematic studies on the usage of the ANTS/DPX assay have not been done; however, we have noted excessively high binding of ANTS to preformed SUV (PS).[89]

Other assays for contents mixing are listed in Table 1A.

## C. The NBD/Rh Assay

Assays for lipid mixing during membrane fusion are often based on resonance energy transfer (RET) between donor and acceptor molecules. The assay we will describe here involves $N$-(7-nitro-2,1,3-benzoxadiazol-4-yl)phosphatidylethanolamine (NBD-PE) as the donor, and $N$-(lissamine rhodamine B sulfonyl) phosphatidylethanolamine (Rh-PE) as the acceptor.[90] In one version of the assay, both molecules are incorporated in the bilayer of one population of vesicles ("labeled" vesicles) at a molar concentration of 0.5 to 1.0% each. At these concentrations, the fluorescence of NBD is transferred to Rh. Thus, fluorescence monitored at the emission maximum of NBD is low. These vesicles are mixed with a population of "unlabeled" vesicles, usually at a 1:1 ratio. Fusion results in the dilution of the probe molecules into the unlabeled membranes. Since RET depends critically on the distance between the fluorophores,[91] NBD fluorescence increases when the average distance between NBD and Rh increases, since the efficiency of RET between the fluorophores is reduced. Fluorescence is monitored at 525 or 530 nm, with the excitation monochromator set at 450 or 475 nm. Light scattering can be simultaneously measured in the second emission channel by using a band-pass filter at 450 nm.

In another version of the assay, NBD-PE and Rh-PE are incorporated (often at 2 mol %) into separate populations of vesicles. The vesicles are initially mixed at a 1:1 ratio. Fusion results in the quenching of NBD fluorescence by Rh because of the proximity of the fluorophores.[92,93] Other types of lipid-mixing assays are listed in Table 1B.

Comparative aspects of the assays described above will be given in Section V.

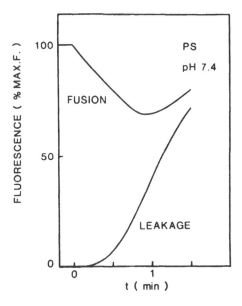

FIGURE 3. Fusion of LUV (PS) in the presence of 5 m*M* Ca$^{2+}$ followed by the ANTS/DPX assay, and the accompanying leakage of contents, monitored by the release of ANTS/DPX complex preencapsulated in the vesicles. (Reprinted with permission from Ellens, H., Bentz, J., and Szoka, F. C., *Biochemistry*, 24, 3099, 1985. Copyright 1985 American Chemical Society.)

## III. CALIBRATION OF ASSAYS AND FUSION KINETICS

Fusion assays have been developed to show whether or not two membranes have fused and whether their aqueous contents have coalesced. The equally important question is how many of the vesicles in a suspension have fused. When the assay responds, it indicates that some fusion has occurred; but to show that this fusion is an average property of the vesicle membranes and, not a rare artifact of the preparation, requires a proper calibration of the assay. In this section we show how the well-behaved fusion assay can reveal not only whether fusion has occurred, but also how much and how fast. With this information, we can find the primary rate constants for vesicle aggregation (which tells us about the physical forces between the membranes) and for membrane fusion per se (which tells us about the molecular mechanism of the fusion process).

How much is 100% fusion? This apparently simple question requires a prescription of the vesicle fusion reaction mechanism, at least implicitly. Here we choose the straightforward mass action kinetic model shown also in Figure 1:

$$
\begin{aligned}
&V_1 + V_1 \overset{C_{11}}{\underset{D_{11}}{\rightleftarrows}} V_2 \qquad \overset{f_{11}}{\rightarrow} F_2 \\[4pt]
&V_1 + V_2 \overset{C_{12}}{\underset{D_{12}}{\rightleftarrows}} V_3 \\[4pt]
&\qquad\qquad\qquad \downarrow f_{11} \\[4pt]
&V_1 + F_2 \overset{C'_{12}}{\underset{D'_{12}}{\rightleftarrows}} V_1F_2 \overset{f_{12}}{\rightarrow} F_3 \\[4pt]
&\qquad\qquad\qquad\quad + \text{ higher order reactions}
\end{aligned}
\qquad (1)
$$

## Table 1A
## FUSION ASSAYS: AQUEOUS CONTENTS MIXING

| Fluorophore | Adjuvant | Mechanism | Inhibitor | Ref. |
|---|---|---|---|---|
| $Tb^{3+}$ | $DPA^{2-}$ (pyridine 2,6-dicarboxylic acid) | $Tb(DPA)_3^{3-}$ chelation complex enhances native $Tb^{3+}$ fluorescence $10^4$-fold | >1 m$M$ Divalent cation and >0.1 m$M$ EDTA competitively chelate $DPA^{2-}$ and $Tb^{3+}$ | Wilschut and Papahadjopoulos[58] and Refs. 59 to 65, 70, 77 to 82, 84, 92, 94, 103, 110, 125, 129, 130, 138, 142 to 145, 156 to 158, 172 to 174, 187 |
| ANTS; (1-aminonaphthalene-3,6,8-trisulfonic acid) | DPX; (*p*-xylylene-*bis*-pyridinium bromide) | Collisional quenching of ANTS by >10 m$M$ DPX | Dilution of DPX into medium reduces collision frequency | Ellens et al.[84] and Refs. 85 to 87, 116, 170 |
| $Co^{2+}$-calcein | EDTA | Calcein-Co chelation complex quenches the fluorescence of calcein; EDTA chelation of $Co^{2+}$ relieves this quenching | >10 m$M$ EDTA in medium competitively chelates $Co^{2+}$; $Ca^{2+}$-calcein complex is slightly quenched | Kendall and MacDonald[135] |
| ABG; (aminobenzoyl glycine) | Phe($NO_2$); (phenylalanine-$NO_2$) | Förster energy transfer from ABG to the Phe ($NO_2$) linked by oligopeptide is relieved by trypsin cleavage | Trypsin inhibitor in media blocks cleavage of peptide | Hoekstra et al.[160] |
| CF-diA; (carboxyfluoroscein diacetate) | Lysosomal esterases | Hydrolysis of both acetates produces highly fluorescent carboxylfluorescein | None | Altstiel and Branton[159] |
| FITC-dextran | None | Relief of self-quenching following fusion | Antifluorescein IgG | Stutzin[194] |

## Table 1B
## FUSION ASSAYS: MEMBRANE MIXING

| Fluorophore | Adjuvant | Mechanism | Anchor | Ref. |
|---|---|---|---|---|
| NBD; (7-nitrobenz-2-oxa-1,3-diazol-4-yl) | Rhodamine, Rh; (Lissamine Rhodamine) | Förster energy transfer from NBD to Rh decreases as lipids are diluted during fusion of labeled membranes with unlabeled membranes | Both fluorophores are conjugated to the amino group of PE | Struck et al.,[90] Hoekstra[93] put NBD-PE and Rh-PE into different liposome populations, and Refs. 84 to 86, 92, 95, 96, 98, 99, 103, 108, 127, 138, 148, 150 to 155, 170, 172, 174, 175, 195 |
| Dansyl; (5-dimethyl-amino-naphthalene-1-sulphonyl) | Rhodamine | Förster energy transfer from dansyl to Rh; fluorophores are incorporated in different vesicle populations | Both fluorophores are conjugated to the amino group of PE | Vanderwerf and Ullman[161] |
| Pyrene | Pyrene | Pyrene eximer fluorescence | Pyrenesulfonyl DPPE; 1-Palmitoyl-2 pyrene-decanoylphosphatidylcholine | Owen[162] Schenkman et al.[177] and Ref. 190 |
| Dansyl | ASPPS [N-(3-sulfopropyl)-4-(p-didecyl aminostyryl) pyridinium] | Förster energy transfer from Dansyl to ASPPS | Dansyl is conjugated to the amino group of PE and ASPPS has two decyl acyl chains | Gibson and Loew[163] and Ref. 139 |
| NBD | Bimane | Förster energy transfer from bimane to NBD | Bimane is conjugated to dipalmitoylphosphatidyl dimethylethanolamine. NBD is conjugated to PE | Pryor et al.[176] |
| NBD | Anthracene | Förster energy transfer from anthracene to NBD | NBD is conjugated to two hexadecyl acyl chains; anthracene is conjugated to cholesterol | Uster and Deamer[128] and Ref. 181 |
| Rhodamine | Fluorescein | Förster energy transfer from fluorescein to rhodamine | Fluorescein isothiocyanate is conjugated to octadecylamine; rhodamine B is conjugated to octadecylamine | Keller et al.[146] |
| | None | Relief of self-quenching fol- | Rhodamine B is conjugated to | Hoekstra et al.[148] and Refs. |

| | | | | |
|---|---|---|---|---|
| | | lowing fusion of labeled and unlabeled vesicles | octadecylamine | 97, 149, 155 |
| Chlorophyll b | Chlorophyll a | Förster energy transfer from chlorophyll b to a | Membrane-bound | Gad and Eytan[164] |
| Trans parinaric acid | None | Photochemical dimerization and relief of self-quenching which depends on membrane concentration | Parinaric acid is conjugated to the 2 position of phospholipids | Morgan et al.[165] Stegmann et al.[95] |
| Diphenylhexatriene (DPH) | Nitro-DPH | Förster energy transfer from DPH to nitro-DPH | DPH and nitro-DPH are esterified to lysophosphatidylcholine and triglyceride analogues | Morgan et al.[166] |
| DPH | None | Measurement of fluorescence lifetimes of DPH | DPH esterified to lysopalmitoyl PC | Parente and Lentz[167] |
| CPS (coumarin) | DABS (dimethylamino-phenylazophenol) | Förster energy transfer from CPS to nonfluorescent JABS | Both CPS and JABS are attached to the acyl chains of PC | Silvius et al.[195] |

## Table 1C
## LEAKAGE ASSAYS: AQUEOUS CONTENTS RELEASE

| Fluorophore | Adjuvant | Inhibitor | Mechanism | Ref. |
|---|---|---|---|---|
| CF (carboxyfluorescein) | — | — | Relief of self-quenching upon release into medium | Weinstein et al.[75] and Refs. 53, 58 to 64, 70, 76, 77, 82, 94, 118, 119, 125, 129, 142, 143, 145, 156, 168, 185, 188 |
| Calcein | — | — | Relief of self-quenching upon release into medium | Allen and Cleland[169] and Refs. 95, 99, 119, 186 |
| ANTS/DPX | — | — | Relief of DPX collisisonal-quenching of ANTS upon dilution into medium | Smolarsky et al.[83] and Refs. 84—88, 116, 170, 171 |
| Tb/DPA | — | — | Loss of fluorescence due to competitive chelation of divalent cations and EDTA in medium; both leakage and influx of medium are monitored | Bentz et al.[78] and Refs. 65, 79, 81, 82, 84, 103, 125, 174 |
| 4-Methyl-umbelliferylphosphate | Alkaline phosphatase | — | Hydrolysis by phosphatase in medium yields fluorescent 4-methyl umbelliferone | Six et al.[183] and Refs. 135, 184 |

where $V_1$ denotes the vesicle monomer, $V_2$ the dimer aggregate, $F_2$ the fused doublet (with merged membranes and aqueous contents), and $V_1F_2$ the aggregate of a monomer and a fused doublet. Higher order reactions do occur, and together with given values of the aggregation ($C_{ij}$), deaggregation ($D_{ij}$), and fusion ($f_{ij}$) rate constants, the appropriate mass action kinetic equations can be solved numerically to give the time-dependent concentrations of each of the constituents of the assay.[64]

The problem is that proper fusion assays can only "see" the fusion products, e.g., $F_2$ and $F_3$. If we think only of dimerization, i.e., formation of $F_2$, we see that there are three rate constants contributing to the kinetics of this process. An intuitive guess is that to fit the three parameters ($C_{11}$, $D_{11}$, and $f_{11}$) based solely on the kinetics of $F_2$ formation will require three independent experiments. Mathematically, a unique curve for $F_2$ over time is predicted for each combination of rate constants. Unfortunately, in the time regime where the higher order fusion reactions are negligible, the differences between many of these curves are experimentally insignificant. Thus, all three parameters cannot be fitted simultaneously. Nevertheless, we have shown that there are three independent experiments, involving different total lipid concentrations and different ratios of labeled (e.g., Tb) to unlabeled (e.g., DPA) vesicles, which, taken together, uniquely specify the primary rate constants, regardless of the value of the higher order rate constants.[64]

The typical experiment begins with vesicle monomers at an initial concentration denoted $X_0 = [V_1]$ at time $t = 0$. Let us suppose that one population, denoted L-type vesicles, contains the fluorophore (e.g., Tb, ANTS, NBD-PE/Rh-PE, or just NBD-PE), while the other population, denoted U-type vesicles, contains the adjuvant (i.e., DPA, DPX, unlabeled vesicles, or Rh-PE, respectively) (Table 1). The "L" and "U" labels are only for bookkeeping purposes, since we assume here that these assays are perfect ("real" situations will be described below) and do not alter the aggregation and fusion characteristics of the vesicles. On the other hand, extensions of the analysis to fusion between vesicles with dissimilar membranes (e.g., fusion of PS liposomes with PA liposomes[94] or the fusion of liposomes with viruses[95]) requires explicit consideration of a fusion reaction pathway more complicated than that shown in Equation 1.[96,97]

The initial fluorescence intensity of the L-type vesicles is denoted $i_0 = q_0[L(t = 0)]$, where $q_0$ is the intensity per mol/$\ell$ of L-type vesicles, and $[L(t = 0)]$ is the initial molar concentration of vesicles (*not* phospholipid concentration). The maximal fluorescence intensity (or quenching, depending on the assay used) attainable in an experiment, $i_f$, is usually taken as the fluorescence obtained if all the vesicles have fused to one continuous membrane and/or all of the aqueous contents have coalesced within the hypothetical megavesicle, with no leakage of contents or influx of medium into the fusing structures. For example, with the Tb/DPA assay, $i_f$ is set by lysing the Tb-vesicles in the presence of 20 $\mu M$ free DPA, which chelates all the Tb to $Tb(DPA)_3^{3-}$. For the NBD/Rh assay, $i_f$ is set by making liposomes containing NBD and Rh at the surface density one would obtain if all the liposomes in the assay had fused together. An early practice of setting $i_f$ by lysing the NBD/Rh liposomes in detergent creates an artificial final state and, worse, the fluorescence intensities of the probes are diminished in the detergent micelles,[85,90] although certain correction factors can be applied.[96,98]

The fusion experiment then measures the relative change in fluorescence given by:

$$I(t) = \frac{i(t) - i_0}{i_f - i_0} \qquad (2)$$

where $i(t)$ is the absolute fluorescence intensity at time $t$. Equation 2 is constructed to show fusion as an increase in $I(t)$, regardless of whether the assay being utilized shows enhanced fluorescence intensity or fluorescence quenching following fusion. Because the intrinsic fluorescence of Tb is insignificant compared with the Tb/DPA complex, the initial intensity, $i_0$, of the Tb-liposomes is essentially zero. In some cases, leakage of liposome contents during storage or fluorescent contaminants may increase $i_0$. For the ANTS/DPX assay, $i_0$ is the fluorescence of the ANTS liposomes and $i_f$ is the intensity after quenching by DPX, calibrated using liposomes with co-encapsulated ANTS and DPX.[84]

## A. Fitting the Rate Constants

Here we demonstrate how the primary kinetic rate constants can be obtained from the experimental fluorescence curve given by Equation 2. For simplicity, we will assume first that fusion occurs without leakage of contents (for the contents-mixing assays), or anomalous probe exchange (for the lipid-mixing assays). These corrections will be treated in Sections B and C below. As shown by Nir et al.[62] and Bentz et al.,[64] the measured fluorescence intensity must also be equal to:

$$I(t) = \sum_{j=1}^{\infty} jB_jF_j(t)/X_0 \qquad (3)$$

where $F_j(t)$ is the concentration of fused j-lets at time t. For example, $F_2(t) = [F_2] + [V_1F_2] + 2[F_2F_2] + $ [all higher order terms containing $F_2$], see Equation 1. $B_j$ is the relative fluorescence factor for each fused j-let. The value of $B_j$ is calculated and depends upon the nature of the fusion assay and the ratio of labeled to unlabeled vesicles, as described in the following pages in Equations 8 and 10. Finding the correct values of the primary rate constants for a given system is simply a matter of fitting the predicted curve from Equation 3 to the experimental curve given by Equation 2.

We have found some accurate approximate equations for the initial changes in fluorescence which simplify this process considerably. Using the Tb/DPA or ANTS/DPX assays with a 1:1 mixture of labeled to unlabeled vesicles, we have found that numerical integration of the mass-action kinetic equations (Equation 1) gives solutions of Equation 3 which are approximated within experimental error ($\pm$ 1%)by:[64]

$$I(t) = A(t)F(t)$$

where

$$A(t) = (1 + 4\hat{C}_{11}X_0t)^{1/4} - 1$$

$$F(t) = 1 - (1 - \exp\{-\hat{f}_{11}t\})/(\hat{f}_{11}t)$$

$$\hat{C}_{11} = C_{11}/(1 + K_r)$$

$$\hat{f}_{11} = f_{11}(1 + K_r)$$

$$K_r = D_{11}/f_{11} \qquad (4)$$

This equation is applicable for the ANTS/DPX assay provided that the encapsulated concentration of DPX is large enough ($\geq$ 90 m$M$;[88]) to completely quench the ANTS in the fused vesicles (see Equation 8 for the values of $B_j$ used.).

This approximate equation is reliable provided that $C_{11}X_0t \leq 0.5$, $f_{11}/C_{11}X_0 \geq 1$ and $D_{11}/f_{11} \leq 4$. The second and third conditions reflect primarily the ranges of values rigorously checked by numerical integration and can probably be relaxed somewhat. The first condition is *quite* serious, because it ensures that most of the observed fluorescence arises from fused doublets and, not the higher order fusion products. The higher order rate constants must be taken into account, since they usually occur to some extent in any experiment, even with rapid-mixing.[99] However, the dimerization rate constants $C_{11}$, $D_{11}$, and $f_{11}$ are of prime interest anyway, since they measure the energy of interaction between two membranes as they approach each other ($C_{11}$ and $D_{11}$) and the rate of local destabilization which leads to fusion ($f_{11}$).[63,64,79]

The fitting procedure requires at least three independent experiments, i.e., one for each parameter. A rigorous description of the procedure is given in Reference 64, while several examples are shown in Bentz et al.[79]

Briefly, the first run requires a very low lipid concentration with a 1:1 ratio of Tb vesicles and DPA vesicles. In the limit of $X_0 \to 0$ the overall fusion kinetics are aggregation rate limited and Equation 4 reduces to

$$I(t) = (1 + 4\hat{C}_{11}X_0t)^{1/4} - 1 \qquad (5)$$

and the fitted fluorescence data give the value of $\hat{C}_{11} = C_{11}/(1+K_r)$. If, by chance, it is the case that the aggregation is irreversible, i.e., $D_{11} \ll f_{11}$, or $K_r \ll 1$, then the value of $C_{11}$ would be immediately known. In general, one cannot assume that vesicle aggregation is irreversible.[79,100-102] It is relevant here to point out that Morris et al.[99] used an apparently different equation to deduce the aggregation rate constants for the $Ca^{2+}$-induced fusion of LUV (PS/PE) with the NBD/Rh assay. However, initially their equation is identical to Equation 5 and both are valid only under aggregation rate limiting conditions. In order to determine the fusion rate constants, as well as the dissociation rate constant $D_{11}$, it is necessary to use either Equation 4-6 or the direct numerical integrations.

The second run is made with a high lipid concentration, e.g., 2 to 3 orders of magnitude higher than the first run, so that the fusion step becomes rate limiting, with a 1:1 ratio of labeled to unlabeled vesicles. When the time course of fluorescence development is fitted by Equation 4, the value of $\hat{f}_{11} = f_{11}(1+K_r)$ is obtained, since $\hat{C}_{11}$ is known. Once again, if $D_{11}/f_{11} \ll 1$, we would immediately obtain the value of the fusion rate constant.

In order to fit rigorously the value of $D_{11}$, and hence obtain $C_{11}$ and $f_{11}$ *per se*, it is necessary to use a 1:9 ratio of labeled to unlabeled vesicles at a lipid concentration such that $\hat{f}_{11}/\hat{C}_{11}X_0 \cong 10$.[64] The reason for this is that within the relevant time regime of $0.5 < C_{11}X_0t$, the predicted fluorescence curves from Equations 1 and 3 for the 1:1 ratio (computed by numerical integration) depends (within experimental error) only upon the values of $\hat{C}_{11}$ and $\hat{f}_{11}$. This is shown in Table 2. For the 1:1 ratio of labeled:unlabeled vesicles, the calculated value of fluorescence, $I(t)$, is relatively indifferent to the exact value of $K_r$, i.e., the differences are experimentally insignificant. However, for the 1:9 ratio, choosing the lipid concentration such that, $\hat{f}_{11}/\hat{C}_{11}X_0 \cong 10$ gives a sufficient discrimination between the various values of $K_r$. Curves such as these can be seen in Bentz et al.[64,79] and Wilschut et al.[103]

We have also found an extension to Equation 4 which is sufficiently accurate to use for fixing the value of $K_r$. When we denote the fraction of labeled:unlabeled vesicles as p, such that p = 0.5 for a 1:1 ratio and 0.1 for a 1:9 ratio, then again:

$$I(t) = \gamma A(t)F(t)$$
where
$$A(t) = (1 + 4\alpha\hat{C}_{11}X_0t)^{1/4} - 1$$

$$\alpha = \begin{cases} 1 \text{ for the 1:1 ratio or p} = 0.5 \\ 2 \text{ for the 1:9 ratio or p} = 0.1 \end{cases}$$

$$\gamma = 1 + \frac{1 + 2K_r}{8 + 2\hat{K}}$$

$$\hat{K} = \hat{f}_{11}/\hat{C}_{11}X_0 \qquad (6)$$

## Table 2
## RELATIVE FLUORESCENCE I(t) FOR 1:1 AND 1:9
## LABELED:UNLABELED MIXTURES

| System (L:U) | $\hat{f}_{11}/\hat{C}_{11}X_0$ | $\hat{C}_{11} X_0 t$ | I(t) $K_r = 0$ | 1 | 2 |
|---|---|---|---|---|---|
| 1:1 | 100 | 0.1 | 0.080 | 0.079 | 0.078 |
|  |  | 0.2 | 0.152 | 0.150 | 0.147 |
|  |  | 0.5 | 0.302 | 0.298 | 0.295 |
|  | 10 | 0.1 | 0.033 | 0.031 | 0.029 |
|  |  | 0.2 | 0.094 | 0.085 | 0.080 |
|  |  | 0.5 | 0.260 | 0.234 | 0.222 |
| 1:9 | 100 | 0.1 | 0.141 | 0.138 | 0.136 |
|  |  | 0.2 | 0.259 | 0.255 | 0.252 |
|  |  | 0.5 | 0.484 | 0.479 | 0.475 |
|  | 10 | 0.1 | 0.059 | 0.056 | 0.052 |
|  |  | 0.2 | 0.164 | 0.149 | 0.136 |
|  |  | 0.5 | 0.425 | 0.385 | 0.352 |

*Note:* Values of I(t) were calculated from Equations 3 and 8 from numerical integration of the mass action equations. The accuracy of Equations 4 and 6 can be checked against these values.[64]

F(t) is defined as before in Equation 4. At low lipid concentrations, $\gamma \rightarrow 1$ and Equation 6 becomes identical to Equation 4, except that it is *also* valid for the 1:9 ratio of labeled to unlabeled vesicles. Fitting the value of $K_r$ simply comes from the $\gamma$ factor.

## B. Correction for Leakage During Fusion

Leakage of aqueous contents during fusion of certain types of vesicles can complicate the fitting of kinetic rate constants in contents-mixing assays. This complication can be overcome by performing a complementary experiment which measures the leakage from vesicles. The proper leakage experiments are described previously in Section II, but the basic point is to measure what happens to the fluorescence of the fused doublet. For example, with the Tb/DPA assay, the fused doublet should contain 1.25 m$M$ Tb$^{3+}$ with 25 m$M$ DPA$^{2-}$ (as 1:1 dilutions of the encapsulated 2.5 m$M$ Tb (citrate) and 50 m$M$ Na dipicolinate from the two populations of vesicles). However, influx of the medium into the fusing vesicles,[82] or leakage of their contents into the medium reduces the amount of Tb(DPA)$_3^{3-}$ complex (due to its dissociation by EDTA and divalent cations) and, hence, the fluorescence. We can correct for this by repeating the fusion experiment with the same concentration of vesicles now containing the Tb/DPA complex (1.25 m$M$ Tb/25 m$M$ DPA, encapsulated during liposome preparation). When the fluorescence curve from the fusion experiment is denoted $I_F(t)$ and the percentage of dissociation (i.e., 100 minus the observed fluorescence in the dissociation experiment) is denoted $I_D(t)$, the total fluorescence which would have resulted if the vesicle fusion had not been leaky, is given rigorously by:[62,64]

$$I(t) = I_F(t) + \delta I_D(t) \tag{7}$$

where $\delta$ is the fraction of the fused liposomes which contain both Tb and DPA, or ANTS and DPX, and must be calculated from the reaction kinetics in general. However, initially $\delta = 0.5$ for a 1:1 ratio of labeled to unlabeled vesicles, or $\delta = 1 - p$, where p is the fraction of labeled vesicles.[64] These initial values are correct until the number of fused triplets becomes significant.

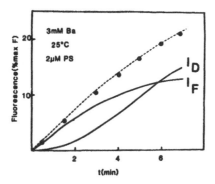

FIGURE 4. Kinetics of intermixing of contents of PS LUV ($I_F$) and dissociation of preencapsulated Tb/DPA complex ($I_D$) in the presence of 3 m$M$ Ba$^{2+}$. The total lipid concentration was 2 $\mu M$, 1 $\mu M$ each of Tb-liposomes and DPA-liposomes for the fusion experiment, and 2 $\mu M$ of Tb/DPA liposomes for the dissociation experiment. The solid circles show the fusion data corrected for dissocation, and the dashed line is the theoretical curve with $C_{11} = 2.8 \times 10^7$ $M^{-1} s^{-1}$ and $X_0 = 2.5 \times 10^{-11} M$. (Reprinted with permission from Bentz, J., Düzgüneş, N., and Nir, S., *Biochemistry*, 24, 1064, 1985. Copyright 1985 American Chemical Society.)

As an example, we consider a 1:1 mixture of Tb and DPA vesicles, i.e., p = 0.5. Suppose that half of the vesicles formed fused doublets and there are no higher order fusion products. If there were no leakage from these fused doublets or any of the unfused vesicles, then $I_D(t)$ = 0 and $I_F(t)$ = 0.25, since only one half of the fused doublets would contain both Tb and DPA, the other half containing either Tb (citrate) or DPA. Thus, I(t) = 0.25. Suppose instead that the fused doublets lysed completely, so that the fusion signal $I_F(t)$ = 0. In this case, $I_D(t)$ = 0.5, since half of the vesicles have lost their contents. But, since δ = 0.5, I(t) = 0.25, as before. Therefore, it is clear that using the corrected fusion fluorescence gives the rate constants for vesicle aggregation and membrane fusion, without specifying whether $f_{11}$ implies mixing of aqueous contents or lysis. That definition is up to the user.[79] However, it must be noted that when the vesicles only leak, without any mixing of aqueous contents, then the rate constants can be obtained only if the leakage per contact is complete, as it was in the example given above. If there is only partial leakage per contact, then one can obtain similar leakage curves with rapid aggregation kinetics and small partial leakage per contact, or with slow aggregation kinetics and large leakage per contact.[64,84,88] When the vesicles show some mixing of contents, obtaining the rate constants is not a problem. Figure 4 shows this procedure for PS LUV induced to fuse by 3 m$M$ Ba$^{2+}$. The solid lines show the original data for fusion ($I_F$) and dissociation ($I_D$) and the solid dots show the corrected value of I(t) given by Equation 7. The dashed curve gives the predicted values based upon a fitted value of $\hat{C}_{11}$ from Equation 5.

For the Tb/DPA, or ANTS/DPX assay, under the appropriate conditions, the $B_j$ coefficients are given by:[64]

$$B_j = (1 - p^{j-1}) \tag{8}$$

where p is the fraction of Tb or ANTS vesicles. This equation holds strictly, provided that in every fused j-let: (i) the formation of the Tb/DPA complex is limited by the available

Tb, and not the DPA, within the vesicles, or (ii) the amount of DPX is sufficiently large to completely quench the ANTS fluorescence. Encapsulating a 10 to 20-fold excess of DPA over Tb[59,62] or at least 90 m$M$ DPX[84,87] is sufficient to meet this condition for fused doublets and triplets. Nir et al.[62] have presented a more general form for $B_j$ for the Tb/DPA assay when higher orders of fusion products are considered. We note that, by definition, $B_1 = 0$.

### C. Calculating the Fluorescence for Lipid Mixing Assays

Once leakage is taken into account for the aqueous contents-mixing assays, the calculation of the $B_j$ coefficients for Equation 3 is straightforward.[62,64] Leakage does not present a problem for the lipid-mixing assays, but the calculation of the expected fluorescence from the fused doublet is not as simple. These assays are based on detecting the surface density of fluorescent lipid probes by Förster energy transfer between a donor and acceptor pair of fluorophores.[91,104-107] Ionotropic phase separation,[93,108,116,179] or intermembrane exchange of the probe molecules during close apposition[109] must either be ruled out or corrected for, in order to calculate the expected fluorescence.

The simplest assumption is to define the fused j-lets as having completely mixed their membrane components, with a homogeneous distribution of the fluorescent probes. Estep and Thompson[105] have shown that diluting anthracene into PC liposomes results in predicted fluorescence intensities, and we have found the same conclusion for NBD-PE and Rh-PE diluted into PS LUV, yielding an $R_0$ (the distance of 50% energy transfer between donor and acceptor probes) of approximately 65 Å.[191] Thus, the Förster energy transfer theory can be used to calculate the $B_j$ coefficients for these probes, provided that the fusion event leads to complete mixing of both the inner and outer monolayers of the fusing membranes.

When the lipid-bound fluorophores can diffuse closer than $R_0/4$ to one another,[107] an approximation by Wolber and Hudson[106] can be used to calculate the ratio of donor quantum yield in the presence and absence of acceptor as:

$$q(v) = 0.6463e^{-4.7497v} + 0.3537e^{-2.0618v} \tag{9}$$

where $v$ = (surface density of acceptor molecules) times $R_0^2$. This approximation is valid for phospholipid-bound fluorophores provided that $R_0 > 30$ Å.

Using the same procedure described for aqueous contents-mixing assays, we can obtain the value of $B_j$ for the lipid-mixing assays. When the labeled vesicles contain *both* probes, the unlabeled vesicles are "empty", and we observe the fluorescence of the donor molecule, then in Equation 2, $i_0 = pX_0q(v)q_0$, and $i_f = X_0q(pv)pq_0$, where p is the fraction of labeled vesicles, v is the number of acceptors/$R_0^2$ in the labeled vesicles, $X_0$ is the molar concentration of vesicles, and $q_0$ is the fluorescence intensity per vesicle in the absence of acceptors. Thus,

$$B_j = \frac{Q_j - q(v)}{q(pv) - q(v)}$$

$$Q_j = \sum_{i=0}^{j-1} \frac{(j-1)!}{i!(j-1-i)!} p^i(1-p)^{j-1-i} q\left(\frac{i+1}{j}v\right) \tag{10}$$

Similar equations can be derived for the case when the acceptor molecules are in the unlabeled vesicles. Unfortunately, due to the nondiscrete nature of fluorescence in lipid-mixing assays, as opposed to the contents-mixing assays, this equation cannot be simplified further, except in the limit of $v \to 0$, i.e., where the acceptor probe densities become vanishingly small. In this case,

$$B_j \to (1 - 1/j) \tag{11}$$

**Table 3**
**B$_j$ FLUORESCENCE COEFFICIENTS FOR**
**1:1 RATIO OF LABELED:UNLABELED**
**VESICLES ($p = 0.5$)**

| | | | B$_j$ | |
| --- | --- | --- | --- | --- |
| | | j = 2 | 3 | 4 |
| Tb/DPA or ANTS/DPX[a] | | 0.5 | 0.75 | 0.875 |
| NBD/Rh[b] | v = 0.1 | 0.5 | 0.667 | 0.751 |
| | 1.0 | 0.5 | 0.723 | 0.824 |
| | 1.285 | 0.5 | 0.750 | 0.862 |

[a] Calculated from Equation 8 which assumes that the Tb is the limiting reagent for the Tb/DPA complex formed or that DPX completely quenches ANTS fluorescence.
[b] Calculated from Equation 10. v is the number of acceptor molecules per R$_0^2$ unit area. When $v \leqslant 0.1$, the limiting expression of Equation 11 is valid.

which is independent of the value of p. Table 3 shows the values of B$_j$ for a few cases. Wilschut et al.[103,174] used Equation 11 for B$_j$ to compare the Tb/DPA and NBD/Rh dilution assays. For the NBD/Rh assay in PS or PC vesicles, taking R$_0$ = 65 Å and the area per lipid molecule as 70 Å$^2$, the values of v = 0.1, 1.0, and 1.285 correspond to 0.17 mol % Rh, 1.7 mol % Rh and 2.1 mol % Rh in the labeled vesicles. For the 1:1 ratio of labeled:unlabeled vesicles B$_2$ = 0.5 for both the Tb/DPA and the NBD/Rh assays. It is interesting to note that using v = 1.285 will align the two assays nearly exactly up to B$_4$. Unfortunately, other factors can influence the fluorescence of these molecules beyond the ideal theory represented by this table.

## IV. THE IDEAL FUSION ASSAY

We describe here the criteria which appear to be necessary to construct a fusion assay. It will become apparent through the course of the review that these criteria are not always sufficient to produce a functioning fusion assay; they serve only to identify the theoretical details.

### A. Assays for the Intermixing of Aqueous Contents
The following criteria must be met for fusion assays monitoring the mixing of aqueous contents between fusing vesicles:

A1. None of the fluorophores, adjuvants, or their products should bind to the membranes. This binding must be checked for each membrane type.
A2. The kinetics of the fluorescence reactions must be fast compared to the kinetics of the fusion event.
A3. Leakage of the probe molecules or their products must reflect leakage of the encapsulated volume. Leakage from intact bilayers, i.e., before fusion has begun or after it has been stopped and the vesicle membranes have reassembled, must be negligible (see Section IV.C. on leakage assays).
A4. The external medium must contain a quencher which prevents the fluorophore and adjuvant from reacting after leakage and thereby mimicking the fusion event.

These criteria have been tested for the Tb/DPA assay and the ANTS/DPX assay. To test binding (A1), empty vesicles are incubated for 1 hr with the fluorophore solution which would be encapsulated for the assay. The vesicles are separated from the incubation mixture by gel filtration through Sephadex® G-75. The fluorescence remaining with the vesicles after detergent lysis (to standardize the conditions) can be compared with the amount of fluorophore which the vesicles would have encapsulated. If the lipid bilayer retains more than a few percent of its encapsulated contents, there is cause for concern.[89]

The next step is testing the kinetics of the fluorescence reaction. Fusion consists of several molecular events (Figure 1), the last of which is the actual reaction between the fluorophore and the adjuvant. If the rate of reaction between the fluorophore and the adjuvant is so slow as to be the rate-limiting step of the entire sequence, then the assay cannot monitor the rat of membrane fusion. To show that the fluorescence reaction is fast enough, it is necessary to show that the observed rate of change of fluorescence during fusion is independent of the concentrations of the encapsulated fluorophore and adjuvant. The reaction rate that *matters* is the rate of fluorescence generation by the reactants at concentrations at which they are encapsulated (e.g., 2.5 m$M$ Tb and 50 m$M$ DPA); but, it is difficult to measure this rate because of inner filter effects at such high concentrations of reactants. Even more important is the fact that the crucial reaction for the assay occurs within the confined volume of two fused liposomes, which is 1 to 10 $\times$ 10$^{-18}$ $\ell$ for liposomes of 0.1 to 0.2 $\mu$m diameter. The bulk phase reaction kinetics in the cuvette are of interest (e.g., mixing 1 $\mu M$ Tb and 20 $\mu M$ DPA[59]), but the assay should be tested under the relevant conditions.

This test has never been applied rigorously to any aqueous contents mixing assay, primarily because changing the encapsulated contents also changes the ionic and osmotic strength of the medium, which can affect fusion kinetics.[65,77,110] Thus, careful and tedious controls would be necessary. However, there is excellent indirect evidence that the kinetics of chelation of the Tb/DPA complex, the collisional quenching of ANTS by DPX and the resonance energy transfer between NBD and Rh are not rate-limiting to the observed fluorescence changes. With PS vesicles, the kinetics of the fluorescence changes increase with increasing Ca$^{2+}$ concentration (Figure 5[51,61,84,90,111]). If any of the fluorescence reactions were rate limiting at low Ca$^{2+}$ concentrations, then it is difficult to imagine how they would be sped up by adding more Ca$^{2+}$; especially for the Tb/DPA assay which would be disrupted if the Ca$^{2+}$ entered the fusing vesicles. The fusion signals clearly arise from the membrane destablization reaction. It *may* be that at the highest Ca$^{2+}$ concentrations, where membrane fusion kinetics appear to reach maximal levels, the rate of Tb and DPA chelation is being seen. Or, it may be that the membrane destabilization step is occurring as rapidly as possible.

The ambient leakage of the fluorophore, adjuvant, or their products from intact vesicles is the next test (A3). The reactants of a reliable assay will show no significant leakage of contents over several days of storage under nitrogen or argon at 4°C. The bimolecular lipid membrane cannot be expected to hold low molecular weight aqueous probes indefinitely.[112] Under experimental conditions, the ambient leakage from intact membranes should be very slow relative to the rates found after membrane destabilization and fusion. Significant corrections for the loss of contents before the experiment begins or extensive leakage during the experiment makes data analysis difficult and of debatable significance.

The final criterion for a reliable assay is that the fluorescence reaction between the fluorophore and the adjuvant, which is taken as the signal of the coalescence of aqueous compartments, should not occur outside the vesicle following the leakage of contents which may accompany fusion. For the Tb/DPA assay, the inclusion of 0.1 m$M$ EDTA in the buffer is necessary to completely inhibit the formation of the Tb/DPA complex in the medium, even though Ca$^{2+}$ will bind to DPA.[59] For the ANTS/DPX assay, the dilution of the DPX concentration below 1 m$M$ is sufficient to eliminate all collisional quenching of ANTS fluorescence. An alternate version of the Co-calcein/EDTA assay tested in our laboratory,

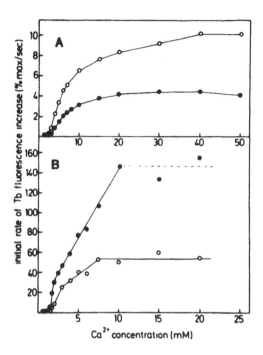

FIGURE 5. Dependence of PS vesicle fusion on the Ca²⁺ concentration. CaCl₂ was injected as a 0.1 or 0.2 *M* solution into a 1:1 mixture of Tb and DPA vesicles. Final millimolar Ca²⁺ concentrations (not corrected for the presence of 0.1 m*M* EDTA in the medium) are indicated: (A) LUV and (B) SUV. Initial rates of the Tb fluorescence increase were determined from the tangents to the fluorescence curves. Lipid concentration: 50 (●) and 200 (○) μ*M* (LUV, panel A) and 10 (○) and 50 (●) μ*M* (SUV, panel B). (Reprinted with permission from Wilschut, J., Düzgüneş, N., Fraley, R., and Papahadjopoulos, D., *Biochemistry*, 19, 6011, 1980. Copyright 1980 American Chemical Society.)

in which Co (citrate) is encapsulated in one population of vesicles and calcein in the other, requires 10 m*M* EDTA in the medium to prevent the Co/calcein complex from forming outside the vesicles. These levels of EDTA are inconvenient if one wishes to examine the divalent cation-induced fusion of liposomes or biological membrane vesicles.

## B. Assays for the Intermixing of Lipids

Most of the assays for lipid mixing listed in Table 1B involve Förster energy transfer, except in the cases of excimer formation and photodimerization, and depend upon the average separation between probe molecules. Here we shall discuss the energy transfer pairs, and for each of the assays consider two configurations: (i) Probe dilution: the fluorescent lipid probes are initially incorporated into one population of vesicles whose fusion with unlabeled (blank or empty) vesicles is reported by the dilution of the probes and the decrease of energy transfer. (ii) Probe mixing: the probes are incorporated into separate vesicle populations whose fusion is reported by probe intermixing and the resulting increase in energy transfer.

These assays must satisfy the following criteria:

B1. The probes must be stably inserted into the membrane phase; i.e., there should be no free monomers in solution and no probe transfer through the aqueous phase.

B2.  The probes in the membrane must be at a sufficiently low surface density so as not to alter the kinetics of vesicle aggregation and fusion.
B3.  The probes must not transfer between bilayers upon intermembrane contact without concomitant bulk phase lipid mixing during membrane fusion.
B4.  The fluorescence quantum efficiency of the probes should not be affected either by cation binding to the probe or the bilayer, or by the process of membrane fusion.

The requirement that the probes remain in the bilayer throughout the course of the experiment (B1) is obvious. Nichols and Pagano[113,114] have exhibited probes that remain in the bilayer and others that exist in sufficient quantities in the aqueous phase to permit observable transport between vesicles. Clearly, this requirement demands only that the critical micelle concentration of the probe be of the same order of magnitude as the phospholipid molecules in the bilayer, i.e., $10^{-14}$ $M$.[115]

Criterion B2 requires that the probe molecules do not alter the stability properties of the membrane. As with criterion A2 for the aqueous-mixing assays, this point can only be established rigorously by repeating the fusion experiment with successively smaller mole fractions of the probe molecules. The fusion kinetics projected from "infinite" probe dilution must be used to choose the "finite" probe concentration which is acceptable. The minimum probe concentration is determined by sufficient resonance energy transfer, a reasonable difference in signal between the unfused vesicles and "completely fused" vesicles (i.e., vesicles used for calibrating the maximal fluorescence signal), the quantum yield of the probe, and the sensitivity of the instrument. The use of fluorescent probes attached to the acyl chains of phospholipids (e.g., $C_{12}$-NBD-PC) should have no effect on the surface properties of the membrane, although these molecules tend to transfer more easily through the aqueous phase than lipids labeled at the head-group.[113,114]

Upon intermembrane contact, the probes are expected to report the flow of the rest of the membrane (B3). While most of the lipid-mixing assays used do not involve lipid exchange when the membranes are separate (B1), the possibility of selective appositional transfer of probe during contact is more difficult to eliminate. Membrane compositions and electrolyte conditions have been found which lead to aggregation and substantial (approx. 40%) transfer of probes (NBD-PE and Rh-PE) without any mixing of aqueous contents or leakage of contents.[84,109,111,116] This observation may be interpreted as either the fusion of outer monolayers, or the selective transfer or probes after contact (Figure 1[103,176]). The "probe dilution" and "probe mixing" versions of the NBD/Rh assay show different rates of fusion (Figure 6). Typically, more extensive probe transfer is observed with the latter version. The problem is that one cannot use the particular assay to show that it is performing its assigned task of following the bulk flow of lipid. This issue will be taken up again when we examine how real assays perform.

The final criterion, B4, is that the quantum efficiency of the probes must not change during the course of the fusion experiments. The need for this is obvious since changes in fluorescence intensity are intended to reflect dilution or mixing of the probes, and not some progressive shift in the polar environment of each molecules. The NBD/Rh assay, which is the most rigorously studied lipid-mixing assay, is affected by $H^+$, $Ca^{2+}$, or $La^{3+}$ binding to PS membranes in which the probe molecules are embedded. However, since the binding reaction is usually rapid with respect to the subsequent fusion reaction and lipid mixing, it is possible to treat this event as an injection offset.[99] The assay is also affected by the transformation of lipids from the lamellar to the hexagonal $H_{II}$ phase transition upon intermembrane contact.[85,117,197] The quantum efficiency of NBD-PE exhibits an increase upon transformation of the membrane into the $H_{II}$ phase. Another factor which affects the fluorescence intensity is the lateral phase separation of NBD-PE during fusion.[108] This sensitivity of NBD fluorescence to phase segregation has been used to analyze whether it is involved in membrane fusion.[93,116,179]

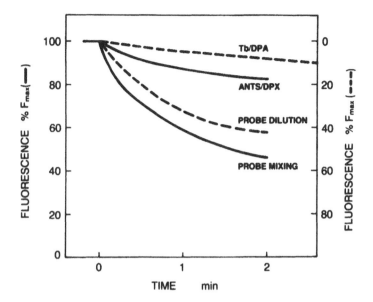

FIGURE 6.  Comparison of the kinetics of fusion of LUV (PS) in 2 m*M* Ca²⁺ obtained by different fusion assays. Lipid concentration: 50 μ*M*. Temperature: 25°C. The fluorescence scale for the solid lines is on the left-hand side and for the dashed lines on the right-hand side. (From Düzgüneş, N., Allen, T. M., Fedor, J., and Papahadjopoulos, D., *Biochemistry,* in press. With permission.)

## C. Assays for the Leakage of Aqueous Contents

Assays for the leakage of internal aqueous contents must also satisfy the first three criteria for contents-mixing assays for fusion (A1-A3): the fluorophores must not bind to the membrane, they must not leak at appreciable rates through intact membranes, and the extent of fluorescence increase must reflect, and preferably equal,[88] the extent of leakage from the destabilized membrane. The most widely used fluorescence assay for leakage involves the encapsulation of CF at a high concentration where its fluorescence is self-quenched. Leakage of the molecule and its dilution into the medium results in the relief of self-quenching and an increase of fluorescence intensity.[53,59,75,76] CF is quite stable inside liposomes unless the pH is reduced below 6, where the molecule begins to be protonated and becomes amphiphilic.[118,119,188] On the other hand, it is important to note that many commercial preparations of this dye, as well as calcein, require additional purification to remove hydrophobic constituents.[89,119,185]

Another sensitive probe of membrane destabilization is the Tb/DPA complex encapsulated in liposomes.[78] The release of this fluorescent complex into the medium containing 0.1 m*M* EDTA and divalent cations results in its dissociation and the loss of fluorescence. Furthermore, the influx of the medium into the fusing liposomes will result in the dissociation of the complex. We have used this assay together with the CF leakage assay to show that fusing PS liposomes first experienced an influx of medium, followed by a release of contents into the medium (Figure 7 and Reference 82).

## V. FUSION OF PHOSPHOLIPID VESICLES AND COMPARISON OF THE ASSAYS

In this section we will give some examples of the diverse fusion behavior of liposomes of varying membrane composition and membrane curvature, discuss the problem of leakage

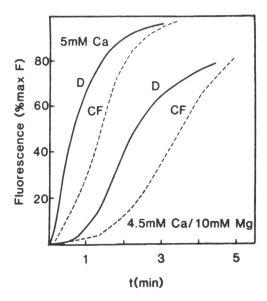

FIGURE 7.   Destabilization of LUV (PS) in 5 m*M*
Ca²⁺ (left pair of curves) and 4.5 m*M* Ca²⁺/10 m*M*
Mg²⁺ (right pair of curves), monitored by the dissociation of Tb/DPA complex (D) or by the release of CF.
The difference between dissociation and CF leakage is
a measure of the influx of medium into the fusing liposomes. (Reprinted with permission from Bentz, J. and
Düzgüneş, N., *Biochemistry*, 24, 5436, 1985. Copyright
1985 American Chemical Society.)

during membrane fusion, and compare the results of the fusion assays described in Section
II.

## A. Control of Membrane Fusion

The diverse behavior of liposomes of different phospholipid composition and bilayer
curvature has been discussed in detail in several recent reviews.[42,44-47,120-124,196] Here we will
give a few illustrations of the factors which control the fusion behavior of liposomes. We
have already shown the kinetics of fusion of PS vesicles in the presence of Ca²⁺ (Figures
2, 3, 5 to 7). Fusion of SUV is considerably faster than LUV and, occurs at a lower threshold
concentration of Ca²⁺.[59] In the case of fusion induced by Mg²⁺, SUV and LUV (PS) behave
very differently. While SUV fuse to a limited extent, LUV are completely resistant to fusion,
although they aggregate (Figure 8[60]). It appears that when fusion of SUV produces vesicles
approximately the size of LUV, the membranes are no longer susceptible to fusion. Thus,
in this vesicle system, an absolute specificity of Ca²⁺ over Mg²⁺ is observed. LUV (PS)
also display considerable variability in their responsiveness to other divalent cations. At
temperatures slightly above the gel-liquid crystalline phase transition of PS (in NaCl), Ba²⁺
induces more rapid and extensive fusion than Ca²⁺ or Sr²⁺ (Figure 9[79,125]).

The presence of phosphatidylcholine (PC) in PS membranes drastically alters the kinetics
and extent of fusion (Figure 10[70,77]). SUV composed of 20 mol % PC in PS undergo a few
rounds of fusion and then stop, with a limited extent of release of internal contents, while
pure PS vesicles collapse after fusion, indicated by the eventual decrease in Tb fluorescence
in Figure 10. As the PC content of PS or phosphatidic acid (PA) vesicles is increased, the
threshold concentration of divalent cations which induce aggregation and fusion increases.[74,77,94,116,126,127,179] LUV composed of a 1:1 mixture of PS and PC become completely

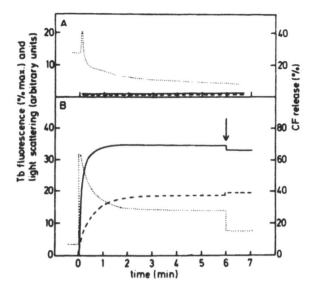

FIGURE 8.   Aggregation (dotted lines) and fusion (solid lines) of PS LUV (A) and SUV (B) in the presence of $Mg^{2+}$. $Mg^{2+}$ was added at t = 0 to a final concentration of 20 m$M$ for LUV and 8 m$M$ for SUV. Fusion was measured by the Tb/DPA assay and aggregation by 90° light scattering. The release of CF is shown by the dashed lines. 10 m$M$ EDTA was added at the arrow. (Reprinted with permission from Wilschut, J., Düzgüneş, N., and Papahadjopoulos, D., *Biochemistry*, 20, 3126, 1981. Copyright 1981 American Chemical Society.)

resistant to fusion in the presence of $Ca^{2+}$.[70] However, when PC is replaced by phosphatidylethanolamine (PE), the vesicles undergo extensive fusion, although at a lower initial rate than PS vesicles.[70,123,128,172]

Other factors which control membrane fusion include membrane fluidity,[129] lateral phase separation of lipids,[120,125,127] amount of divalent cation-bound per anionic lipid,[77,78,82] dehydration of the membrane surface,[70,93,130] and osmotic pressure gradients.[110,131-133] Cytosolic $Ca^{2+}$-binding proteins and polyamines can increase the rate of fusion and decrease the threshold $Ca^{2+}$ concentration for fusion of certain phospholipid membranes.[142-145,187] Viral envelope glycoproteins confer very different fusion requirements to the membranes in which they are embedded.[49] These aspects of membranes fusion have been reviewed in detail.[42,45,47,48]

## B. Comparison of the Assays

While Table 1 contains a considerable number of membrane fusion and leakage assays, there have been very few systematic comparisons. The PS vesicles (LUV and SUV) have become the standard upon which new assays are tested, simply because they have been most intensively studied with respect to fusion kinetics and assays. Vesicle fusion depends upon many factors and a proper comparison between assays requires that they be examined under identical conditions, e.g., vesicle preparation techniques and lipids. We will limit our discussion to those assays for which this has been done.

Comparison of the kinetics of fusion of LUV (PS) detected by the ANTS/DPX assay and the Tb/DPA assay indicates minor but significant differences (Figure 11, see also Figure 6[84,111]). It is not known at present whether this is due to differences in the nature of the reactions which the assays monitor (collisional quenching for the ANTS/DPX assay, complex formation for the Tb/DPA assay), differences in the size distribution of vesicles prepared in the various solutions, or differences in the sensitivity of the reactants to divalent cations.

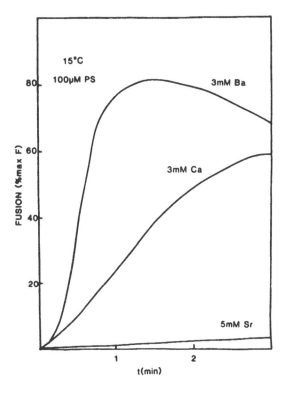

FIGURE 9.   Fusion of LUV (PS) in the presence of 3 m$M$ Ca$^{2+}$, 3 m$M$ Ba$^{2+}$, or 5 m$M$ Sr$^{2+}$ at 15°C. Fusion was monitored by the Tb/DPA assay. (Reprinted with permission from Bentz, J., Düzgüneş, N., and Nir, S., *Biochemistry*, 24, 1064, 1985. Copyright 1985 American Chemical Society.)

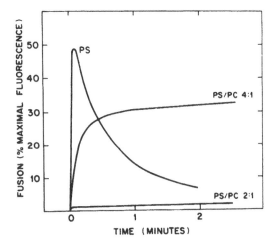

FIGURE 10.   The effect of membrane composition on the kinetics of fusion of SUV in the presence of 3 m$M$ Ca$^{2+}$ in 100 m$M$ NaCl, pH 7.4. Fusion was followed by the Tb/DPA assay. (From Düzgüneş, N., Nir, S., Wilschut, J., Bentz, J., Newton, C., Portis, A., and Papahadjopoulos, D., *J. Membr. Biol.*, 59, 115, 1981. With permission.)

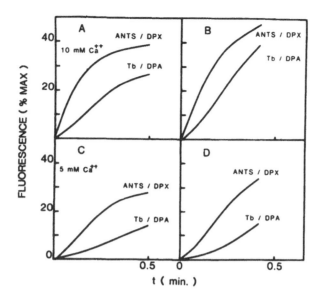

FIGURE 11.  Kinetics of fusion of PS LUV in 5 m$M$ (C and D) and 10 m$M$ (A and B) Ca$^{2+}$, monitored by the ANTS/DPX and Tb/DPA assays. Panels B and D show the fusion curves corrected for leakage (ANTS/DPX) or dissociation (Tb/DPA) (see Section III.B. and Equation 7). The ANTS/DPX curves are given in units of percent maximum quenching (i.e., 100 minus the measured fluorescence curve). (Reprinted with permission from Ellens, H., Bentz, J., and Szoko, F. C., *Biochemistry*, 24, 3099, 1985. Copyright 1985 American Chemical Society.)

Nevertheless, the rates of fusion differ by no more than a factor of 2. In the case of fusion induced by transition metal ions such as Mn$^{2+}$ or Ni$^{2+}$, the Tb/DPA assay produces a very small fluorescence signal for mixing of contents and a rapid dissociation of the preencapsulated Tb/DPA complex (Figure 12). The ANTS/DPX assay, however, shows extensive fusion.[82] We believe this can be attributed to a rapid influx of the transition metal ions into the PS liposomes (as also observed for Co$^{2+}$ and SUV [PS][134]), more rapid than Ca$^{2+}$ or Mg$^{2+}$, thereby preventing the formation of the Tb/DPA complex, but not interfering with ANTS fluorescence or its quenching by DPX. In such cases, it might be advantageous to have recourse to yet a third assay for the mixing of aqueous contents.

The observations of Kendall and MacDonald[135] that SUV (PS) in the presence of Ca$^{2+}$ lyse without coalescence of the encapsulated contents (using the Co-calcein/EDTA assay) are in contradiction to the results of the Tb/DPA, Tb-release, and CF-release assays. The authors have not explained the possible reasons for this discrepancy. We have analyzed their assay and other assays in detail to understand why such differences arise. One aspect which we examined was the amount of fluorophore bound to the vesicle membranes, using preformed SUV (PS) incubated with the reactants used in the assays. Tb and DPA associated with the membrane minimally (3.9 and 0.41% of the amount encapsulated by an SUV, assuming an internal volume of 0.2 μℓ/μmol) whereas excessive quantities of Co/calcein bound to the vesicles (564 to 700% of the amount encapsulated inside an SUV[89]). Thus, it appears that most of the leakage observed from SUV (PS) using the Co-calcein/EDTA assay arises from the bound fluorophores as they detach from the outer surface of the vesicle when Ca$^{2+}$ destabilizes the membrane. Furthermore, the "encapsulated" (i.e., bound plus encapsulated) Co/calcein is very permeable, even before the onset of the fusion reaction. We also found that ANTS bound excessively to preformed SUV (PS), at about 100% of the

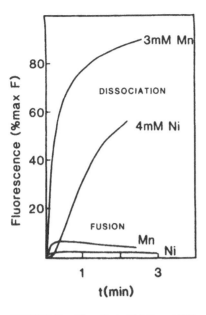

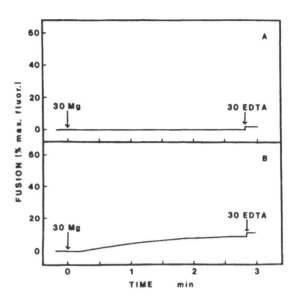

FIGURE 12. The effect of $Mn^{2+}$ and $Ni^{2+}$ on the Tb/DPA fusion signal (lower pair of curves) and the dissociation signal (upper pair of curves). The very small fusion signals imply that either very little mixing of contents occurs, or that the influx of the transition metals into the fusing liposomes prevents Tb/DPA chelation. Using the ANTS/DPX assay, it was determined that the latter explanation is correct. (Reprinted with permission from Bentz, J. and Düzgüneş, N., *Biochemistry*, 24, 5436, 1985. Copyright 1985 American Chemical Society.)

FIGURE 13. The effect of $Mg^{2+}$ (30 m$M$) on LUV (PG) monitored by the Tb/DPA assay (A) and the NBD/Rh probe dilution assay (B). (From Rosenberg, J., Düzgüneş, N., and Kayalar, C., *Biochim. Biophys. Acta*, 735, 173, 1983. With permission.)

encapsulated volume. On the other hand, with LUV (PS) we found that *none* of the assays (Tb/DPA, ANTS/DPX, or Co-calcein/EDTA) showed excessive binding of any constituents. Nevertheless, the Co-calcein/EDTA assay behaved quite differently from both the Tb/DPA and ANTS/DPX assays. For example, it produced essentially identical fluorescence signals for both 5 and 10 m$M$ $Ca^{2+}$ (compare with Figures 3, 5, and 11), and these signals could not be interpreted as monitoring the mixing of contents.[89] We can only conclude that the Co/calcein assay has yet to be proven as a reliable vesicle fusion assay.

Reliable lipid mixing and contents-mixing assays provide complementary information about the fusion behavior of liposomes. However, in some cases artifacts may arise. Here we give examples of both cases.

When the Tb/DPA assay and the lipid probe dilution assay are used to detect fusion of phosphatidylglycerol (PG) vesicles in the presence of $Mg^{2+}$, different results are obtained (Figure 13[136]). The results have been explained by the possibility that the outer monolayers of the vesicles mix during aggregation, although the bilayer is not destabilized to give rise to communication between the internal contents of the vesicles. Lipid mixing in the absence of either contents mixing or leakage has also been observed in the case of LUV (cholesteryl hemisuccinate (CHEMS)/PE; 3:7) in the presence of low concentrations of $Ca^{2+}$ or $Mg^{2+}$ (Figure 14[84]). At higher concentrations, contents mixing is also observed, but lipid mixing appears to proceed at a faster rate. Addition of polylysine to phosphatidate (PA)/PE vesicles results in lipid probe intermixing in the absence of significant contents mixing or leakage.[137]

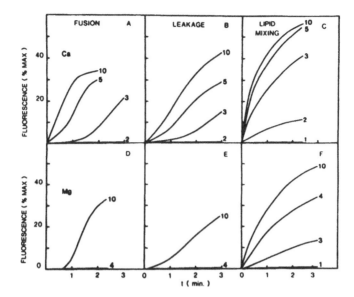

FIGURE 14. Fusion, leakage, and lipid mixing in PE/cholesteryl hemisuc-cinate (713) liposomes in the presence of $Ca^{2+}$ (upper panels, A to C) or $Mg^{2+}$ (lower panels, D to F). Fusion was monitored by the ANTS/DPX assay, leakage by the dilution of preencapsulated ANTS/DPX, and the lipid mixing by the probe dilution method using NBD/Rh. Note the extensive lipid mixing in 4 m$M$ Mg, where there is no measurable leakage or mixing of contents. The ratio of labeled (ANTS or NBD/Rh) to unlabeled vesicles (DPX or empty) was 1:9. (Reprinted with permission from Ellens, H., Bentz, J., and Szoka, F. C., *Biochemistry*, 24, 3099, 1985. Copyright 1985 American Chemical Society.)

Ababei and Hildenbrand[138] reported that the NBD/Rh probe dilution assay showed more $Ca^{2+}$-induced fusion of LUV (PS) than did the Tb/DPA assay. Wilschut et al.[103] quantitated this result by finding that the fusion rate constants ($f_{11}$) measured by the probe dilution assay were an order of magnitude higher than those obtained using the Tb/DPA assay. It was also found that the aggregation rate constants are the same for both assays at 15°C, but higher for the probe dilution assay at 35°C, indicating that lipid mixing can occur during reversible vesicle aggregation. Pryor et al.[176] concluded that the reversible aggregation of SUV (dimyristoyl PE) is also accompanied by probe mixing. Ababei and Hildenbrand[138] found that a higher concentration of $Ca^{2+}$ was required to initiate fusion for their LUV (PS) than had other studies.[59,61,84,129] This descrepancy may be attributed to the different sources of PS and vesicle size distributions.

With SUV (PS), lipid mixing detected by the probe-mixing assay[93] appears to take place at the same rate as contents mixing (Figure 15[61,92]). This difference between the SUV and the LUV (where lipid mixing can occur before mixing of contents) agrees with the kinetic analysis of these systems. Under the conditions used, the $Ca^{2+}$-induced fusion of the SUV (PS) is rate limited by the aggregation step (i.e., $C_{11}$) and that of the LUV (PS) is partially dependent upon the membrane fusion kinetics (i.e., $f_{11}$[63,64]). Seen in this light, it is not surprising that the contents mixing and the lipid-mixing assays agree with respect to the SUV (PS) and can show different kinetics with the LUV (PS).

However, recent results on LUV have indicated that the probe-mixing assay (in which NBD-PE and Rh-PE are placed in separate vesicle populations) gives considerably faster "fusion" rates than the probe-dilution method (in which the probes are initially placed in one population of vesicles) (Figure 6[109,111]). For example, the initial rate of probe mixing

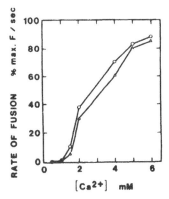

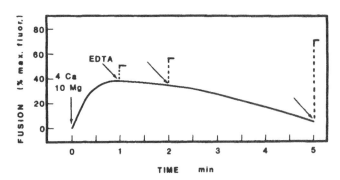

FIGURE 15.    The initial rate of fusion of SUV (PS) as a function of Ca²⁺ concentration, monitored by the NBD/Rh probe-mixing method (circles) or by the Tb/DPA assay (triangles). The lipid concentration was 50 μM (From Wilschut, J., Düzgüneş, N., Hong, K., Hoekstra, D., and Papahadjopoulos, D., *Biochim. Biophys. Acta*, 734, 309, 1983. With permission.)

FIGURE 16.    Fusion of LUV (PS) in the presence of 4 m*M* Ca²⁺/10 m*M* Mg²⁺ monitored by the NBD/Rh probe dilution method. 14 m*M* EDTA was added at the times indicated by the arrows. (From Rosenberg, J., Düzgüneş, N., and Kayalar, C., *Biochim. Biophys. Acta*, 735, 173, 1983. With permission.)

is an order of magnitude faster than probe dilution in the case of LUV (PS/PE) in the presence of 5 m*M* Ca²⁺. In fact, significant probe mixing occurs in the absence of any probe dilution or mixing of contents in certain types of vesicles, such as LUV (PS) in the presence of Mg²⁺, or LUV (PS/PC) in the presence of Ca²⁺ or Mg²⁺. Thus, without careful controls, the probe-mixing assay can respond to many other events, other than simple membrane fusion. In fact, it has been proposed to use it strictly as an assay of aggregation when no fusion occurs (cf. Gibson and Loew[139]).

One final note on the lipid-mixing assays, as well as the contents-mixing assays, is that massive aggregation of the vesicles can lead to erroneous estimates of fusion. In Figure 16, we show an example of the effect of vesicle aggregation on NBD fluorescence. Stopping the fusion reaction with EDTA at different times reveals a more accurate level of fluorescence than the time-course curve, which declines as the number of fluorescent particles exposed to the excitation beam decreases as a result of massive vesicle aggregation. A similar deception can occur with the ANTS/DPX assay where loss of fluorescent intensity could be interpreted as fusion. On the other hand, it is important to note that stopping Ca²⁺ induced fusion with excess EDTA is known to cause an additional transient destabilization.[59,116]

## C. Fusion vs. Lysis

One of the most common features of studies dealing with membrane fusion is a discussion, however brief, of whether the fusion event is accompanied by leakage of encapsulated contents, and if so, whether the kinetics of this leakage is sufficient to identify the event as "lysis", as opposed to "fusion with some leakage" (Figure 1). This issue has provoked a long-lived discourse whose origins can be found in the first assay for mixing of aqueous contents developed by Ingolia and Koshland.[140] This assay depended on the luminescence of firefly extract, encapsulated in one population of asolectin vesicles, upon interaction with Mg²⁺ and ATP, encapsulated in the other population. Some mixing of contents was found, but only 3% of the maximum attainable luminescence could be recovered during the fusion process. The conclusion was that only a few percent of the liposomes actually fused. Holz

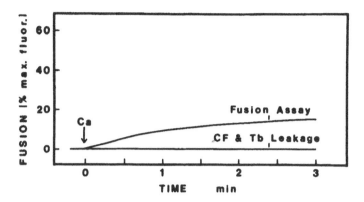

FIGURE 17.    Fusion of LUV (PG) in the presence of 30 m$M$ Ca$^{2+}$, monitored by the Tb/DPA assay. The leakage of vesicle contents was monitored by CF release or Tb release. (From Rosenberg, J., Düzgüneş, N., and Kayalar, C., *Biochim. Biophys. Acta*, 735, 173, 1983. With permission.)

and Stratford[141] repeated these experiments using purified firefly luciferase with Mg$^{2+}$ in one population of asolectin liposomes and ATP in the other population, with the membrane-permeable luciferin present in the medium as a co-factor for the reaction. They also found only about 3 to 5% of the maximal luminescence and concluded that the liposomes had leaked their contents. What really happens in not known, because there have been no follow-up reports on the luciferase assay in other types of liposomes, or on the fusion of asolectin liposomes using other aqueous contents-mixing assays.

Membrane fusion in liposome systems is not necessarily a leaky process. Fusion of PS LUV in 2 m$M$ Ca$^{2+}$ is nonleaky for the first 2 min (Figure 2B). Even at higher Ca$^{2+}$ concentrations, the early stages of fusion, i.e., when only dimers and trimers are present, are not accompanied by leakage.[59-64] Similar results have been obtained with PS/PE LUV.[70,99] No leakage of either CF or Tb is observed during Ca$^{2+}$-induced fusion of LUV composed of PG (CL) (Figure 17[136]). The internal aqueous contents of LUV made of *Bacillus subtilis* cardiolipin and dioleoyl PC are completely retained during fusion induced by Ca$^{2+}$ at 10°C; leakage is observed at higher temperatures (Figure 18[61]). Even SUV (PS) can show fusion with little leakage and no collapse for several hours when the ionic strength is low (Figure 19[65]). Compared with Figure 10, the critical difference in this case is that further aggregation of the vesicles, beyond dimers and trimers, is inhibited by the strong electrostatic repulsion between the fused vesicles, which increases sharply with decreased ionic strength and increased particle size.[178]

It has become obvious over time that the issue of vesicle leakage is inextricably bound to the extent of aggregation and its role in inducing vesicle collapse. Our present interpretation of the fusion of LUV (PS) induced by Ca$^{2+}$, for example, is that leakage of contents occurs after the fusion event and is primarily due to a collapse of the fused products. After all, the equilibrium structures are anhydrous cochleates.[51] Other interpretations have been advanced which consider the ways in which the fusion assays may be leading us astray. It is worthwhile to briefly review these possibilities.

The first is that the Tb/DPA complex might be forming outside the aggregating vesicles due to locally high concentrations of the released contents.[59,61,99,182] This has been examined by terminating the fusion reaction by excess EDTA in the case of Ca$^{2+}$-induced fusion of PS vesicles.[59] If the fluorescence is generated by complex formation outside the vesicles, addition to EDTA should dissociate the complex and quench the fluorescence. If the fluorescent complex is inside the fusing vesicles, EDTA should have no effect on the fluorescence. Figure 20 shows that EDTA addition at the early stages of Ca$^{2+}$-induced fusion of

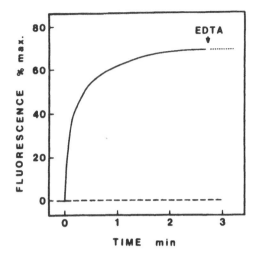

FIGURE 18. Fusion (Tb/DPA assay, solid line) and release of contents (CF, dashed line) of LUV composed of an equimolar mixture of cardiolipin from *Bacillus subtilis* and dioleoyl PC, at 10°C. (From Wilschut, J., Düzgüneş, N., Hong, K., Hoekstra, D., and Papahadjopoulos, D., *Biochim. Biophys. Acta*, 734, 309, 1983. With permission.)

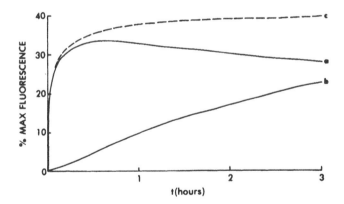

FIGURE 19. Fusion over hours of SUV(PS) in the presence of 20 m$M$ Na$^+$ and 1 m$M$ Ca$^{2+}$, added at t = 0. Curve a shows the intermixing of aqueous contents (Tb and DPA), curve b the dissociation of Tb/DPA complex, and curve c the corrected fusion obtained by adding one half of the dissociation values to curve a (see Equation 7). (From Nir, S., Düzgüneş, N., and Bentz, J., *Biochim. Biophys. Acta*, 735, 160, 1983. With permission.)

PS vesicles results in the arrest of the Tb fluorescence signal. This indicates that the Tb/DPA complex is exclusively within the vesicles. EDTA addition at later times causes a considerable drop in the fluorescence, especially in the case of LUV (Figures 20A and B). This effect is thought to arise from the disruption of some of the vesicles by EDTA, causing the release of their contents and not from the mere entry of EDTA into the vesicle interior.[59] The effect of EDTA is more pronounced at higher Ca$^{2+}$ concentrations (Figures 20C and D). Another experiment indicating that the contents do not interact outside the vesicles is shown in Figure 20E. The relative drop in fluorescence after EDTA addition is the same

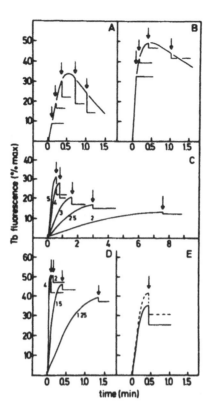

FIGURE 20. Termination of $Ca^{2+}$-induced fusion of PS vesicles by the addition of excess EDTA. Fusion was monitored by the Tb/DPA assay. (A) LUV; 5 m$M$ $Ca^{2+}$; (B) SUV; 1.5 m$M$ $Ca^{2+}$; (C) LUV; various concentrations of $Ca^{2+}$ in m$M$; (D) SUV; various concentrations of $Ca^{2+}$; (E) LUV; 10 m$M$ $Ca^{2+}$ alone (dashed line) or 20 m$M$ $Ca^{2+}$ with 10 m$M$ EDTA. (Reprinted with permission from Wilschut, J., Düzgüneş, N., Fraley, R., and Papahadjopoulos, D., *Biochemistry*, 19, 6011. Copyright 1980 American Chemical Society.)

whether fusion is initiated with $Ca^{2+}$ alone or excess $Ca^{2+}$ in the presence of EDTA. The high EDTA concentration in the latter case would have prevented the formation of the Tb/DPA complex in the vicinity of the aggregated vesicles.

A second way in which the assays could show mixing of contents, while the vesicles have lysed, is if large aggregates have formed, and within the aggregate, the leaked contents are protected from the external medium.[61,78] That this is not the case, at least for vesicle systems we have examined, has been proven in two ways. First, the kinetics of Tb/DPA complex formation and ANTS/DPX quenching have been shown to be rate limited by vesicle aggregation (at sufficiently low lipid concentrations) for many different systems.[59,60,62-64,79,84-86,103,116,136,174] This can only imply that the vesicles fuse upon contact, whatever happens to the higher order aggregates. Second, the assays do show clear cases of lysis. With LUV (PS) and certain $Ca^{2+}$ and $Mg^{2+}$ mixtures, Wilschut et al.[61] found very little Tb/DPA fluorescence despite extensive leakage from aggregated vesicles. Likewise, with LUV (PE/CHEMS) at low pH, Bentz et al.[170] used the ANTS/DPX assay to show contact-mediated lysis and no fusion. These same liposomes show fusion with $Ca^{2+}$ or $Mg^{2+}$.[84] Thus, there is no evidence that large aggregates are obscuring the interpretation of the aqueous content-mixing assays.

A third possible interpretation of the LUV (PS) fusion event involving lysis has been proposed by Rand et al.[40] They present evidence that some apposed vesicles rupture, leading to lysis, while other apposed vesicles fuse. The data, however, are somewhat ambiguous in that the evidence of rupture of apposed vesicles was based upon observation of very large, multilamellar vesicles (1 to 10 μm). Also, the possibility that (invisible) small vesicles could be causing the apparent rupture event cannot be excluded. Other data included rapid-freezing freeze-fracture studies of LUV (PS) fusing in the presence of $Ca^{2+}$, which revealed the formation of flattened areas of interbilayer contact between the vesicles within 100 msec.[40] At a later time (30 sec) the vesicles appeared collapsed, with highly curved ridges. Unfortunately, it is impossible to compare these observations directly with kinetic measurements with fluorescence assays. The electron microscopy was performed with vesicles in a medium of very low ionic strength (2 m$M$ TES), under which conditions $Ca^{2+}$ binding to PS is much stronger than in the presence of 100 m$M$ NaCl.[65,77,78] The amount of $Ca^{2+}$-bound/PS under these conditions is several-fold above the threshold value required for the fusion reaction [78,82]. Thus, the vesicles are in a highly fusion-prone state as soon as they aggregate. The high vesicle and $Ca^{2+}$ concentrations, and the low ionic strength drive the reaction quickly to the state of vesicle collapse.[59]

The conditions used for the fluorescence assays allow fusion to proceed in a more controlled manner. All the biophysical evidence indicates that collapse and release of vesicle contents is a process secondary to communication between the contents, when the experiments are performed in physiological ionic strength and low vesicle concentrations where the fusion of dimers and trimers predominate and before higher-order aggregates form. Nevertheless, the fluorometric data cannot exclude the possibility that the $Ca^{2+}$-induced fusion of LUV (PS) involves some lytic events, just as there is no direct proof that these lytic events occur. Our strongest evidence that the assays presented can be properly interpreted is that they will show the existence of all possible interactions between vesicles, from aggregation without destabilization, to fusion, to contact-induced lysis.

One final comment on the usage of aqueous contents-mixing assays is warranted. Clearly, the mixing of aqueous contents (under the appropriate kinetic conditions) implies a straightforward fusion event in the biological sense; whereas the lipid mixing assays can only monitor the motion of the membrane constituents. Nevertheless, statements to the effect that "aqueous contents mixing assays fail (or are of limited value) when there is a leakage," often appear in the literature. This is plainly untrue. The sole task of the Tb/DPA or ANTS/DPX assay is to report on the mixing of vesicular contents and their leakage. If there is lysis and that is what the assay reports, then it has succeeded in providing a valuable piece of information about the membrane destabilization. Under conditions of extensive leakage, the lipid-mixing assays can be used to ascertain whether mixing of membrane components occurs. However, lipid mixing need not imply membrane fusion. A paradigm of this problem is PE-containing liposomes incubated at temperatures above their hexagonal $H_{II}$ phase transition temperature. Upon intervesicular contact, there is rapid lysis and lipid mixing.[85,86,170] As the final product is closer to inverted lipid tubes than vesicles, this event should probably not be called membrane fusion.

## VI. APPLICATIONS TO BIOLOGICAL MEMBRANES

It is obvious that the fusion of pure phospholipid vesicles is not expected to model every aspect of fusion events in biological membranes. Use of pure phospholipids enables us to construct viable fusion assays with simple vesicle systems and to understand the physicochemical properties of particular phospholipids with respect to their participation in the membrane-fusion reaction.[42,46,196] In this section we give several examples of the successful application of fluorescence assays to membrane fusion in biological membrane systems.

Keller et al.[146] used octadecyl-rhodamine (R-18) and octadecanoyl-aminofluorescein (F-18) after insertion of the probes into separate populations of cells, to study virus-induced fusion of cultured cells. Holowka and Baird[147] provided a careful description of the partitioning of these probes into cellular vesicles. Hoekstra et al.[148] inserted R-18 into membranes at a self-quenching concentration and followed the dilution of the probe into unlabeled membranes. Using this method, they studied the kinetics of fusion of influenza virus with liposomes, Sendai virus with erythrocyte ghosts, and Sendai virus-induced fusion of ghosts. Hoekstra et al.[149] investigated the pH, temperature, and protein requirements of Sendai virus-erythrocyte ghost fusion, again utilizing the R-18 probe. Nir et al.[97] performed a mass action kinetic analysis of this fusion process and derived an expression for the number of virus particles which have fused with a single cell. They concluded that the number of virus particles which fuse with a single ghost is limited to 100 to 200, and that provided the number of particles per cell is below 100, most of the cell-associated viruses fuse within 2 min. Citovsky et al.[180] studied the fusion of Sendai virus with liposomes using R-18 and found that fusion with PC/cholesterol liposomes was inhibited by the same agents which block fusion with biological membranes. This technique was also used to demonstrate the low pH-induced fusion of influenza virus with cultured cells.[192] The R-18 probe has also been inserted into neutrophil secretory vesicles to monitor their fusion with target membranes.[152]

The lipid probe dilution method described in Section II.A has been used for detecting fusion of labeled liposomes with chromaffin granule ghosts, influenza virus, bacterial membrane vesicles, and neutrophil secretory granules. Bental et al.[150] found that cardiolipin CL/PC/cholesterol liposomes could fuse with chromaffin granule ghosts in the absence of $Ca^{2+}$. Stegmann et al.[95] studied the pH-dependent fusion of influenza virus with liposomes and found that the pH optimum is around 5, and that negatively charged phospholipids produce more efficient fusion, which was also quite leaky. Nir et al.[96] analyzed the kinetics of fusion between influenza virus and LUV (CL), and concluded that the virus fuses with several liposomes, and that at pH 5 and 37°C all the virus particles are involved in fusion. Bacterial membrane vesicles were also found to fuse with LUV (CL/dioleoyl PC) at mildly acidic pH.[151] Meers et al.[152] monitored the fusion of secretory granules isolated from human neutrophils with PA/PE liposomes, to study the role of synexin-like proteins from neutrophil cytosol and arachidonic acid in membrane fusion. Morris and Bradley[153] labeled chromaffin granules with NBD-PE and Rh-PE by incubating them with SUV composed of the probe lipids. Labeling of the granule membrane could be followed by the relief of self-quenching of the fluorophores.

The lipid probes NBD-PE and Rh-PE were incorporated into reconstituted membranes containing viral envelope proteins, and their dilution into unlabeled target membranes was followed to assay for the fusion activity of the reconstituted proteins. Eidelman et al.[154] found that reconstituted membranes incorporating the G-protein from vesicular stomatitis virus fuse at pH below 5 with target membranes containing negatively charged lipids. Harmsen et al.[155] monitored the fusion of reconstituted Sendai virus envelopes with erythrocytes. They found the kinetics of fusion to be different for reconstituted membranes and intact virions. Citovsky and Loyter[156] monitored the fusion of reconstituted Sendai virus with liposomes and erythrocyte membrane vesicles, by the dequenching of the fluorescence of NBD-PE incorporated into the reconstituted membrane. They reported that fusion of reconstituted envelopes with PC/cholesterol vesicles was nonleaky, as monitored by the release of CF from the vesicles. Incorporation of virus receptors, i.e., sialoglycolipids or sialoglycoproteins, into the target membrane rendered the fusion process leaky. Van Meer et al.[189] showed that MDCK and BHK cells infected with influenza virus fuse with NBD/

Rh-labeled liposomes at low pH, presumably due to the expression of the viral fusion protein in the plasma membrane.

The Tb/DPA assay has also been used to study fusion of biological membranes. Hoekstra et al.[157,158] encapsulated Tb citrate and Na dipicolinate in separate populations of erythrocyte ghosts, and followed the kinetics of fusion induced by Ca-phosphate. Citovsky and Loyter[156] monitored the intermixing of aqueous contents of reconstituted Sendai virus envelopes and target liposomes using the Tb/DPA assay.

Altstiel and Branton[159] developed an assay to study the fusion of isolated coated vesicles with lysosomes, utilizing the esterase activity of the lysosomes to generate CF from non-fluorescent carboxydiacetylfluorescein, initially entrapped inside the coated vesicles. Their results indicated that the clathrin coat inhibited fusion, which required $Ca^{2+}$, and that the surface proteins of the "uncoated" vesicles were not involved in this fusion reaction.

The studies outlined above clearly indicate the applicability and usefulness of fluorescence assays in understanding the molecular mechanisms of membrane fusion in biological systems. The next generation of experiments is likely to elucidate the molecular control of exocytosis, by utilizing secretory vesicles and inside-out plasma membranes, and of secondary lysosome formation, by the use of isolated endosomes and primary lysosomes. It may be possible to insert fluorescent probes into biological membrane vesicles using phospholipid exchange proteins, in addition to currently used techniques such as R-18 insertion. Alternatively, fluorogenic substrates for the indigenous enzymes of certain secretory vesicles, such as those of polymorphonuclear leukocytes, may be useful for following aqueous contents-mixing, if they can be encapsulated in phagocytic vacuoles. Fluorescence techniques for detecting and analyzing membrane fusion will be an integral part of the discovery of molecules and mechanisms which control intracellular membrane-fusion phenomena, both in vitro and *in situ*.

## ACKNOWLEDGMENTS

We thank Dr. D. Papahadjopoulos for his continued advice and enthusiasm, Dr. H. Ellens for critically reading the manuscript, and Ms. A. Mazel and R. Antonucci for assistance with the preparation of the manuscript. This work was supported by NIH Grants GM28117, GM 31506, and a Grant-in-Aid from the American Heart Association with funds contributed in part by the California affiliate.

## REFERENCES

1. **Epel, D. and Vacquier, V. D.,** Membrane fusion events during invertebrate fertilization, *Cell Surf. Rev.,* 5, 1, 1978.
2. **Bedford, J. M. and Cooper, G. W.,** Membrane fusion events in the fertilization of vertebrate eggs, *Cell Surf. Rev.,* 5, 65, 1978.
3. **Shapiro, B. M.,** Molecular aspects of sperm-egg fusion, *Ciba Found. Symp.,* 103, 86, 1984.
4. **Palade, G.,** Intracellular aspects of the process of protein synthesis, *Science,* 189, 347, 1975.
5. **Meldolesi, J., Borgese, N., De Camilli, P., and Ceccarelli, B.,** Cytoplasmic membranes and the secretory process, *Cell Surf. Rev.,* 5, 509, 1978.
6. **Orci, L. and Perrelet, A.,** Ultrastructural aspects of exocytotic membrane fusion, *Cell Surf. Rev.,* 5, 629, 1978.
7. **Pollard, H. B., Pazoles, C. J., Creutz, C. E., and Zinder, O.,** The chromaffin granule and possible mechanisms of exocytosis, *Int. Rev. Cytol.,* 58, 159, 1979.
8. **Henson, P. M., Ginsberg, M. H., and Morrison, D. C.,** Mechanisms of mediator release by inflammatory cells, *Cell Surf. Rev.,* 5, 407, 1978.

9. **White, J. G.,** The secretory process in platelets, in *Cell Biology of the Secretory Process,* Cantin, M., Ed., S. Kargar, Basel, 1984, 546.

10. **Silverstein, S. C., Steinman, R. M., and Cohn, Z. A.,** Endocytosis, *Annu. Rev. Biochem.,* 46, 669, 1977.

11. **Bretscher, M. S.,** Endocytosis: relation to capping and locomotion, *Science,* 224, 681, 1984.

12. **Goldstein, J. L., Anderson, R. G. W., and Brown, M. S.,** Coated pits, coated vesicles, and receptor-mediated endocytosis, *Nature (London),* 279, 679, 1979.

13. **Pastan, I. H. and Willingham, M. C.,** The pathway of endocytosis, in *Endocytosis,* Pastan, I. and Willingham, M. C., Eds., Plenum Press, New York, 1985, 1.

14. **Pearse, B. M. F. and Bretscher, M. S.,** Membrane recycling by coated vesicles, *Annu. Rev. Biochem.,* 50, 85, 1981.

15. **Helenius, A., Mellman, I., Wall, D., and Hubbard, A.,** Endosomes, *Trends Biochem. Sci.,* 8, 245, 1983.

16. **Geuze, H. J., Slot, J. W., Strous, G. J. A. M., Lodish, H. F., and Schwartz, A. L.,** Intracellular site of asialoglycoprotein receptor-ligand uncoupling: double-label immunoelectron microscopy during receptor-mediated endocytosis, *Cell,* 32, 277, 1983.

17. **Farquhar, M. G.,** Traffic of products and membranes through the Golgi complex, in *Transport of Macromolecules in Cellular Systems,* Silverstein, S. C., Ed., Dahlem Konferenzen, Berlin, 1978, 341.

18. **Kelly, R. B.,** Pathways of protein secretion in eukaryotes, *Science,* 230, 25, 1985.

19. **Choppin, P. W. and Compans, R. W.,** Reproduction of paramyxoviruses, in *Comprehensive Virology,* Vol. 4, Fraenkel-Conrat, H. and Wagner, R., Eds., Plenum Press, New York, 1975, 95.

20. **Poste, G. and Pasternak, C. A.,** Virus-induced cell fusion, *Cell Surf. Rev.,* 5, 305, 1978.

21. **Helenius, A. and Marsh, M.,** Endocytosis of enveloped animal viruses, *Ciba Found. Symp.,* 92, 59, 1982.

22. **Marsh, M.,** The entry of enveloped viruses into cells by endocytosis, *Biochem. J.,* 218, 1, 1984.

23. **Papadimitriou, J. M.,** Macrophage fusion in vivo and in vitro: a review, *Cell Surf. Rev.,* 5, 181, 1978.

24. **Allan, D., Limbrick, A. R., Thomas, P., and Westerman, M. P.,** Release of spectrin-free spicules on reoxygenation of sickled erythrocytes, *Nature (London),* 295, 612, 1982.

25. **Baggiolini, M. and Dewald, B.,** Exocytosis by neutrophils, in *Regulation of Leukocyte Function,* Snyderman, R., Ed., Plenum Press, New York, 1984, 221.

26. **Goldstein, I. M.,** Neutrophil degranulation, in *Regulation of Leukocyte Function,* Snyderman, R., Ed., Plenum Press, New York, 1984, 189.

27. **Klatzmann, D., Barre-Sinoussi, F., Nugeyre, M. T., Dauguet, C., Vilmer, E., Griscelli, C., Brun-Vezinet, F., Rousioux, C., Gluckman, J. C., Chermann, J. -C., and Montagnier, L.,** Selective tropism of lymphoadenopathy associated virus (LAV) for helper-inducer T lymphotyces, *Science,* 225, 59, 1984.

28. **Satir, B. H., Schooley, C., and Satir, P.,** Membrane fusion in a model system: mucocyst secretion in Tetrahymena, *J. Cell Biol.,* 56, 153, 1973.

29. **Satir, B. H. and Oberg, S. G.,** Paramecium fusion rosettes: possible function as $Ca^{2+}$ gates, *Science,* 199, 536, 1978.

30. **Ahkong, Q. F., Fisher, D., Tampion, W., and Lucy, J. A.,** Mechanisms of cell fusion, *Nature (London),* 253, 194, 1975.

31. **Chi, E. Y., Lagunoff, D., and Koehler, J. K.,** Freeze-fracture study of mast cell secretion, *Proc. Natl. Acad. Sci. U.S.A.,* 73, 2823, 1976.

32. **De Camilli, P., Peluchetti, D., and Meldolesi, J.,** Dynamic changes of the luminal plasmalemma in stimulated parotid acinar cells. A freeze-fracture study, *J. Cell. Biol.,* 70, 59, 1976.

33. **Lawson, D., Raff, M. C., Gomperts, B., Fewtrell, C., and Gilula, N. B.,** Molecular events during membrane fusion: a study of exocytosis in rat peritoneal mast cells, *J. Cell Biol.,* 72, 242, 1977.

34. **Orci, L., Perrelet, A., and Friend, D. S.,** Freeze-fracture of membrane fusions during exocytosis in pancreatic B-cells, *J. Cell Biol.,* 75, 23, 1977.

35. **Chandler, D. E. and Heuser, J. E.,** Membrane fusion during secretion: cortical granule exocytosis in sea urchin eggs as studied by quick-freezing and freeze-fracture, *J. Cell Biol.,* 83, 91, 1979.

36. **Chandler, D. E. and Heuser, J. E.,** Arrest of membrane fusion events in mast cells by quick-freezing, *J. Cell Biol.,* 86, 666, 1980.

37. **Heuser, J. E., Reese, T. S., Dennis, M. J., Jan, Y., Jan, L., and Evans, L.,** Synaptic vesicle exocytosis captured by quick freezing and correlated with quantal transmitter release, *J. Cell Biol.,* 81, 275, 1979.

38. **Heuser, J. E. and Reese, T. S.,** Structural changes after transmitter release at the frog neuromuscular junction, *J. Cell Biol.,* 88, 564, 1981.

39. **Miller, D. C. and Dahl, G. P.,** Early events in calcium-induced liposome fusion, *Biochim. Biophys. Acta,* 689, 165, 1982.

40. **Rand, R. P., Kachar, B., and Reese, T. S.,** Dynamic morphology of calcium-induced interactions between phosphatidylserine vesicles, *Biophys. J.,* 47, 483, 1985.

41. **Ornberg, R. L. and Reese, T. S.**, Beginning of exocytosis captured by rapid freezing of Limulus amebocytes, *J. Cell Biol.*, 90, 40, 1981.

42. **Düzgüneş, N.**, Membrane fusion, *Sub-cell. Biochem.*, 11, 195, 1985.

43. **Papahadjopoulos, D.**, Calcium-induced phase changes and fusion in natural and model membranes, *Cell Surf. Rev.*, 5, 765, 1978.

44. **Papahadjopoulos, D., Poste, G., and Vail, W. J.**, Studies on membrane fusion with natural and model membranes, *Methods Membr. Biol.*, 10, 1, 1979.

45. **Nir, S., Bentz, J., Wilschut, J., and Düzgüneş, N.**, Aggregation and fusion of phospholipid vesicles, *Prog. Surf. Sci.*, 13, 1, 1983.

46. **Düzgüneş, N., Wilschut, J., and Papahadjopoulos, D.**, Control of membrane fusion by divalent cations, phospholipid head-groups, and proteins, in *Physical Methods on Biological Membranes and their Model Systems*, Conti, F., Blumberg, W. E., de Gier, J., and Pocchiari, F., Eds., Plenum Press, New York, 1985, 193.

47. **Düzgüneş, N., Hong, K., Baldwin, P. A., Bentz, J., Nir, S., and Papahadjopoulos, D.**, Fusion of phospholipid vesicles induced by divalent cations and protons: modulation by phase transitions, free fatty acids, monovalent cations and polyamines, in *Cell Fusion*, Sowers, A. E., Ed., Plenum Press, New York, 1987, 241.

48. **Hong, K., Düzgüneş, N., Meers, P. R., and Papahadjopoulos, D.**, Protein modulation of liposome fusion, in *Cell Fusion*, Sowers, A. E., Ed., Plenum Press, New York, 1987, 269.

49. **White, J., Kielian, M., and Helenius, A.**, Membrane fusion proteins of enveloped animal viruses, *Q. Rev. Biophys.*, 16, 151, 1983.

50. **Pagano, R. E. and Weinstein, J. N.**, Interactions of liposomes with mammalian cells, *Annu. Rev. Biophys. Bioeng.*, 7, 435, 1978.

51. **Papahadjopoulos, D., Vail, W. J., Jacobson, K., and Poste, G.**, Cochleate lipid cylinders: formation by fusion of unilamellar lipid vesicles, *Biochim. Biophys. Acta*, 394, 483, 1975.

52. **Newton, C., Pangborn, W., Nir, S., and Papahadjopoulos, D.**, Specificity of $Ca^{2+}$ and $Mg^{2+}$ binding to phosphatidylserine vesicles and resultant phase changes of bilayer membrane structure, *Biochim. Biophys. Acta*, 506, 281, 1978.

53. **Portis, A., Newton, C., Pangborn, W., and Papahadjopoulos, D.**, Studies on the mechanism of membrane fusion: evidence for an intermembrane $Ca^{2+}$-phospholipid complex, synergism with $Mg^{2+}$, and inhibition by spectrin, *Biochemistry*, 18, 780, 1979.

54. **Papahadjopoulos, D. and Bangham, A. D.**, Biophysical properties of phospholipids. II. Permeability of phosphatidylserine liquid crystals to univalent ions, *Biochim. Biophys. Acta*, 126, 185, 1966.

55. **Papahadjopoulos, D. and Ohki, S.**, Stability of asymmetric phospholipid membranes, *Science*, 164, 1075, 1969.

56. **Papahadjopoulos, D., Vail, W. J., Newton, C., Nir, S., Jacobson, K., Poste, G., and Lazo, R.**, Studies on membrane fusion. III. The role of calcium-induced phase changes, *Biochim. Biophys. Acta*, 465, 579, 1977.

57. **Ginsberg, L.**, Does calcium cause fusion or lysis of unilamellar lipid vesicles? *Nature (London)*, 275, 758, 1978.

58. **Wilschut, J. and Papahadjopoulos, D.**, $Ca^{2+}$-induced fusion of phospholipid vesicles monitored by mixing of aqueous contents, *Nature (London)*, 281, 690, 1979.

59. **Wilschut, J., Düzgüneş, N., Fraley, R., and Papahadjopoulos, D.**, Studies on the mechanism of membrane fusion: kinetics of $Ca^{2+}$-induced fusion of phosphatidylserine vesicles followed by a new assay for mixing of aqueous vesicle contents, *Biochemistry*, 19, 6011, 1980.

60. **Wilschut, J., Düzgüneş, N., and Papahadjopoulos, D.**, Calcium/magnesium specificity in membrane fusion: Kinetics of aggregation and fusion of phosphatidylserine vesicles and the role of bilayer curvature, *Biochemistry*, 20, 3126, 1981.

61. **Wilschut, J., Düzgüneş, N., Hong, K., Hoekstra, D., and Papahadjopoulos, D.**, Retention of aqueous contents during divalent cation-induced fusion of phospholipid vesicles, *Biochim. Biophys. Acta*, 734, 309, 1983.

62. **Nir, S., Bentz, J., Wilschut, J.**, Mass action kinetics of phosphatidylserine vesicle fusion as monitored by coalescence of internal vesicle volumes, *Biochemistry*, 19, 6030, 1980.

63. **Nir, S., Wilschut, J., and Bentz, J.**, The rate of fusion of phospholipid vesicles and the role of bilayer curvature, *Biochim. Biophys. Acta*, 688, 275, 1982.

64. **Bentz, J., Nir, S., and Wilschut, J.**, Mass action kinetics of vesicle aggregation and fusion, *Colloids Surf.*, 6, 333, 1983.

65. **Nir, S., Düzgüneş, N., and Bentz, J.**, Binding of monovalent cations to phosphatidylserine and modulation of $Ca^{2+}$- and $Mg^{2+}$-induced vesicle fusion, *Biochim. Biophys. Acta*, 735, 160, 1983.

66. **Grenthe, I.**, Stability relationships among the rare earth dipicolinates, *J. Am. Chem. Soc.*, 83, 360, 1961.

67. **Barela, T. D. and Sherry, A. D.**, A simple, one-step fluorometric method for determination of nanomolar concentrations of terbium, *Anal. Biochem.*, 71, 351, 1976.

68. **Thomas, D. D., Carlsen, W. F., and Stryer, L.,** Fluorescence energy transfer in the rapid diffusion limit, *Proc. Natl. Acad. Sci. U.S.A.,* 75, 5746, 1978.
69. **Szoka, F. and Papahadjopoulos, D.,** Procedure for preparation of liposomes with large internal aqueous space and high capture by reverse-phase evaporation, *Proc. Natl. Acad. Sci. U.S.A.,* 75, 4194, 1978.
70. **Düzgüneş, N., Wilschut, J., Fraley, R., and Papahadjopoulos, D.,** Studies on the mechanism of membrane fusion: Role of head-group composition in calcium- and magnesium-induced fusion of mixed phospholipid vesicles, *Biochim. Biophys. Acta,* 642, 182, 1981.
71. **Düzgüneş, N., Wilschut, J., Hong, K., Fraley, R., Perry, C., Friend, D. S., James, T. L., and Papahadjopoulos, D.,** Physicochemical characterization of large unilamellar vesicles prepared by reverse-phase evaporation, *Biochim. Biophys. Acta,* 732, 289, 1983.
72. **Papahadjopoulos, D. and Watkins, J. C.,** Phospholipid model membranes. II. Permeability properties of hydrated liquid crystals, *Biochim. Biophys. Acta,* 135, 630, 1967.
73. **Huang, C. C.,** Studies on phosphatidylcholine vesicles: formation and physical characteristics, *Biochemistry,* 8, 344, 1969.
74. **Düzgüneş, N. and Ohki, S.,** Calcium-induced interaction of phospholipid vesicles with bilayer lipid membranes, *Biochim. Biophys. Acta,* 467, 301, 1977.
75. **Weinstein, J. N., Yoshikami, S., Henkart, P., Blumenthal, R., and Hagins, W. A.,** Liposome-cell interaction: transfer and intracellular release of a trapped fluorescent marker, *Science,* 195, 489, 1977.
76. **Blumenthal, R., Weinstein, J. N., Sharrow, S. O., and Henkart, P.,** Liposome-lymphocyte interactions: saturable sites for transfer and intracellular release of liposome contents, *Proc. Natl. Acad. Sci. U.S.A.,* 74, 5603, 1977.
77. **Düzgüneş, N., Nir, S., Wilschut, J., Bentz, J., Newton, C., Portis, A., and Papahadjopoulos, D.,** Calcium- and magnesium-induced fusion of mixed phosphatidylserine/phosphatidylcholine vesicles: effect of ion binding, *J. Membr. Biol.,* 59, 115, 1981.
78. **Bentz, J., Düzgüneş, N., and Nir, S.,** Kinetics of divalent cation induced fusion of phosphatidylserine vesicles: correlation between fusogenic capacities and binding affinities, *Biochemistry,* 22, 3320, 1983.
79. **Bentz, J., Düzgüneş, N., and Nir, S.,** Temperature dependence of divalent cation induced fusion of phosphatidylserine liposomes: evaluation of the kinetic rate constants, *Biochemistry,* 24, 1064, 1985.
80. **Hoekstra, D., Düzgüneş, N., and Wilschut, J.,** Agglutination and fusion of globoside GL-4 containing phospholipid vesicles mediated by lectins and $Ca^{2+}$, *Biochemistry,* 24, 565, 1985.
81. **de Kroon, A. I. P. M., van Hoogevest, P., Guerts van Kessel, W. S. M., and de Kruijff, B.,** Influence of glycophorin incorporation on $Ca^{2+}$-induced fusion of phosphatidylserine vesicles, *Biochemistry,* 24, 6382, 1985.
82. **Bentz, J. and Düzgüneş, N.,** Fusogenic capacities of divalent cations and the effect of liposome size, *Biochemistry,* 24, 5436, 1985.
83. **Smolarsky, M., Teitelbaum, D., Sela, M., and Gitler, C.,** A simple fluorescent method to determine complement-mediated liposome immune lysis, *J. Immunol. Methods,* 15, 255, 1977.
84. **Ellens, H., Bentz, J., and Szoka, F. C.,** $H^{+}$- and $Ca^{2+}$-induced fusion and destabilization of liposomes, *Biochemistry,* 24, 3099, 1985.
85. **Ellens, H., Bentz, J., and Szoka, F. C.,** Destabilization of phosphatidylethanolamine liposomes at the hexagonal phase transition temperature, *Biochemistry,* 25, 285, 1986.
86. **Ellens, H., Bentz, J., and Szoka, F. C.,** Fusion of phosphatidylethanolamine containing liposomes and the mechanism of the $L_{\alpha}$-$H_{II}$ phase transition, *Biochemistry,* 25, 4141, 1986.
87. **Düzgüneş, N., Straubinger, R. M., Baldwin, P. A., Friend, D. S., and Papahadjopoulos, D.,** Proton-induced fusion of oleic acid-phosphatidylethanolamine liposomes, *Biochemistry,* 24, 3091, 1985.
88. **Ellens, H., Bentz, J., and Szoka, F. C.,** pH-Induced destabilization of phosphatidylethanolamine-containing liposomes. Role of bilayer contact, *Biochemistry,* 23, 1532, 1984.
89. **Bentz, J., Alford, D. R., and Düzgüneş, N.,** Fluorescence assays for mixing of aqueous contents during vesicle fusion, in preparation.
90. **Struck, D. K., Hoekstra, D., and Pagano, R. E.,** Use of resonance energy tranfer to monitor membrane fusion, *Biochemistry,* 20, 4093, 1981.
91. **Förster, T.,** Zwischenmolekulare energiewanderung und Fluoreszenze, *Ann. Phys. (Leipzig),* 2, 55, 1948.
92. **Hoekstra, D.,** Kinetics of intermixing of lipids and mixing of aqueous contents during vesicle fusion, *Biochim. Biophys. Acta,* 692, 171, 1982.
93. **Hoekstra, D.,** Role of lipid phase separations and membrane hydration in phospholipid vesicle fusion, *Biochemistry,* 21, 2833, 1982.
94. **Sundler, R., Düzgüneş, N., and Papahadjopoulos, D.,** Control of membrane fusion by phospholipid head groups. II. The role of phosphatidylethanolamine in mixtures with phosphatidate and phosphatidyl-inositol, *Biochim. Biophys. Acta,* 649, 751, 1981.
95. **Stegmann, T., Hoekstra, D., Scherphof, G., and Wilschut, J.,** Kinetics of pH-dependent fusion between influenza virus and liposomes, *Biochemistry,* 24, 3107, 1985.

96. **Nir, S., Stegmann, T., and Wilschut, J.,** Fusion of influenza virus with cardiolipin liposomes at low pH: mass action analysis of kinetics and extent, *Biochemistry,* 25, 257, 1986.

97. **Nir, S., Klappe, K., and Hoekstra, D.,** Kinetics and extent of fusion between Sendai virus and erythrocyte ghosts: application of a mass action kinetic model, *Biochemistry,* 25, 2155, 1986.

98. **Tanaka, Y. and Schroit, A. J.,** Insertion of fluorescent phosphatidylserine into the plasma membrane of red blood cells. Recognition by autologous macrophages, *J. Biol. Chem.,* 258, 11335, 1983.

99. **Morris, S. J., Gibson, C. C., Smith, P. D., Greif, P. C., Stirk, C. W., Bradley, D., Haynes, D. H., and Blumenthal, R.,** Rapid kinetics of Ca$^{2+}$-induced fusion of phosphatidylserine/phosphatidylethanolamine vesicles: the effect of bilayer curvature on leakage, *J. Biol. Chem.,* 260, 4122, 1985.

100. **Bentz, J. and Nir, S.,** Aggregation of colloidal particles modeled as a dynamical process, *Proc. Natl. Acad. Sci. U.S.A.,* 78, 1634, 1981.

101. **Bentz, J. and Nir, S.,** Mass action kinetics and equilibria of reversible aggregation, *J. Chem. Soc. Faraday Trans. 1,* 77, 1249, 1981.

102. **Day, E. P., Kwok, A. Y. W., Hark, S. K., Ho, J. T., Vail, W. J., Bentz, J., and Nir, S.,** Reversibility of sodium induced aggregation of sonicated phosphatidylserine vesicles, *Proc. Natl. Acad. Sci. U.S.A.,* 77, 4026, 1980.

103. **Wilschut, J., Scholma, J., Bental, M., Hoekstra, D., and Nir, S.,** Ca$^{2+}$-induced fusion of phosphatidylserine vesicles: mass action kinetic analysis of membrane lipid mixing and aqueous contents mixing, *Biochim. Biophys. Acta,* 821, 45, 1985.

104. **Tweet, A. G., Bellamy, W. D., and Gaines, G. L.,** Fluorescence quenching and energy transfer in monomolecular films containing chlorophyll, *J. Chem. Phys.,* 41, 2068, 1964.

105. **Estep, T. N. and Thompson, T. E.,** Energy transfer in lipid bilayers, *Biophys. J.,* 26, 195, 1979.

106. **Wolber, P. K. and Hudson, B. S.,** An analytic solution to the Förster energy transfer problem in two dimensions, *Biophys. J.,* 28, 197, 1979.

107. **Snyder, B. and Freire, E.,** Fluorescence energy transfer in two dimensions, *Biophys. J.,* 40, 137, 1982.

108. **Hoekstra, D.,** Fluorescence method for measuring the kinetics of Ca$^{2+}$-induced phase separations in phosphatidylserine-containing lipid vesicles, *Biochemistry,* 21, 1055, 1982.

109. **Allen, T. M. and Düzgüneş, N.,** Fusion assays: lipid exchange, aggregation, lipid mixing, or coalescence of contents?, *Biophys. J.,* 47, 169a, 1985.

110. **Ohki, S.,** Effects of divalent cations, temperature, osmotic pressure gradient and vesicle curvature on phosphatidylserine vesicle fusion, *J. Membr. Biol.,* 77, 265, 1984.

111. **Düzgüneş, N., Allen, T. M., Fedor, J., and Papahadjopoulos, D.,** Lipid mixing during membrane aggregation and fusion: why fusion assays disagree, *Biochemistry,* in press.

112. **Andersen, O. S.,** Permeability properties of unmodified lipid bilayer membranes, in *Membrane Transport in Biology,* Vol. 1, Tosteson, D. C., Ed., Springer Verlag, New York, 1978, 369.

113. **Nichols, J. W. and Pagano, R. E.,** Kinetics of soluble lipid monomer diffusion between vesicles, *Biochemistry,* 20, 2783, 1981.

114. **Nichols, J. W. and Pagano, R. E.,** Use of resonance energy transfer to study the kinetics of amphiphile transfer between vesicles, *Biochemistry,* 21, 1720, 1982.

115. **Tanford, C.,** The Hydrophobic Effect: formation of Micelles and Biological Membranes, *Wiley-Interscience,* New York, 1980.

116. **Leventis, R., Gagné, J., Fuller, N., Rand, R. P., and Silvius, J. R.,** Divalent cation-induced fusion and lipid lateral segregation in phosphatidylcholine-phosphatidic acid vesicles, *Biochemistry,* 25, 6978, 1986.

117. **Hong, K., Baldwin, P. A., Allen, T. M., and Papahadjopoulos, D.,** Fluorometric detection of the bilayer to hexagonal phase transition in liposomes, *Biochemistry,* submitted.

118. **Szoka, F. C., Jacobson, K., and Papahadjopoulos, D.,** The use of aqueous space markers to determine the mechanism of interaction between phospholipid vesicles and cells, *Biochim. Biophys. Acta,* 551, 295, 1979.

119. **Straubinger, R. M., Hong, K., Friend, D. S., and Papahadjopoulos, D.,** Endocytosis of liposomes and intracellular fate of encapsulated molecules: encounter with a low pH compartment after internalization in coated vesicles, *Cell,* 32, 1069, 1983.

120. **Düzgüneş, N. and Papahadjopoulos, D.,** Ionotropic effects on phospholipid membranes: calcium-magnesium specificity in binding, fluidity, and fusion, in *Membrane Fluidity in Biology,* Vol. 2, Aloia, R. C., Ed., Academic Press, New York, 1983, 183.

121. **Sundler, R.,** Role of phospholipid head group structure and polarity in the control of membrane fusion, *Biomembranes,* 12, 563, 1984.

122. **Wilschut, J. and Hoekstra, D.,** Membrane fusion: from liposomes to biological membranes, *Trends Biochem. Sci.,* 9, 479, 1984.

123. **Düzgüneş, N., Hong, K., and Papahadjopoulos, D.**, Membrane fusion: the involvement of phospholipids, proteins, and calcium binding, in *Calcium-binding Proteins: Structure and Function*, Siegel, F. L., Carafoli, E., Kretsinger, R. H., MacLennan, D. H., and Wasserman, R. H., Eds., Elsevier/North Holland, New York, 1980, 17.

124. **Blumenthal, R.**, Membrane fusion, *Curr. Top. Membr. Transp.*, 29, 203, 1987.

125. **Düzgüneş, N., Paiement, J., Freeman, K. B., Lopez, N. G., Wilschut, J., and Papahadjopoulos, D.**, Modulation of membrane fusion by ionotropic and thermotropic phase transitions, *Biochemistry*, 23, 3486, 1984.

126. **Ohki, S., Düzgüneş, N., and Leonards, K.**, Phospholipid vesicle aggregation: effect of monovalent and divalent ions, *Biochemistry*, 21, 2127, 1982.

127. **Silvius, J. R. and Gagné, J.**, Calcium-induced fusion and lateral phase separations in phosphatidylcholine-phosphatidylserine vesicles. Correlation by calorimetric and fusion measurements, *Biochemistry*, 23, 3241, 1984.

128. **Uster, P. S. and Deamer, D. W.**, Fusion competence of phosphatidylserine-containing liposomes quantitatively measured by a fluorescence resonance energy transfer assay, *Arch. Biochem. Biophys.*, 209, 385, 1981.

129. **Wilschut, J., Düzgüneş, N., Hoekstra, D., and Papahadjopoulos, D.**, Modulation of membrane fusion by membrane fluidity: temperature dependence of divalent cation-induced fusion of phosphatidylserine vesicles, *Biochemistry*, 24, 8, 1985.

130. **Ohki, S.**, A mechanism of divalent-ion induced phosphatidylserine membrane fusion, *Biochim. Biophys. Acta*, 689, 1, 1982.

131. **Miller, C., Arvan, P., Telford, J. N., and Racker, E.**, $Ca^{2+}$-induced fusion of proteoliposomes: dependence on transmembrane osmotic gradient, *J. Membr. Biol.*, 30, 271, 1976.

132. **Cohen, F. S., Akabas, M. H., Zimmerberg, J., and Finkelstein, A.**, Parameters affecting the fusion of unilamellar phospholipid vesicles with planar bilayer membranes, *J. Cell. Biol.*, 98, 1054, 1984.

133. **Fisher, L. R. and Parker, H. S.**, Osmotic control of bilayer fusion, *Biophys. J.*, 46, 253, 1984.

134. **Morris, S. J., Bradley, D., and Blumenthal, R.**, The use of cobalt ions as a collisional quencher to probe surface charge and stability of fluroescently labeled bilayer vesicles, *Biochim. Biophys. Acta*, 818, 365, 1985.

135. **Kendall, D. A. and MacDonald, R. C.**, A fluorescence assay to monitor vesicle fusion and lysis, *J. Biol. Chem.*, 257, 13892, 1982.

136. **Rosenberg, J., Düzgüneş, N., and Kayalar, C.**, Comparison of two liposome fusion assays monitoring the intermixing of aqueous contents and of membrane components, *Biochim. Biophys. Acta*, 735, 173, 1983.

137. **Bondeson, J. and Sundler, R.**, Lysine peptides induce lipid intermixing but not fusion between phosphatidic acid-containing vesicles, *FEBS Lett.*, 190, 283, 1985.

138. **Ababei, L. and Hildenbrand, K.**, Kinetics of calcium-induced mixing of lipids and aqueous contents of large unilamellar phosphatidylserine vesicles, *Chem. Phys. Lipids*, 35, 39, 1985.

139. **Gibson, G. A. and Loew, L. M.**, Application of Förster resonance energy transfer to interactions between cell and lipid vesicle surfaces, *Biochem. Biophys. Res. Commun.*, 88, 141, 1979.

140. **Ingolia, T. D. and Koshland, D. E.**, The role of calcium in fusion of artificial vesicles, *J. Biol. Chem.*, 253, 3821, 1978.

141. **Holz, R. W. and Stratford, C. A.**, Effects of divalent ions on vesicle-vesicle fusion studied by a new luminescence assay for fusion, *J. Membr. Biol.*, 46, 331, 1979.

142. **Hong, K., Düzgüneş, N., and Papahadjopoulos, D.**, Role of synexin in membrane fusion: enhancement of calcium-dependent fusion of phospholipid vesicles, *J. Biol. Chem.*, 256, 3641, 1981.

143. **Hong, K., Düzgüneş, N., and Papahadjopoulos, D.**, Modulation of membrane fusion by calcium-binding proteins, *Biophys. J.*, 37, 297, 1982.

144. **Hong, K., Schuber, F., and Papahadjopoulos, D.**, Polyamines: biological modulators of membrane fusion, *Biochim. Biophys. Acta*, 732, 469, 1983.

145. **Schuber, F., Hong, K., Düzgüneş, N., and Papahadjopoulos, D.**, Polyamines as modulators of membrane fusion: aggregation and fusion of liposomes, *Biochemistry*, 22, 6134, 1983.

146. **Keller, P. M., Person, S., and Snipes, W.**, A fluorescence enhancement assay of cell fusion, *J. Cell Sci.*, 28, 167, 1977.

147. **Holowka, D. and Baird, B.**, Structural studies on the membrane-bound immunoglobulin E-receptor complex. I. Characterization of large plasma membrane vesicles from rat basophilic leukemia cells and insertion of amphipathic fluorescent probes, *Biochemistry*, 22, 3466, 1983.

148. **Hoekstra, D., de Boer, T., Klappe, K., and Wilschut, J.**, Fluorescence method for measuring the kinetics of fusion between biological membranes, *Biochemistry*, 23, 5675, 1984.

149. **Hoekstra, D., Klappe, K., de Boer, T., and Wilschut, J.**, Characterization of the fusogenic properties of Sendai virus: kinetics of fusion with erythrocyte membranes, *Biochemistry*, 24, 4739, 1985.

150. **Bental, M., Lelkes, P. I., Scholma, J., Hoekstra, D., and Wilschut, J.,** Ca$^{2+}$-independent, protein-mediated fusion of chromaffin granule ghosts with liposomes, *Biochim. Biophys. Acta,* 774, 296, 1984.

151. **Driessen, A. J. M., Hoekstra, D., Scherphof, G., Kalicharan, R. D., and Wilschut, J.,** Low pH-induced fusion of liposomes with membrane vesicles derived from *Bacillus subtilis, J. Biol. Chem.,* 260, 10880, 1985.

152. **Meers, P., Ernst, J., Düzgüneş, N., Hong, K., Fedor, J., Goldstein, I. M., and Papahadjopoulos, D.,** Synexin-like proteins from human polymorphonuclear leukocytes. Identification of granule-aggregating and membrane-fusing activities, *J. Biol. Chem.,* 262, 7850, 1987.

153. **Morris, S. J. and Bradley, D.,** Calcium-promoted fusion of isolated chromaffin granules detected by resonance energy transfer between labeled lipids embedded in the membrane bilayer, *Biochemistry,* 23, 4642, 1984.

154. **Eidelman, O., Schlegel, R., Tralka, T. S., and Blumenthal, R.,** pH-Dependent fusion induced by vesicular stomatitis virus glycoprotein reconstituted into phospholipid vesicles, *J. Biol. Chem.,* 259, 4622, 1984.

155. **Harmsen, M. C., Wilschut, J., Scherphof, G., Hulstaert, C., and Hoekstra, D.,** Reconstitution and fusogenic properties of Sendai virus envelopes, *Eur. J. Biochem.,* 149, 591, 1985.

156. **Citovsky, V. and Loyter, A.,** Fusion of Sendai virions or reconstituted Sendai virus envelopes with liposomes or erythrocyte membranes lacking virus receptors, *J. Biol. Chem.,* 260, 12072, 1985.

157. **Hoekstra, D., Wilschut, J., and Scherphof, G.,** Kinetics of calcium phosphate-induced fusion of human erythrocyte ghosts monitored by mixing of aqueous contents, *Biochim. Biophys. Acta,* 732, 327, 1983.

158. **Hoekstra, D., Wilschut, J., and Scherphof, G.,** Fusion of erythrocyte ghosts induced by calcium phosphate: kinetic characteristics and the role of Ca$^{2+}$, phosphate, and calcium-phosphate complexes, *Eur. J. Biochem.,* 146, 131, 1985.

159. **Altstiel, L. and Branton, D.,** Fusion of coated vesicles with lysosomes: measurement with a fluorescence assay, *Cell,* 32, 921, 1983.

160. **Hoekstra, D., Yaron, A., Carmel, A., and Scherphof, G.,** Fusion of phospholipid vesicles containing a trypsin-sensitive fluorogenic substrate and trypsin, *FEBS Lett.,* 106, 176, 1979.

161. **Vanderwerf, P. and Ullman, E. F.,** Monitoring of phospholipid vesicle fusion by fluorescence energy transfer between membrane-bound dye labels, *Biochim. Biophys. Acta,* 596, 302, 1980.

162. **Owen, C. S.,** A membrane-bound fluorescent probe to detect phospholipid vesicle-cell fusion, *J. Membr. Biol.,* 54, 13, 1980.

163. **Gibson, G. A. and Loew, L. M.,** Phospholipid vesicle fusion monitored by fluorescence enrgy transfer, *Biochem. Biophys. Res. Commun.,* 88, 135, 1979.

164. **Gad, A. E. and Eytan, G. D.,** Chlorophylls as probes for membrane fusion, *Biochim. Biophys. Acta,* 727, 170, 1983.

165. **Morgan, C. G., Williamsen, H., Fuller, S., and Hudson, B.,** Mellitin induces fusion of unilamellar phospholipid vesicles, *Biochim. Biophys. Acta,* 732, 668, 1983.

166. **Morgan, C. G., Thomas, E. W., and Yianni, Y. P.,** The use of fluorescence energy transfer to distinguish between poly(ethyleneglycol)-induced aggregation and fusion of phospholipid vesicles, *Biochim. Biophys. Acta,* 728, 356, 1983.

167. **Parente, R. A. and Lentz, B. R.,** Fusion and phase separation monitored by lifetime changes of a fluorescent phospholipid probe, *Biochemistry,* 25, 1021, 1986.

168. **Hagins, W. A. and Yoshikami, S.,** Intracellular transmission of visual excitation in photoreceptors: electrical effects of chelating agents introduced into rods by vesicle fusion, in *Vertebrate Photoreception,* Barlow, H. B. and Fatt, P., Eds., Academic Press, London, 1977, 97.

169. **Allen, T. M. and Cleland, L. G.,** Serum-induced leakage of liposome contents, *Biochim. Biophys. Acta,* 597, 418, 1980.

170. **Bentz, J., Ellens, H., Lai, M.-Z., and Szoka, F. C., Jr.,** On the correlation between H$_{\Pi}$ phase and the contact-induced destabilization of phosphatidylethanolamine-containing membranes, *Proc. Natl. Acad. Sci. U.S.A.,* 82, 5742, 1985.

171. **Nayar, R. and Schroit, A. J.,** Generation of pH-sensitive liposomes: use of large unilamellar vesicles containing *N*-succinyldioleoylphosphatidylethanolamine, *Biochemistry,* 24, 5967, 1985.

172. **Silvius, J. R. and Gagné, J.,** Lipid phase behavior and calcium-induced fusion of phosphatidylethanolamine-phosphatidylserine vesicles. Calorimetric and fusion studies, *Biochemistry,* 23, 3232, 1984.

173. **Sundler, R. and Wijkander, J.,** Protein-mediated intermembrane contact specifically enhances Ca$^{2+}$-induced fusion of phosphatidate-containing membranes, *Biochim. Biophys. Acta,* 730, 391, 1983.

174. **Wilschut, J., Nir, S., Scholma, J., and Hoekstra, D.,** Kinetics of Ca$^{2+}$-induced fusion of cardiolipin-phosphatidylcholine vesicles: correlation between vesicle aggregation, bilayer destabilization and fusion, *Biochemistry,* 24, 4630, 1985.

175. **Bondeson, J., Wijkander, J., and Sundler, R.,** Proton-induced membrane fusion. Role of phospholipid composition and protein-mediated intermembrane contact, *Biochim. Biophys. Acta,* 777, 21, 1984.

176. **Pryor, C., Bridge, M., and Loew, L. M.**, Aggregation, lipid exchange, and metastable phases of dimyristoylphosphatidylethanolamine vesicles, *Biochemistry*, 24, 2203, 1985.
177. **Schenkman, S., Araujo, P. S., Dijkman, R., Quina, F. H., and Chaimovich, H.**, Effects of temperature and lipid composition on the serum albumin-induced aggregation and fusion of small unilamellar vesicles, *Biochim. Biophys. Acta*, 649, 633, 1981.
178. **Nir, S. and Bentz, J.**, On the forces between phospholipid bilayers, *J. Colloid Interface Sci.*, 65, 399, 1978.
179. **Graham, I., Gagné, J., and Silvius, J. R.**, Kinetics and thermodynamics of calcium-induced lateral phase separations in phosphatidic acid containing bilayers, *Biochemistry*, 24, 7123, 1985.
180. **Citovsky, V., Blumenthal, R., and Loyter, A.**, Fusion of Sendai virions with phosphatidylcholine-cholesterol liposomes reflects the viral activity required for fusion with biological membranes, *FEBS Lett.*, 193, 135, 1985.
181. **Uster, P. S. and Deamer, D. W.**, pH-Dependent fusion of liposomes using titratable polycations, *Biochemistry*, 24, 1, 1985.
182. **Rand, R. P. and Parsegian, V. A.**, Physical force considerations in model and biological membranes, *Can. J. Biochem. Cell Biol.*, 62, 752, 1984.
183. **Six, H. R., Young, W. W., Vemura, K., and Kinsky, S. C.**, Effect of antibody-complement on multiple vs. single compartment liposomes. Application of a fluorometric assay for following changes in liposomal permeability, *Biochemistry*, 13, 4050, 1974.
184. **Tagesson, C., Magnusson, K. E., and Stendahl, O.**, Physicochemical consequences of opsonization: perturbation of liposomal membranes by *Salmonella typhimurium* 395 MS opsonized with IgG antibodies, *J. Immunol.*, 119, 609, 1977.
185. **Ralston, E., Hjelmeland, L. M., Klausner, R. D., Weinstein, J. N., and Blumenthal, R.**, Carboxyfluorescein as a probe for liposome-cell interactions. Effects of impurities and purification of the dye, *Biochim. Biophys. Acta*, 649, 133, 1981.
186. **Straubinger, R. M., Düzgüneş, N., and Papahadjopoulos, D.**, pH-Sensitive liposomes mediate cytoplasmic delivery of encapsulated macromolecules, *FEBS Lett.*, 179, 148, 1985.
187. **Meers, P., Hong, K., Bentz, J., and Papahadjopoulos, D.**, Spermine as a modulator of membrane fusion: Interaction with acidic phospholipids, *Biochemistry*, 25, 3109, 1986.
188. **Barbet, J., Machy, P., Truneh, A., and Leserman, L. D.**, Weak acid-induced release of liposome-encapsulated carboxyfluorescein, *Biochim. Biophys. Acta*, 772, 347, 1984.
189. **Van Meer, G., Davoust, J., and Simons, K.**, Parameters affecting low-pH-mediated fusion of liposomes with the plasma membrane of cells infected with influenza virus, *Biochemistry*, 24, 3593, 1985.
190. **Morris, S. J., Bradley, D., Gibson, C. C., Smith, P. D., and Blumenthal, R.**, Use of membrane-associated fluorescence probes to monitor fusion of bilayer vesicles: application to rapid kinetics using pyrene excimer/monomer fluorescence, in *Spectroscopic Membrane Probes*, Loew, L. M., Ed., CRC Press, Boca Raton, Fla., in press.
191. **Düzgüneş, N. and Bentz, J.**, unpublished data.
192. **Düzgüneş, N., White, J., and Fedor, J.**, unpublished data.
193. **Bentz, J., Alford, D., Cohen, J., and Düzgüneş, N.**, $La^{3+}$ induced fusion of phosphatidylserine liposomes: close approach, intermembrane intermediates and electrostatic surface potential, *Biophys. J.*, submitted.
194. **Stutzin, A.**, A fluorescence assay for monitoring and analyzing fusion of biological membrane vesicles in vitro, *FEBS Lett.*, 197, 274, 1986.
195. **Silvius, J., Leventis, R., Brown P. M., and Zvokerman, M.**, Novel fluorescent phospholipids for assays of lipid mixing between membranes, *Biochemistry*, 26, 4279, 1987.
196. **Bentz, J. and Ellens, H.**, Membrane fusion: kinetics and mechanisms, *Colloids Surf.*, in press.
197. **Bentz, J., Ellens, H., and Szoka, F. C.**, Destabilization of phosphatidylethanolamine-containing liposomes: hexagonal phase and asymmetric membranes, *Biochemistry*, 26, 2105, 1987.

Chapter 7

USE OF MEMBRANE-ASSOCIATED FLUORESCENCE PROBES TO
MONITOR FUSION OF BILAYER VESICLES: APPLICATION TO RAPID
KINETICS USING PYRENE EXCIMER/MONOMER FLUORESCENCE*

**Stephen J. Morris, Diane Bradley, Carter C. Gibson, Paul D. Smith, and
Robert Blumenthal**

TABLE OF CONTENTS

*   All computer programs and documentation for the data acquisition system described herein are available from
    C. C. Gibson, BEIB, DRS, Bldg. 13 Room 3W13, NIH, Bethesda, Md 20892.

# I. INTRODUCTION

**A. Scope of this Review**

The fusion of membrane compartments within living cells is well-understood to underlie the basic mechanisms of transfer, processing, release, and reuptake of materials.[1-6,12] The fusion processes can be modeled using various preparations of membrane bilayers composed of phospholipids incorporating various other biological components (proteins, ganglio-sides).[5,6,13] Unilamellar phospholipid bilayer vesicles are one of the most popular preparations of membrane bilayers because of their relative ease of preparation and the possibility of incorporating materials into both the bilayer and the sealed core space. These particles can be manufactured in large quantities in very narrowly monodisperse sizes from 0.025 $\mu$m to several hundred microns in diameter.[14]

Since this review is written in a conceptual vein, we will not present a comprehensive summary of data in the literature on membrane fusion and on fluorescent probes to measure fusion. For this we refer the reader to a number of reviews, which have come out in the past years. For instance, the volume on membrane fusion edited by Poste and Nicholson[1] is still an excellent compendium of facts and hypotheses about membrane fusion. A collection of more recent reviews on exocytosis and secretion can be found in Hand and Oliver,[2] and viral fusion has recently been reviewed by White et al.[3] A series of reviews on cell fusion can be found in a recent Ciba Foundation Symposium volume,[4] and more recent reviews on membrane fusion include those by Düzgüneş,[5] and by Blumenthal.[6] What we hope to achieve here is to provide the reader with methodology to monitor his or her particular fusion system using fluorescent probes and to delineate what is fact or artifact in those measurements. We believe that fluorescence and rapid kinetics will be of great use in attempting to understand the mechanism of membrane fusion.

Many studies have been carried out on the interaction of lipid vesicles with cells.[19,20] The aim of those studies was to use lipid vesicles as vehicles to introduce biologically active substances into cells. In spite of indications that material was transferred from vesicles to cells, there was no clear demonstration of fusion nor quantitation of the interaction. Most of the quantitation in these studies was done by labeling the liposome with aqueous or lipid markers and measuring the amount of cell-associated material. However, molecules from liposomes can become cell-associated by other mechanisms than fusion, such as stable adsorption, endocytosis, and lipid exchange.[19] An assay for fusion of lipid vesicles with cells was developed in which transfer of contents can be visualized.[21,22] This assay is based on self-quenching properties of the water-soluble dye carboxyfluorescein and monitors transfer of contents of liposomes to cells. Based on measurements of the amounts of material transferred to cells, the quantity of putative fusion was much less than originally estimated, and the process was saturable.[22] However, even using that assay the amount of fusion was overestimated. There was considerable cell-induced leakage of liposome contents.[23] Moreover, fluorescence photobleaching showed that fluorescent lipid probes incorporated into lipid vesicles remained immobilized on the cell surface.[23,24] This, taken together with studies on energy transfer between lipid probes incorporated into liposomes,[28] indicated that only a very small portion of the liposome-cell interaction was membrane fusion. Transfer of contents seems to take place by some other, still undefined, mechanism.[20,24] These considerations necessitate a cautious approach to the analysis of these (liposome) systems before identifying the process as fusion. This situation where core mixing could be demonstrated without lipid mixing was dubbed "appositional transfer".[25]

## B. Definitions of Membrane-Fusion Assays

The process of merging two bilayers from two separate membranes into a single continuous bilayer involves a number of steps as shown in Figure 1 a to d. Intermediate steps, such as those shown in Figures 1b and 1c have been designated as membrane fusion in the literature.[12] We define membrane fusion as the process by which the two bilayers, however closely apposed, become one bilayer (Figure 1d). At some point in the discussion we will need to refer to partial fusion of bilayers as the process by which the single leaflet of two bilayers becomes confluent leaving the other leaflets unmixed (Figure 1c.)[7-11]

We define membrane-mixing assays as those in which the probes are associated with the membrane bilayer in an irreversible manner (Figure 2, positions a, b, and c). Core-mixing assays are defined as assays in which the signal for fusion is the interaction of soluble materials which have originally been sealed in the lumens of separate populations of vesicles (Figure 2, position d).

Assay of the fusion of vesicles can rely on either mixing of soluble materials trapped in the aqueous core spaces of the vesicles or mixing of the membrane components. This review will deal primarily with membrane-mixing assays which utilize fluorescent-labeled phospholipids as probes. These have become the most variegated, as well as the easiest to manipulate, method for assaying membrane fusion. The core-mixing assays will be the subject of a separate chapter of this monograph.[15] We also will concentrate the discussion upon fusion of vesicular structures of originally equal size to each other. This is not necessarily the usual situation in biology where a small vesicle fuses to a large, almost planar membrane or buds off from it. However, there are several of the assays which employ only one probe or only one population of labeled vesicles fusing with an unlabeled population of vesicles which could be adapted for use with planar membranes,[16] or very large vesicles,[17,18] or cells.[19-25]

## C. Application of Fusion Assays to Stopped-Flow Mixing

It is not the purpose of this review to extensively present the case for time-dependent resolution of fluorescent signal changes. However, if the possibility of extracting such data is available, many interesting experimental questions can be answered concerning the factors (ions, lipid headgroups, proteins) which determine apposition, destabilization, and the fusion event itself. Fusion kinetics has been analyzed according to the mass action scheme:[13,15,64]

$$V_1 + V_1 \underset{D_{11}}{\overset{C_{11}}{\rightleftarrows}} V_2 \overset{f_{11}}{\to} F_2 \tag{1}$$

a.

b.

Aggregation
(No lipid mixing. No core mixing.)

c.

Possible Fusion Intermediate
(Outer bilayer leaflet free to mix.
Inner leaflets unmixed. No core
mixing.)

d.

Fusion
(Both leaflets free to mix. Core
spaces contiguous.)

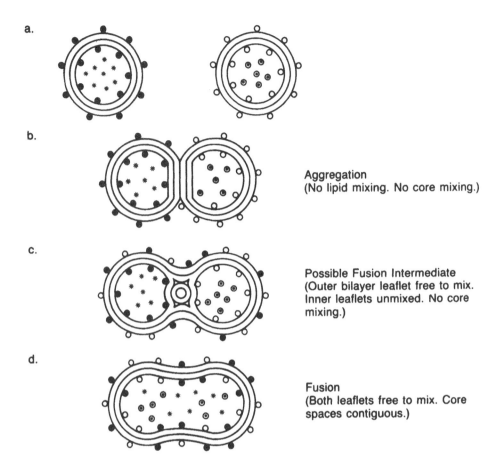

FIGURE 1.   Definitions of membrane fusion. Separate bilayer vesicles (a), aggregate (b), go through one or more intermediate stages which may allow mixing of some, but not all, phospholipids (c), and finally fuse (d). Only at the last step is there contact between the core spaces of the original particles as well as unrestricted lateral diffusion of all phospholipids. The possible intermediate drawn in (c) is the $H_{II}$ phase.[7-11] It would allow mixing of the outer monolayer phospholipids while keeping the inner monolayer lipids and core spaces separate.

where $V_1$ denotes the original vesicle, $V_2$ the dimer aggregate, and $F_2$ the fused dimer, when the bilayers have merged and the aqueous contents mixed. Extraction of the rate constants for aggregation ($C_{11}$), dissociation ($D_{11}$), and fusion ($f_{11}$), requires at least three independent experiments, one for each parameter.[15] At very low lipid concentration, aggregation rate constants can be obtained since the overall fusion kinetics are aggregation-rate limited. Fusion becomes rate-limited at high lipid concentration. Hand-mixing techniques can resolve events on a time scale of seconds. Thus, there are cases where fusion proceeds slow enough to be resolved by hand-mixing techniques even at high lipid concentrations.[13,15,64] For instance, we have found that polylysine/low pH-induced fusion of PS:PE vesicles shows second-order (collision-dependent) aggregation, while fusion proceeds at much slower, non-second order rates.[83] However, in some cases the fusion reactions can proceed so rapidly that hand-mixing of the reactants will not resolve the initial aggregation and fusion processes.[26]

Assays for fusion would be most useful if they were amenable to stopped-flow techniques. It should be noted that the change in probe property to be measured must run 10 to 100-fold faster than the reaction to be followed. Thus, mixing due to lateral diffusion of lipid probes in the bilayer or mixing due to diffusion of aqueous markers between vesicle lumens

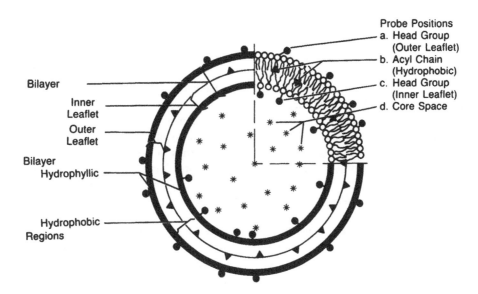

FIGURE 2. Possibilities for placement of probes in bilayers. Probe groups may be attached to the polar headgroups of the phospholipids (a,c) or to the acyl chains (b). Soluble probes may be placed in the aqueous core space (d).

**Table 1**
**POSSIBLE EXTRANEOUS INFORMATION REPORTED BY MEMBRANE-BOUND AND SOLUBLE (CORE-SPACE) PROBES USED FOR FUSION ASSAYS**

| Position | Possible extraneous information reported |
| --- | --- |
| a. Outer (hydrophilic, headgroup) surface of the bilayer | Vesicle-vesicle adhesion; partial fusion; exchange of phospholipids between bilayers; binding of solutes; surface charge or potential changes; lateral phase separation and other phase changes; Flip-flops (*trans*-bilayer exchange) |
| b. Middle (hydrophobic acyl-chain) region of the bilayer | Lateral phase separation and other phase changes |
| c. Inner (hydrophilic, headgroup) surface of the bilayer | Flip-flops; lateral phase separation and other pahse changes; surface charge or potential changes |
| d. Core (solution) space | Leakiness of the membrane; interactions with inner-membrane constituents |

*Note:* See Figure 2 for details of the probe positions.

should not be rate-limiting. Probe reaction times are usually not rate-limiting, unless enzyme reactions or slow dissociation of complexes are involved.

We will stress experimental methodologies, especially those for making rapid kinetic measurements on these systems. These techniques can be applied to either core- or membrane-mixing assays, although the discussion will center on the latter. At the time this article was written core-mixing assays had yet to be adapted to rapid-mixing, although there is no intrinsic reason why this cannot be done. In some cases to be discussed below, core-mixing assays are unsuitable for probing fusion of biological storage vesicles, which lose their ability to fuse when disrupted during the insertion of materials into their lumens.[27] Conversely, membrane-mixing assays may report artifacts such as lipid transfer or exchange (see Table 1).

We have previously adapted to stopped-flow mixing[26] a membrane-mixing assay which depends upon the changes in resonance energy transfer (RET) seen upon fusion of vesicles incorporating headgroup-labeled 7-nitro-1,2,3-benzoxydiazole-4-amino (NBD)- and rhoda-mine- phosphatidylethanolamine (NBD-PE and Rh-PE) fluorophores.[28,31] This assay system has several problems which limit its use. For example, $Ca^{2+}$ and other divalent cations are often used to initiate membrane fusion. However, both of these probes, and especially NBD-PE are prone to reporting artifactual processes due to the interactions of divalent cations[26,27,29,30] and polyamines[83] which directly affect probe quantum yield.

Our laboratory has recently succeeded in perfecting a membrane-mixing assay based on the ratio of excimer to monomer fluorescence (E/M ratio) of chain-labeled pyrene-phos-phatidylcholine (PC). Use of such a ratio of the properties of a single fluorophore solves many of the problems of other membrane-mixing assays involving resonance energy transfer (RET) or other spectral properties. Discussions of these problems and a detailed discussion of the new assay will form the conclusion of this review.

### D. Possible Artifacts Encountered in Fluorescence Measurements

Most fusion assays rely upon a change in the strength of the fluorescent signal by quenching or resonance energy transfer. However, such signal changes, reflecting changes in quantum yield, can come from a variety of processes which have nothing to do with the fusion process, but can mimic the expected change in fluorescence. Other processes which may change fluorescence levels are outlined in Table 1 and discussed below.

Membrane-mixing assays depend upon the fluid mosaic nature of the bilayer to allow mixing of the probe molecules. However, it is well-understood that other properties of fluid mosaic bilayers can change the properties of fluorescent probes associated with them. Examples are changes in headgroup tilt due to binding of a divalent cation, changes in headgroup local environment due to changes in surface potential or dielectric constant, lateral phase separation of the surrounding lipids, and changes in the average motion of the probe secondary to changes in the rigidity of the acyl chains.

Most possibilities can be checked with suitable controls. The experimenter should always be aware of the alternate explanations and control for them when necessary. A nonexhaustive list of possible artifacts follows.

#### 1. Changes in the Microenvironment Surrounding the Probe

An example is the pH-dependent changes in fluorescein quantum yield, which can be used as a pH indicator for its local environment.[32] Changes in quantum efficiency due to solvent effects are extensively discussed in Reference 33. Probes such as dansyl, diphenyl-hexatriene (DPH), and pyrene are highly quenched when attached to the headgroups of phospholipids, but have good quantum yields when coupled to the ends of the acyl chains. Changes in quantum efficiency due to possible changes in local potential or surface charge are discussed in relation to NBD/calcium in Section IV.A.

These types of artifacts are easily revealed by control experiments in which only probe-containing vesicles are subjected to the catalytic event, such as addition of calcium (see Section IV.A, Figure 9). In assays involving the use of two different fluorescent probes such as RET assays, each probe must be checked separately.

#### 2. Exchange of Probe from One Membrane to Another

This may take place in the absence of any fusion.[34-36] Since most assays presently in use involve dilution of one or both probes, this can be a very insidious problem. A typical control would be to mix probe-containing vesicles with unlabeled vesicles in the absence of (or at subthreshold concentrations of) the fusion catalyst. Unfortunately, this type of control will not reveal exchange which is based upon transfer after aggregation,[37] a situation which may become clear by the separation of the kinetics of the two processes, or by the use of asymmetrically labeled vesicles with probe molecules on the inner leaflet only (see Section III.C.3).

*3. Reduction in Fluorescence Due to Photobleaching*

This is a straightforward problem which can be checked in a number of ways: for example, by placing probe-containing vesicles in a fluorimeter and noting any time-dependent change in fluorescence levels, first with the excitation shutter closed, and, then after an equal time with the shutter open. Photobleaching is often a near linear (0-order) process. If the bleaching is slow compared to the process being studied it can be ignored. Alternately, one can simply substract the appropriate ramp from the data.

*4. Reduction in Fluorescence Due to Innerfilter Effects*[38]

This will be especially troublesome in systems containing larger (>1000 Å) vesicles or as smaller vesicles aggregate. If innerfilter effects due to turbidity become too bothersome it is possible to use "front face" measurements.[38] Another useful trick is to use "crossed" polaroid filters.[26,28] Place one filter in front of the cuvette and the second in the opposite orientation in front of the fluorescent photomultiplier. Since scattered light is highly polarized in the plane of excitation and fluorescence from nonhindered fluorophores is not, this will remove most of the scattered light while reducing the fluorescence signal by about half. Grating monochrometers highly polarize light; therefore, selecting the excitation component which is parallel to the monochrometer grating will reduce excitation intensity losses.

The obvious answer to these problems is to choose a system which is not subject to untoward changes under the experimental conditions. While this is not always possible, an assay recently developed in our laboratory specifically designed to avoid most of these artifacts will be discussed in Section IV, which should help our readers to understand many of these problems. It should be noted that many of these artifacts are not important if quantative information (fusion kinetics and/or quantitation of extent of fusion) is not needed.

## II. CORE-MIXING ASSAYS

General strategies for core-mixing assays are presented in Figures 3 and 4. These assays are discussed extensively in Reference 15. Each type of assay offers certain advantages and is subject to certain flaws. The experimenter should be aware of both in choosing an assay for any particular application.

### A. Advantages

1.  When properly controlled, a core-mixing assay reports only when positive connections have been established between the lumens of the vesicles which are large enough to allow the passage of the encapsulated probes. Depending upon the hydrated size of the probe chosen, these passages can be a few angstroms to about 0.2 nm in diameter. Other intermediate interactions between particles (exchange of lipids, contact, adhesion, mixing of membrane lipids, etc.) are excluded. If the experimenter is clever, he can set up the core-mixing assay with a suitable membrane-mixing assay to simultaneously report both fusion and other activities, e.g., if the fusion seen by membrane-mixing is a leaky or a tight process.[15]
2.  Core-mixing assays remove the need of perturbing the structural properties of the bilayer with bulky probes.

### B. Disadvantages

1.  The experimenter must control for leakage of probe to prevent artifactual results. This often requires antiprobe antibodies[26] or, in some cases, chelators,[15] which may upset the catalyst concentrations and change the surface properties by removing bound-

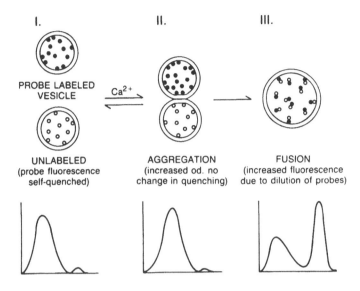

**FUSION OF CORE SPACES DETECTED BY RESONANCE ENERGY TRANSFER**

FIGURE 3. Fusion of core spaces detected by resonance energy transfer. Either the donor or the acceptor is incorporated into the core space of two sets of vesicles. Aggregation is detected as before. Fusion allows the donor and acceptors to mix and RET occurs. Since RET is proportional to the inverse sixth power of the distance between donor and acceptor, this experimental arrangement is only practical at millimolar concentrations of probes. This problem is overcome if the donor and acceptor form a complex, such as $Tb^{2+}/DPA$. For other core-mixing assays see Reference 15.

charged components. On the other hand, Düzgüneş and Bentz[15] point out that core-mixing assays report both fusion *and* leakage. The latter process might provide valuable information about membrane destabilization.

2.  The soluble components must be placed within the vesicles. This creates problems for preformed particles such as biological vesicles, where perturbations such as lysis and resealing may destroy biological functions or create artifacts. For example, lysis and resealing of chromaffin granules destroy their ability to fuse in the presence of $Ca^{2+}$ at pH 7.4.[27]

3.  Depending on the assay, one must have the ingredients in separate vesicle populations, which requires preparing two sets of labeled vesicles, each of which must retain other activities (enzyme activity, storage of material, and fusibility). Many membrane-mixing assays require labeling only one of the two vesicle populations, even when two probes are involved.

## C. Applications to Biological Membranes

The $Tb^{3+}$/dipicolinic acid (DPA) (Figure 3) assay depends upon the chelation of the trivalent cation by citrate.[39] Assuming that one can reseal the probe mixture within biological membrane structures, problems might still be encountered. For example, many membranes have fairly high-affinity binding sites for divalent cations. In the case of the chromaffin granule, these are of high enough activity that they will effectively compete with the citrate for binding the $Tb^{3+}$,[40] thus greatly reducing or abolishing any potential signal, or perhaps distorting the apparent fusion kinetics by the off-time for the $Tb^{3+}$/membrane-site complex. Most of the core-mixing assays involving biological membranes have depended upon relief

**FUSION OF CORE SPACES DETECTED BY
RELIEF OF FLUORESCENCE SELF-QUENCHING**

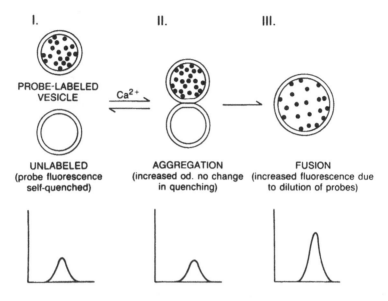

FIGURE 4.   Fusion of core spaces detected by relief of fluorescence self-quenching. Probe is incorporated into the core space of one vesicle population at self-quenching concentrations. These vesicles are mixed with unlabeled vesicles. Aggregation of vesicles can be detected by changes in light scattering or optical density. Fusion of the vesicles will dilute the probe in a greater volume, relieving the self-quenching and increasing the fluorescent emission.

of self-quenching of a fluorophore (see Section I.B. and Figure 4). An aqueous probe, such as carboxyfluorescein[21,22] or calcein[43] is incorporated into the core of the vesicles. The latter would be preferable, since it lacks the tremendous pH-dependent changes in quantum yield seen with fluorescein.[32] The carboxyfluorescein self-quenching technique has been used extensively to monitor leakage of contents during fusion.[15,26] Presuming no leakage takes place, relief of self-quenching will occur when a probe-containing vesicle fuses with a nonprobe-containing vesicle. As mentioned in Section I.B., care must be taken with the interpretation of carboxyfluorescein self-quenching data as a probe for fusion, since mechanisms such as "appositional transfer" can report false positives.[23,24] Hydrophobic impurities in the commercial preparation of carboxyfluorescein have produced misleading results; they can be removed by passage through a Sephadex® LH20 column.[41] Stutzin[44] demonstrated pH-dependent fusion of lysed, resealed chromaffin granules containing fluorescein isothyocyanate, deactivated (FITC)-dextran. This larger probe is less likely to transfer spontaneously across the bilayers without fusion.

## III. MEMBRANE-MIXING ASSAYS

### A. Overview of Membrane-Mixing Assays

As noted above, this review will focus on membrane-mixing assays. An extensive list of the available assays can be found in Reference 15. We will concentrate on the use of phospholipid-based probes as being the easiest to interpret, since they are least likely to exchange spontaneously between membranes. However, all the principles noted here could be applied to the use of derivatized fatty acids, glycolipids, and proteins as probes. These could be tremendously useful for specific problems or applications. The ability of certain mono- and divalent cations to induce aggregation or fusion of given lipid mixtures has been extensively reviewed.[13] These reviews should be consulted for general principles for choosing lipid mixtures, particularly with respect to headgroup specificity. For example, $Mg^{2+}$ will not fuse pure, large, PS vesicles.[42]

### 1. Resonance Energy Transfer (RET)

This has proven to be the most popular approach to membrane mixing due to the availability of high-extinction coefficient probes.[28,31,46] The assay can be set up either with both probes in the same membrane before fusion or in separate sets of vesicles, and one can either follow donor-quenching or RET. However, the latter configuration can give rise to artifactual changes in signal due to vesicle aggregation.[15]

### 2. Dilution of a Self-Quenched Probe

The principle of this assay is the same as that described for core mixing in Section II except that the diffusion/dilution would be limited to the two-dimensional plane of the bilayer. A probe-containing phospholipid is incorporated into the bilayer of one set of vesicles at a high enough concentration to be self-quenching. Aggregation produces no change in fluorescence levels, but fusion with a nonprobe-containing vesicle would dilute probe-containing phospholipids, resulting in an increased fluorescence signal. Like the double-probe configuration of the RET assay (Figure 5a), this assay requires only one set of membranes to be labeled. Citovsky and Loyter[49] have used this approach with reconstituted Sendai virus envelopes. Fusion with liposomes and cells was monitored by dequenching of NBD-PE fluorescence which was incorporated in the membranes during reconstitution at about 10 mol % lipid. An example of this type of assay is presented in Figure 9.

An assay based upon incorporation of octadecylrhodamine (R18) at 9 mol % has been described by Hoekstra et al.[47] The probe is quenched in the labeled vesicle, and upon fusion with biological membranes or liposomes it becomes dequenched. Agents such as phenylmethylsulfonyl fluoride and dithiothreitol, which block fusion of Sendai virus, but do not affect binding to cells, inhibit R18 dequenching,[54] indicating that the probe does not exchange spontaneously. Since in general membrane fluidity increases with increasing fatty acid concentration,[59] it is possible that the bulk properties of the viral bilayer have changed with such a large percentage of fatty acid. However, the studies indicate that labeling with R18 does not affect viral fusion markedly.[54]

### 3. Quenching of a Fluorescent Membrane Probe by a Phospholipid-Based Quencher

The quenching molecule (for example, brominated phospholipid[60] or spin-label[61] could either be incorporated with the fluorescent probe and dequenching followed, or the probes could be placed in separate vesicle populations. Lapme and Nelsetuen[61] used this approach to study the stopped-flow kinetics of myelin basic protein-induced vesicle fusion.

### 4. Pyrene Excimer/Monomer Ratio

Pyrene not only has a distinctive emission fluorescence spectrum but can, under certain conditions, produce a second broad emission peak which is red-shifted about 50 nm from the monomer spectrum. This excimer fluorescence is dependent upon the distance between fluorophores. Utilization of this property for a membrane-fusion assay is diagramed in Figure 7. A detailed explanation of this assay will be found in Section IV of this review. Using the ratio of two fluorescent properties of the same probe greatly reduces many of the problems discussed above. It should be noted that this same approach could be used in a core-mixing assay. However, the low quantum yields of pyrene in a hydrophilic environment would require very high concentrations of probe.

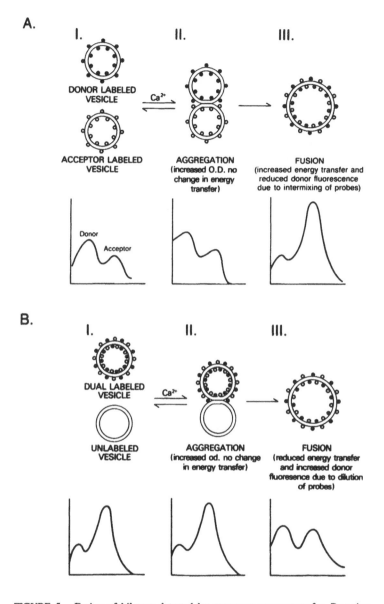

FIGURE 5. Fusion of bilayers detected by resonance energy transfer. By using fluorescence-labeled phospholipids which are free to diffuse only in the plane of the bilayer, it becomes possible to achieve the necessary local concentrations of donors or acceptors without heavily loading the membrane with probe-bearing lipids. This is especially true of probes with large $R_o$s such as NBD/Rh (see Reference 15 for a list of energy transfer pairs). A. Placement of donor and acceptor in separate populations. This approach is similar to that outlined in Figure 4. Aggregation is reported by the change in apparent optical density (turbidity) with little if any change in RET. Energy transfer and donor-quenching are seen when the bilayers fuse and the probes are free to diffuse in the same plane. Either signal can be followed. B. Placing both donor and acceptor in the same membrane gives donor-quenching and energy transfer at the beginning of the experiment. When these vesicles fuse with unlabeled vesicles, average distances between probes increase and RET is reduced while donor fluorescence increases. This arrangement requires that only one set of vesicles needs to be labeled. It is possible to manipulate the concentrations of probes in the vesicles of either experimental situation to weight the signal changes in favor of early events or late events in the overall fusion process. For example, if one wishes to track early events in situation (A), one uses relatively high concentrations of probe.

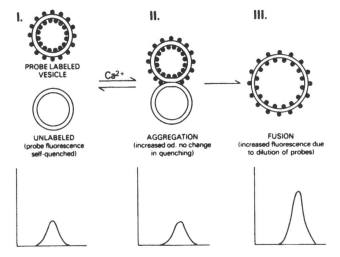

FIGURE 6.   Fusion of bilayers detected by relief of fluorescence self-quenching. Probe is incorporated into the bilayer of one vesicle population at self-quenching concentrations. These vesicles are mixed with unlabeled vesicles. Aggregation of vesicles can be detected by changes in light scattering or optical density. Fusion of the vesicles will give the probe a greater area in which to diffuse laterally, relieving the self-quenching and increasing the fluorescent emission.

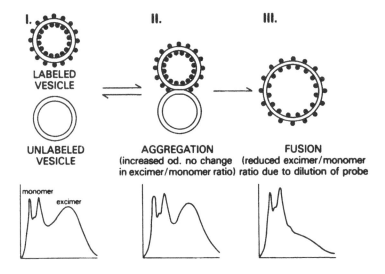

FIGURE 7.   Fusion of bilayers detected by changes in the pyrene monomer/excimer ratio. Chain-labeled pyrene-PE is incorporated into one set of vesicles at a concentration high enough to see appreciable excimer formation (see Figures 11 and 13). These vesicles are mixed with unlabeled vesicles. Aggregation of vesicles can be detected by changes in light scattering or optical density. Fusion of the vesicles will give the probe a greater area in which to diffuse laterally, reducing the number of excimers formed, and, thus reducing the E/M ratio. Changes in monomer and excimer fluorescence emission are recorded (simultaneously if possible) and the extent of fusion judged by the reduction of the ratio (see Figure 13).

*5. Changes in Fluorescence Lifetimes*

Parente and Lentz[30] have reported changes in fluorescence lifetimes of DPH-PC during fusion. Due to the time required to make the lifetime measurement, the assay is not amenable to stopped-flow mixing, but rapid kinetics could conceivably be analyzed by quench-flow methodologies.

**B. General Rules for Choosing Probe Systems**

Below we have enumerated general guidelines to follow for choosing a probe or probe pair. Obviously no probe can meet all the requirements listed, and some trade-off is necessary. Also, there may be cogent reasons for choosing a certain probe precisely because it breaks one of the rules and thereby give the investigator specific information, for instance about lipid phase changes:

1.  The probe(s) should be stable (not subject to photobleaching, chemical reactions, free radical formation).
2.  A probe should not interact extensively with components of the bilayer. It should report its bulk concentration properties in the bilayer.
3.  The probe(s) should not interact with the solvent or solute.
4.  Probes should have high-quantum efficiencies and extinction coefficients (for highest sensitivity).
5.  For RET assays the probes should have sufficient spectral overlap to produce good transfer without having so much overlap that results are ambiguous. A good example of this problem can be found in the NBD to Rh RET assay, where NBD emission sufficiently overlaps the rhodamine emission, so that increases in NBD emission due to presence of divalent cations (Figure 8) can look like transfer of energy into the rhodamine spectrum. This particular problem may be controlled by following donor-quenching after having first ascertained that no untoward artifactual changes will bias these results. It should be noted in passing that the acceptor need not itself be a fluorophore to quench the donor.[62]
6.  Natural fluorescence: as noted above for core-mixing assays, it is often possible to exploit the natural fluorescence of proteins, or other membrane constituents as probes. For example, quenching of tryptophan fluorescence in a membrane protein contained in one vesicle population by an appropriate probe (DANSYL-PE) in a second population could signal fusion of the bilayers.

**C. Incorporation of Phospholipid Probes into Membranes**

Methods for placing lipid probes are extensively covered in the other chapters. The methods may be divided into those in which the probes are incorporated as the vesicles are formed and those systems in which the probes are incorporated into already existing vesicles. We will mainly discuss incorporation of phospholipid probes. To a great extent, the type of vesicle to be made will determine the labeling procedure to be used. For any vesicles which will be formed from lipids in organic solvents or formed from lipids dried down from organic solvents, the fluorescent phospholipid probe(s) can be introduced into the organic solvent. For labeling vesicles asymmetrically on the outer leaflet, labeling biological membranes, and inserting probes attached to proteins or other molecules which could not stand up to the organic solvent, other methods are used.

*1. Incorporation of Probes During the Vesicle Preparation Procedure*

Methods for making various types of phospholipid vesicles are extensively covered in three volumes of *Liposome Technology*.[14] Vesicle sizes can vary from the small unilamellar vesicles formed by sonication of phospholipid dispersions to very large vesicles formed by

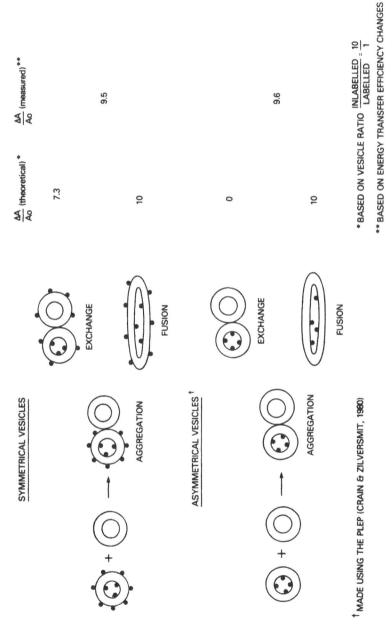

EVIDENCE FOR VESICLE FUSION

† MADE USING THE PLEP (CRAIN & ZILVERSMIT, 1980)

* BASED ON VESICLE RATIO $\frac{INLABELLED}{LABELLED} = \frac{10}{1}$

** BASED ON ENERGY TRANSFER EFFICIENCY CHANGES

FIGURE 8.   Predicted and observed RET changes for polylysine-induced fusion of symmetrical and asymmetrical PS:PC SUV. The SUVs were composed of PS:PC:NBD-PE:Rh-PE (49:49:1:1). The asymmetrical SUVs were prepared using the PLEP in as described in the text. To induce fusion polylysine (20,000 daltons) was incubated with PS:PC SUV (50 $\mu$g/m$\ell$) at a 1:1 lysine:serine ratio for 15 min at room temperature. (see Reference 56 for more details). The ratio of fluorescently labeled SUV to unlabeled SUV was 10:1. The measured $\Delta A/A_0$ was derived from energy transfer changes between NBD-PE and Rh-PE according to Equation 2. The predicted $\Delta A/A_0$ for exchange involves outer monolayer mixing only, whereas the predicted value for fusion involves both monolayers. Probe headgroups are marked by closed dots on the inner and outer vesicle membrane surfaces.

extrusion. For studying fusion of biological membranes, reconstitution has been a useful appraoch.[48,49] The protein, lipid, and probes are dissolved in detergent, and proteoliposomes are formed by removing detergent using a variety of methods, including: dilution, dialysis, and gel chromatography.[50] Most preparation methods produce heterogeneous dispersions of vesicles. However, homogeneous populations can be obtained by gel chromatography[51] or high-performance liquid chromatography.[52]

### 2. Incorporation by Passive Exchange

It is sometimes possible to label preformed vesicles, viruses, or biological membranes by collision with probe dissolved in an organic solvent such as ethanol. This approach has been used extensively to incorporate lipid probes into plasma membranes of intact cells for lateral mobility studies.[53] As noted above, Hoekstra and co-workers[47] showed that the fatty acid octadecylrhodamine can be irreversibly incorporated into viruses and chromaffin granules. The unincorporated material is separated by column chromatography. The probe is quenched in the labeled vesicle, and upon fusion with biological membranes or liposomes it becomes dequenched[54] (Figure 6). Acyl chain-labeled probes can be incorporated into cell membranes by liposome exchange.[55] However, those probes have limited use for fusion studies.

### 3. Preparation of Asymmetrically Labeled Vesicles

Recently, using phospholipid exchange proteins, an assay for fusion was developed for vesicles with fluorescent probes present only in the inner monolayer of the vesicle membrane,[56] (see Rh-PE, Figure 8). SUVs containing the two fluorescent probes, NBD-PE and on the inner monolayer only were prepared by using the nonspecific phospholipid exchange protein (PLEP) isolated as described by Crain and Zilversmit.[57] Nichols and Pagano[36] studied protein-mediated lipid transfer between vesicles using this PLEP. Fluorescently labeled SUVs were mixed with a tenfold excess of large unlabeled vesicles composed of phosphatidylserine (PS):PC (1:1). The REVs served as an acceptor for the fluorescently labeled lipids on the outer monolayer of the SUVs and a source of replacement lipid for PS and PC. The mixtures were incubated for 24 hr on a rotating shaker at 37°C and samples taken periodically to follow the progress of the exchange. The two populations of vesicles were separated by centrifugation in an airfuge (120,000 × $g_{av}$) for 5 min causing the REVs to pellet and leaving the SUVs in the supernatant. The energy transfer efficiency (ETE) was measured in the SUV portion, and the concentrations of the two probes were determined by direct excitation in detergent-dispersed aliquots. The ETE of vesicles incubated with PLEP remained identical to the initial values but the amount of NBD-PE and Rh-PE was reduced to about 50% of the initial values as would be expected if the outer monolayer lipids were replaced and the inner monolayer remained unperturbed. The change in surface area upon fusion of symmetrical or asymmetrical vesicles can be calcualed from the equation derived by Fung and Stryer[31] where ETE is related to probe surface density:

$$(\Delta\,A/A_0)\ =\ [\ln(F_i/F_{max})/\ln(F_x/F_{max})] - 1 \qquad (2)$$

where $F_i$ and $F_x$ are the initial and observed fluorescence at 530 nm respectively. $F_{max}$ is the NBD-PE fluorescence when the probes are at infinite dilution, $\Delta A$ = change in area/probe, $A_0$ = initial area/probe. The outer:inner surface area of SUV used in these calculations was 2:1.[57] Figure 8 shows predicted and observed results for polylysine-induced lipid mixing of PS:PC (1:1) SUV with symmetrical and asymmetrical probe distribution when labeled and unlabeled vesicles were mixed in a ratio of 1:10.[57] Since most other lipid-mixing processes (monomer exchange, inverted micellar intermediates) will disproportionately involve the outer monolayer or result in random mixing of both monolayers (detergent effects), this assay makes it possible to clearly establish that ordered mixing of both monolayers occurred.

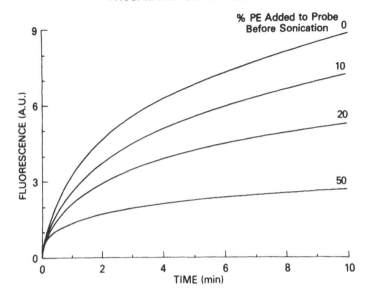

**INHIBITION OF PROBE INCORPORATION
INTO CHROMAFFIN GRANULES BY
PHOSPHATIDYL ETHANOLAMINE**

FIGURE 9. Effect of increasing concentrations of phosphatidylethanolamine on the ability of small unilamellar vesicles containing NBD-PE to fuse to chromaffin granules. Small unilamellar vesicles were sonicated from mixtures of NBD-PE, containing increasing amounts of PE. These vesicles were mixed with isolated chromaffin granules and the incorporation of the probe into the granules was monitored by relief of self-quenching of the probe. The data were normalized to total amount of available probe. Increasing the mol percent PE decreased the ability of the vesicles to fuse to the granule membrane. See text and Reference 27 for details.

This method will prove to be extremely useful for incorporating nonexchangeable fluorescent lipid probes into biological membranes (secretory granules, viruses).

### 4. Fusion of Target Membranes with Probe-Containing Vesicles

As an alternative to phospholipid exchange proteins, it may be possible to fuse probe-containing vesicles to the target vesicles. This approach has the advantage of labeling both sides of the bilayer. However, unless very low concentrations of probe lipids will suffice, this requires that the probe vesicles contain high concentrations of the probe-labeled lipids. Such vesicles have been successfully employed to label the membranes of chromaffin granules.[27,63] One study demonstrated that sonicating dispersions of 100 mol % NBD-PE or Rh-PE resulted in formation of vesicles as indicated by entrapment of [14]C-sucrose.[27] Vesicle-encapsulated colloidal gold appeared in the lumen of chromaffin granules after fusion as shown by thin-section electronmicroscopy.[63] As shown diagrammatically in Figure 6, when the probe is in high concentration it will be subject to self-quenching. Therefore, when the probe-containing vesicles fuse with nonprobe-containing membranes, the probe molecules will separate by lateral diffusion, self-quenching will decrease, and the quantum yield will increase. This can be shown by following the time-dependent increase in fluorescence after mixing of the two populations (Figure 9), and allows for immediate checking of incorporation. As can be seen, lipid mixture becomes an important variable; as the probe-PE was diluted with unlabeled PE, incorporation rate and final amplitude decreased.

### D. Advantages and Disadvantages of Membrane-Mixing Assays

One can choose essentially stable nonexchangeable probes. This greatly reduces the need for extensive controls to examine artifactual interactions. Probes can be chosen which are essentially inert and require no special precautions in their use. Thus one can avoid the need for chelators antibodies or other special materials in the medium which can bias the results.

Membrane-mixing assays can report events involved in prefusion activities (e.g., priming of the membranes by catalysts, initial contact, and partial fusion). It has been noted above that this type of information may promote misinterperetations (see Section I.D) especially if the kinetics of prefusion events, such as aggregation or ''partial'' fusion are different than fusion as defined by complete mixing of bilayers.

As has been noted in Section III.C, one can choose the position of the probe with great accuracy. By knowing the relationship between probe distances and the strength of the spectral property to be measured one can accurately place the site of the interaction within a few angstroms. Thus, as will be described, one can judiciously choose placement of more than one probe and by altering the measurement technique (wavelength, etc.), perhaps detect several different activities in the same experimental situation (perhaps at the same time).

Certain assays require placement of only one probe in one vesicle population. This is of great advantage is assaying effects of a membrane-bound component, such as a protein, on the fusion process or upon interaction of two different types of membranes with each other.

Disadvantages of membrane-mixing assays are the generation of a multitude of artifactual positive results due to, for example, prefusion interactions, partial fusion, exchange of lipids, and aggregation (see Table 1 and Reference 15).

## IV. USE OF PYRENE MONOMER/EXCIMER FLUORESCENCE RATIOS TO MONITOR MEMBRANE FUSION

### A. Introduction

In a previous study using resonance energy transfer from NBD-PE to Rh-PE to assess the kinetics of $Ca^{2+}$-promoted fusion of bilayers, we encountered an artifactually fast process which distorted the initial change in RET due to vesicle-vesicle fusion. This artifact is almost completely due to a direct action of divalent cations on the quantum yield of the NBD probe.[26,27,29,30] This can be seen clearly in Figure 10. Mixing a sample of NBD-PE labeled PS:PE membranes with various $Ca^{2+}$-concentrations produced a rapid, pseudofirst-order process which ran many times faster than aggregation of the vesicles, as measured by turbidity changes. The amplitude of this change was about 130% of the initial NBD fluorescence. The half-times for the initial rise were at least tenfold faster than aggregation and approached 100-fold at 5 m$M$ $Ca^{2+}$. A good description of the data for $Ca^{2+}$-concentrations between 1 and 3 m$M$ was an exponential.[26] Above this concentration, the initial rise took place within the mixing time and the amplitudes were calculated from the difference between initial and saturation readings. We have observed similar results from NBD-labeled vesicles mixed with $Mg^{2+}$ (not shown). The basis for this interaction will not be considered here.

In order to side-step this problem, we turned to the use of pyrene as a probe. The strategy for this assay has been outlined in Figure 7. The ability of pyrenes to form excimers with a characteristic shift in the fluorescence spectrum (Figure 11) has been used to explore a variety of membrane properties, including lateral mobility[66,67] exchange of phospholipids between apoprotein complexes,[68] and exchange between phospholipid bilayers.[34] Two reports use changes in the ratio of excimer and monomer fluorescence (E/M ratio) to demonstrate membrane fusion. Owen[69] demonstrated the fusion of headgroup-labeled pyrene phosphatidylcholine vesicles to cell plasma membranes and Shenkman, et al.[70] using pyrene chain-labeled PE as a probe, showed that bovine serum albumin would slowly fuse phosphatidylcholine vesicles. Neither group attempted continuous measurements nor resolution of the fusion kinetic processes.

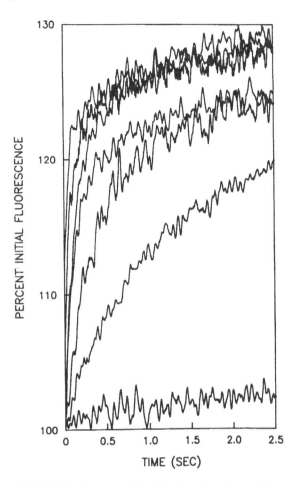

FIGURE 10.    Progress of the $Ca^{2+}$-dependent changes in fluorescence of NBD-PE incorporated into PS:PE (1:1) small unilamellar vesicles. The NBD-PE was incorporated at 0.67 mol %. Lipid = 70 μg/mℓ. NBD-PE:PS:PE = 1:75:75. Temperature = 25°C. Excitation/emission, 475/530 nm. The total fluorescence increase for all experiments except 1.0 m$M$ $Ca^{2+}$ was about 130% of the initial reading. Curves from left to right are for 5.0, 4.0, 3.0, 2.5, 2.0, 1.5, and 1.0 m$M$ $CaCl_2$.

Pyrene-PC (3-palmitoyl-2-(1-pyrenedecanoyl)-L-phosphatidylcholine) was purchased from Molecular Probes, Junction City, Oregon. *E. coli* phosphatidylethanolamine (PE), bovine brain phosphatidylserine (PS), egg phosphatidylcholine (PC), and NBD-PE were purchased from Avanti Polar Lipids, Birmingham, Alabama. All lipids were stored at −20°C in chloroform in sealed ampules until used. PS, PE, and PC were used immediately upon opening. The sonicated suspensions were kept no longer than one week at 0°C. Pyrene-PC and NBD-PE were stored at 1.0 mg/mℓ concentration in chloroform at −20°C for up to one month before discarding the solution. No significant difference in results between freshly opened probe and month old solutions were noted. Other chemicals used were of reagent grade.

Pyrene is very sensitive to quenching by water; this is easily judged by the decrease in quantum yield and disappearance of the fine structure in the emission spectrum between the main monomer peaks at 378 and 398 nm. For this reason, preliminary experiments established that headgroup-labeled pyrene phospholipid was quite unsuitable and explains why Owen[69]

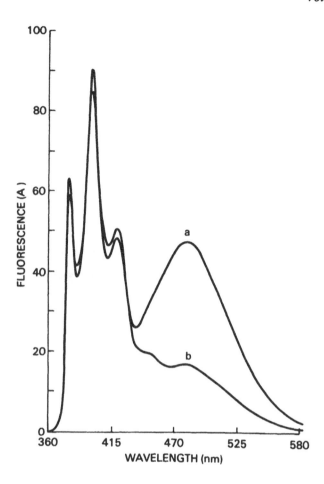

FIGURE 11.    Fusion of pyrene PC-labeled PS:PE SUV with unlabeled SUV reduces the excimer/monomer ratio. Preparation of pyrene-PC-labeled small unilamellar vesicles (SUVs) from PS:PE (1:1) (mol:mol) in 145 m$M$ NaCl, 10 m$M$ HEPES, pH 7.4 was performed as described previously.[26,27] The pyrene-PC stock solution was added to the lipid mixture in chloroform. Pyrene-PC-labeled SUV (5 mol % pyrene PC) were mixed with unlabeled SUV in a ratio of 1:2. Total lipid concentration = 100 μg/mℓ. Excitation 346 nm. Excitation and emission slits set at 2 nm. Temperature = 25°C. Monomer and excimer emission spectrum of pyrene PC (a) before and (b) after the addition of 4 m$M$ CaCl$_2$. Before the calcium is added one sees the typical monomer peaks at 378, 398, and 410 nm, plus the large broad peak of excimer fluorescence with a peak at about 475 nm. After the addition of 4 m$M$ CaCl$_2$, the monomer peaks are increased slightly while the excimer peak is greatly reduced. The E/M ratios before and after the addition were 0.556 and 0.185, respectively, giving a total change of 66.7% of the original ratio. This is exactly the change predicted when diluting the probe by a factor of 2 (cf. Figure 13 and the text).

was required to make his SUVs from >70% probe-labeled PC before analyzable changes in E/M ratio could be obtained. Coupling the pyrene to the phospholipid chain through sulfonyl chloride also produced vesicles with highly quenched spectra. We presume this is due to the hydration of the sulfate within the bilayer or, more likely, to the sulfate dragging the pyrene back to the surface of the bilayer where it would be quenched by water. Some commercially available preparations of these probes proved to be unreliable. However,

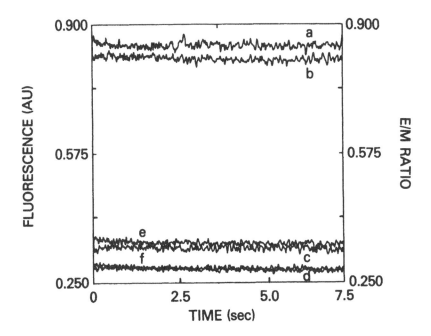

FIGURE 12.    Excimer and monomer fluorescence and E/M ratio changes in the presence of divalent cations. Lines a and b are monomer fluorescence, lines c and d, excimer fluorescence, and lines e and f are E/M ratios of pyrene-PC-labeled PS:PE vesicles in the presence of $Ca^{2+}$ and $Mg^{2+}$, respectively. [Lipid] = 50 μg/mℓ. Five percent pyrene-PE. $[Ca^{2+}]$ = 4.0 m$M$. $[Mg^{2+}]$ = 8.0 m$M$.

excellent results were obtained using the pyrene-PC supplied by Molecular Probes. Pyrene is also very sensitive to quenching by oxygen.[71] No attempt was made to exclude oxygen from the experiments described here. If suitable apparatus were available, this probably would increase the sensitivity of the assay.

A major problem with phospholipid probes used to report fusion events is the possibility that exchange may be taking place in the absence of fusion. Recent reports show that phospholipids chain-labeled with ampipathic fluorophores can exchange between small unilamellar vesicles in the absence of any fusion (see I.D), whereas headgroup-labeled phospholipids seem not to exchange. The spontaneous exchange seems to occur via an aqueous compartment rather than collisions between vesicles. Half-times for exchange increase with increasing acyl chain length and are only slightly affected by the head group, with the exception of diglyceride, which exchanges 20-fold more slowly. Roseman and Thompson[34] observed only a very slow exchange of chain labeled pyrene-PC, with a half-time of several hours. However, Galla and Hartman,[66] using stopped-flow mixing of chain-labeled pyrene-PC incorporated into DPPC SUVs and mixed against unlabeled DPPC SUVs, measured a 25% change in E/M ratio, with a half-time of 11 sec at 29°C, as well as a slower process ($t_{1/2}$ = 6 hr). Rapid exchange of pyrene-PC would greatly distort any change in E/ M ratio due to mixing by fusion.

We reinvestigated this point using chain-labeled pyrene-PC incorporated into either PC or PS:PE SUVs mixed with unlabeled SUVs in the absence of $Ca^{2+}$ or $Mg^{2+}$ (results similar to Figure 12e, f). Results for both membrane types were identical. For hand-mixed experiments, exchange of label was virtually null after 14 hr at 0°C or 4 hr at 25°C. Stopped-flow experiments showed no changes in aggregation or E/M ratio for the time frame 5 to 15,000 msec. We conclude that the chain-labeled pyrene-PC shows no significant exchange in either neutral PC or negatively charged PS:PE bilayers. It is possible that Galla and Hartman[66] were using a probe containing a pyrene-labeled contaminant, such as a fatty acid, which would have a relatively short exchange time.[35] Other possibilities are discussed by Roseman and Thompson.[34]

## B. Steady-State Measurements of Fusion

Steady-state (before vs. after) measurements of fluorescence changes can be used to gain a qualitative idea of whether or not fusion has taken place. Also, it is possible to quantitate the amplitude (extent) of the fusion, if care is taken to be sure that the system has reached equilibrium and appropriate factors (such as quenching by added detergents, etc.) are considered. Quantitation of fusion amplitudes using RET is discussed in References 15 and 28.

The pyrene PC steady-state fluorescent measurements were made in $5 \times 5$ mm cuvettes in a Perkin Elmer® model 650-10S spectrofluorimeter fitted with a thermostated cuvette holder. Crossed polaroid filters were employed to reduce scattering artifacts. Samples were scanned, after which small volumes of concentrated solutions of $CaCl_2$ or $MgCl_2$ were added to the cuvette, thoroughly mixed, and the changes in excimer fluorescence (350/475 nm) were followed. Samples were scanned a second time after no further change in excimer fluorescence was detected. The E/M ratio was calculated as the ratio of the fluorescence intensities at 398 and 475 nm (Figure 11).

The spectrum of pyrene-PC containing SUVs is shown in Figure 11. Measurement of the E/M ratio of PS:PE (1:1) SUVs containing various concentrations of pyrene-PE up to 6 mol %, confirmed the expected result that the ratio was linear with increasing probe concentration (Figure 13).[70] One would also predict that as labeled vesicles fuse with unlabeled vesicles, dilution of the probe should be linear with mixing of membrane contents. Figure 11 also shows the expected reductions in the E/M ratio of 0.50 and 0.80 when labeled and unlabeled vesicles were mixed in ratios of 1:1 and 1:5 and fused with 3.0 m$M$ $CaCl_2$. Similar results (not shown) were obtained using $Mg^{2+}$.

## C. Kinetic Measurements of Initial Fusion Events

Stopped-flow measurements are easily done with rather old equipment if proper experimental caution is exercised. We have found that the fusing lipid mixes adhere to the stopped-flow cuvette windows requiring frequent cleaning with a mild detergent solutions (Windex® is excellent) followed by scrupulous removal of the detergent not to destroy the lipid vesicles. Otherwise the usual cautions apply, especially in obtaining the light intensity readings needed to convert the transmittance data to optical density units by the Lambert-Beer law.

Stopped-flow measurements of changes in the apparent optical density (turbidity), monomer fluorescence, and excimer fluorescence were made simultaneously using a modified version of the apparatus described previously[26] which is diagramed in Figure 14. The machine consists of an Aminco® argon lamp, and Aminco®-Chance single-beam monochrometer, and an Aminoc-Morrow® thermostated stopped-flow drive and cuvette. This configuration limits us to two-channel mixing. (Other configurations with multichannel mixers are also readily available.) Mixing time for this machine, including the delays introduced by the computer software, is about 3 msec. The fastest half-times for the studies reported here are about 200 msec, implying errors of less than 1.5% for this delay. Changes in transmittance are measured by a phototube mounted in line with the excitation beam. Fluorescent light is excluded by a Schott® low-pass UG11 filter. The monomer and excimer fluorescence are detected using two photomultipliers mounted at right angles to the beam. Excitation is at 347 nm, emission wavelengths are selected using bandpass interference filters (Corion® 35-3219-QPT1-2, $\lambda_{max} = 398$ nm; Corion® P10-488-S2584-D110, $\lambda_{max} = 488$ nm).

Unless the volume of the cuvette is rather large (not the case for most stopped-flow machines), one will need to average the data from several experiments to improve the signal to noise ratio. One may use an averaging oscilloscope, but will also eventually wish to digitalize the data for analysis by modern curve-fitting programs. Also, the averaging is

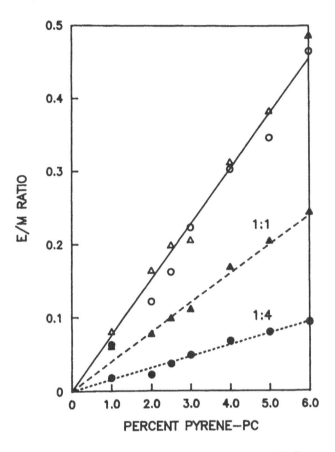

FIGURE 13.   E/M ratios as a function of mol percent pyrene-PC. Pyene-PC was incorporated into PS:PE SUVs at the stated concentrations. Labeled and unlabeled vesicles were mixed in the ratios indicated below and the E/M ratio was determined at 346/398 and 346/475, nm, respectively. Some experiments were observed continuously for an hour with no change in either absolute fluorescence intensities or E/M values, showing that the probe is quite resistant to photobleaching and that no spontaneous exchange of label occurs. A small volume of concentrated $Ca^{2+}$ or $Mg^{2+}$ was added, the excimer fluorescence was monitored until it stabilized, and the readings were repeated. Temperature was 25°C for all experiments. Open symbols and closed symbols represent, respectively, readings before and after addition of 4 m$M$ $CaCl^{2+}$. Circles represent labeled:unlabeled SUV 1:1, and triangles labeled:unlabeled SUV 1:4.

best done off-line, since artifacts often appear in individual records which should be rejected. An excellent method for doing this is via a lab computer with analog to digital convertors (ADCs) and decent storage capacity.

Our acquisition and averaging software uses the strategy outlined above. The output photomultiplier currents are converted to voltages, amplified to appropriate gains, and connected to a Digital Equipment Corp.® AR11, 10-bit precision, multiplexed, analog to digital convertor. Data acquisition and storage is controlled by a DEC PDP11/34 computer.* Data are collected as a series of points (usually 500) on one time base (2 msec/point) followed by shifting to a second, slower time base for an additional series of points (maximum 1024 points). Both the time bases and number of points are programmable. The data acquisition

---

*   All computer programs and documentation for the data acquisition system are available from C. C. Gibson, BEIB, DRS, Bldg. 13 Room 3W13, NIH, Bethesda, Md 20892.

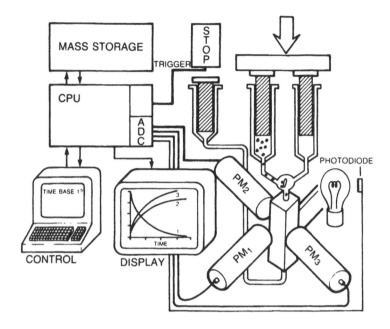

FIGURE 14. Configuration of the stopped-flow-rapid-mixing spectrometer. The three photomultiplier (PM) tubes allow for simultaneously gathering data for change in sample transmittance as well as the monomer and excimer fluorescence. *Note:* the photodiode channel is used to monitor fluctuations in the incident beam. With the installation of the six-pole, 60 Hz notch filters, the AC noise has virtually disappeared from the data, requring no data correction for short-term drift. The lamp is quite stable requiring no correction for drift over the course of an experiment (usually less than 10 sec).

system stores this information, as well as the gains on all channels, the temperature of the stopped-flow cuvette, the excitation and emission wavelengths, experiment name, file name, date, and time of creation. Each point is displayed in real time as it is acquired on a Tektronix® model 603 display to facilitate setting the gains and offsets for each data channel. Twenty to 25 individual experiments are stored, averaged, and transmitted to the NIH DEC10 facility for further reduction and curve fitting, using the MLAB interaction monitor.[72]

### D. Calculations of Stopped-Flow Data

Pyrene excimer/monomer ratios were calculated directly from the fluorescence emission intensities at 398 and 488 nm, respectively. Changes in optical transmittance were converted to optical densities and both types of data were fitted to second-order processes according to:[26,73,74]

$$\% \text{ E/M (or \% absorbance)} = A_1 k_1 t / (1 + k_1 t) + A_2 k_2 t / (1 + k_2 t) \tag{3}$$

where $A_1$ and $A_2$ are the % amplitudes ($A_1 + A_2 = 100\%$) and $k_1$ and $k_2$ are the rate constants. The rate constant $k_1$ is then equivalent to $2C_{11}X_0$, where $X_0$ is the total lipid concentration[64] and $C_{11}$ is defined in Equation 1. The first term on the right-hand side of Equation 3 is an exact solution of Equation 1 for the dimeric case with $(f_{11}/C_{11}X_0) > 100$ (aggregation rate-limiting) and $D_{11} \rightarrow 0$ (irreversible reaction).[74] The second term on the right-hand side of Equation 3 represents "dumping" of all higher-order aggregates and fused species.

For a fixed number of pyrene-PCs evenly distributed among all vesicles and evenly mixed within each vesicle, the steady-state value of E/M should be linearly proportional to the molar concentration of pyrene-PC.[34,70] The true E/M ratio is given by:[34]

$$E/M = CE_{max}/C_h M_{max} \tag{4}$$

where C is the number of pyrene-PCs per total phospholipid molecules, $E_{max}$ = excimer intensity at inifinite C, $M_{max}$ = monomer intensity at C = 0 and $C_h$ is the half-value concentration. In considering the kinetics of phospholipid transfer, Roseman and Thompson[34] point out that for a continually changing distribution of lipids, the true E/M ratio will only equal the observed ratio at time 0 and $t \rightarrow \infty$. At other times the observed ratio will be the number-averaged sum of all the species. Since translational diffusion times for lipids are on the order of $10^{-8}$ cm$^2 \cdot$ sec$^{-1}$,[67] the mixing time for the phospholipids for two 250 Å vesicles after continuity of the bilayers is established is < 100 μsec and the relaxation of the E/M ratio due to lipid mixing will not influence half-times in the >100 msec range. Each fusion product ($V_i$) will yield a characteristic $(E/M)_i$ ratio according to the chain reaction:

$$\overset{\displaystyle V_1 \quad\quad V_1 \quad\quad\; V_1 \quad\quad\quad\; V_1}{V_1 \rightarrow V_2 \rightarrow \;..... \rightarrow V_i\; ....... \rightarrow V_n} \tag{5}$$

If we can assume that fusion produces no change in the lifetimes of the monomer and excimer species, that there is no difference in the reaction rates between labeled and unlabeled vesicles, and no mixing of lipids takes place by exchange, the observed E/M ratio will be equal to the number-averaged sum of the ratios of all the multimeric species according to:

$$(E/M)_{obs} = \sum X_i (E/M)_i \tag{6}$$

where $X_i$ is the fraction of pyrene molecules in an i-mer. Eidelman[84] has shown that Equation 6 is correct provided that $(E/M)_i \ll 1$. Thus, we can use the ratio formed from the recorded values of excimer and monomer fluorescence to follow the membrane mixing. Furthermore, the amplitudes for the formation of dimers and small oligomers should be predictable from Equation 6.

### E. Results of Stopped-Flow Measurements of Divalent Cation-Induced Vesicles Fusion

As discussed above, using resonance energy transfer from NBD-PE to Rho-PE to assess the kinetics of Ca$^{2+}$-promoted fusion of bilayers, we encountered an artifactually fast process which distorted the initial change in RET due to vesicle-vesicle fusion. This artifact is almost completely due to a direct action of divalent cations on the quantum yield of the NBD probe (Figure 9). No change in the E/M ratio is seen upon Ca$^{2+}$ or Mg$^{2+}$ addition to samples containing only pyrene-PC labeled vesicles, although both the excimer and monomer fluorescence change (Figure 12). Thus, the E/M ratio is insensitive to this type of artifact. Experiments were carried out using both Ca$^{2+}$ and Mg$^{2+}$ as the catalyst to fuse mixtures of pyrene-PC labeled and unlabeled SUVs. Initial rates and amplitudes for aggregation and fusion were calculated by directly fitting second-order expressions to the data as described previously[26,74] (Figure 15). The rate of a second-order process should be linearly dependent upon substrate concentration. Figure 16 demonstrates that both Mg$^{2+}$-dependent aggregation and fusion of this lipid mixture is second-order to 200 μg/mℓ lipid. Furthermore, fusion is aggregation rate-limited. These results are qualitatively identical to those reported previously for Ca$^{2+}$ using the NBD-PE/Rh-PE energy transfer assay. However, in the previous case, the raw data contained an artifactual fast process which required much data manipulation to remove. The pyrene-PC assay requires no such contortions.

The fusion rate constant seems to be consistently higher as compared to the aggregation

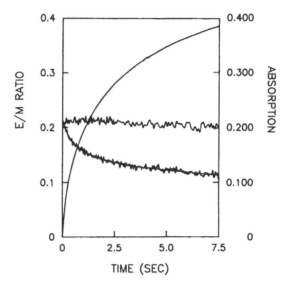

FIGURE 15.    Time-dependence of initial rates of $Mg^{2+}$-promoted aggregation and fusion of PS:PE (1:1) SUV. [Lipid] = 140 µg/mℓ. [$Mg^{2+}$] = 12 m$M$. Pyrene-labeled:unlabeled SUV = 1:1. Smooth line represents the fit of Equation 3 for percent absorbance or percent E/M. The rate for the first process ($k_1$) was 1.68 sec$^{-1}$ and 1.52 sec$^{-1}$ for absorption and E/M, respectively. $A_1$ for E/M was a half of the total change in E/M. The horizontal line is a control experiment with no unlabeled SUV. Pyrene-labeled SUV (100 µg/mℓ lipid) was mixed with 12 m$M$ MgCl$^{2+}$.

rate constant (Figures 16 and 17). This discrepancy could be due to the fact that vesicle size changes are not correctly expressed as percent absorbance in Equation 3, especially if the amplitude $A_1$ of the particles varies with time as the vesicle dimer undergoes deformations.

The results of Figures 11 and 13 show that the expected total change in fluorescence for complete fusion of any mixture of pyrene-labeled and unlabeled vesicles can be calculated on the basis of a simple linear mixing rule. It should also be possible to calculate the expected amplitude of any step of a stochastic model of vesicle fusion from Equation 6. In the present case, starting with equal amounts of labeled and unlabeled vesicles, one should see half of the expected total change in the E/M ratio occur in the first round of fusion. The average change in amplitude for the first second-order process of the data of Figure 13 was 0.235 ± 0.002 (SEM) of the total change, which is almost exactly the expected value of 0.25 of the inital E/M ratio. The total change in E/M ratio is again almost exactly half the value of the initial ratio as predicted from Figure 13.

We can use this type of argument to produce a second line of evidence showing that the changes in E/M ratio due to dilution of the pyrene probes in PS:PE SUVs in the presence of divalent cations occur through fusion of the membranes rather than phospholipid exchange. Because of its small diameter and high radius of curvature, a small unilamellar vesicle will have about 65% of its lipids located in the outer monolayer and 35% in the inner monolayer. If $Ca^{2+}$ or $Mg^{2+}$ were promoting exchange of outer leaflet lipids and the flip-flop rates were low compared to this process for a 1:1 mixture of labeled and unlabeled vesicles, one would predict that the overall amplitude of the change in E/M ratio would be 0.65 × 0.5 ≈ 0.33 rather than the observed 0.5, and in the unlikely event that this should have aggregation rate-limited second order kinetics, the initial change in ratio should be half of this prediction

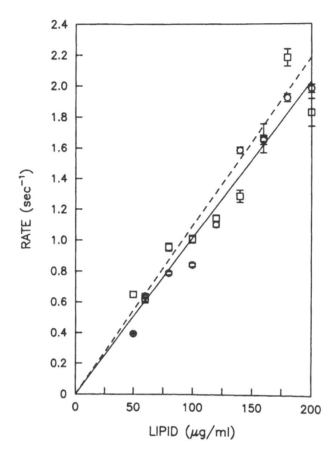

FIGURE 16.    Rate constants for aggregation and fusion. PS:PE (1:1) SUV; $[Mg^{2+}]$ = 12.0 m$M$. Temperature 25°C. Circles represent absorbance measurements (aggregation) and squares (E/M) measurements (fusion). Rate constants ($k_1$) were obtained by fitting the data to Equation 3.

or, 0.17 compared to the observed 0.25. Thus, the amplitudes of the changes are consistent with fusion rather than exchange.

### F. Conclusions about Mechanisms of Fusion

Divalent cations aggregate phosphatidylserine (PS) bilayers; however, there is great selectivity for the further interactions which lead to fusion. For example, magnesium ions will not fuse large PS vesicles;[42] however, diluting the PS with PE removes this selectivity to fusion.[75] Using the pyrene E/M ratio assay, Figure 17 shows that initial rates for monomer to dimer aggregation and initial fusion are identical, thus $Mg^{2+}$ promotes fusion with the same aggregation rate limited kinetics we previously reported for $Ca^{2+}$ fusion of SUVs.[26] Figure 17 shows the $Ca^{2+}$ and $Mg^{2+}$-dependence of initial rates of aggregation and fusion. The threshold for $Ca^{2+}$ (about 1.5 m$M$) is about threefold lower than that for $Mg^{2+}$ (about 4.5 m$M$) and rates for $Ca^{2+}$ run two- to threefold faster than $Mg^{2+}$.

In the absence of aggregation, binding of $Ca^{2+}$ to PS vesicles gives rise to a Ca-PS complex with a 1:1 stoichiometry and binding constant of 12 $M^{-1}$.[76] However, under conditions in which membranes aggregate, there is a dramatic discontinuity in the binding curve with a change to a 1:2 stoichiometry for the complex.[77] The discontinuity has been shown with pure PS, PS:PE(1:1), and PS/galactosylcerebroside (10:1), but, not with PS/PC 1:1. The higher affinity 1:2 complex is interpreted as the formation of a new type of cooperative

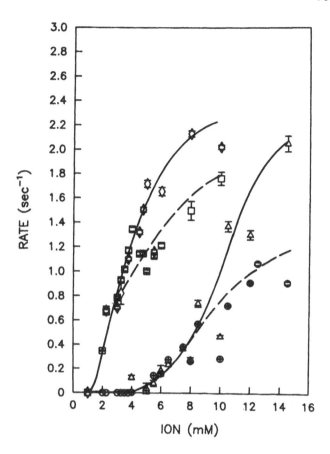

FIGURE 17. Dependence of aggregation and fusion on divalent cations. The curves on the left represent rate constants ($k_1$) derived from absorbance measurements (aggregation) (squares) and (E/M) measurements (fusion) (diamonds) in the presence of $Ca^{2+}$. The curves on the right represent rate constants ($k_1$) derived from absorbance measurements (aggregation) (triangles) and (E/M) measurements (fusion) (circles) in the presence of $Mg^{2+}$, obtained by fitting the data to Equation 3. PS:PE (1:1) SUV. [Lipid] = 80 μg/mℓ.

binding site (possibly the anhydrous complex) between closely apposed bilayer membranes. $Mg^{2+}$ binds to PS with an affinity of 8 $M^{-1}$,[76] but does not form a 1:2 intermembrane dehydration complex. However, PS/PE vesicles will fuse simply by charge neutralization of the PS by $Mg^{2+}$, similar to the way pure PE vesicles (formed by titrating the amino group to a negative charge at pHs $> 9$,[79]) fuse by lowering the pH to neutral.[80] The PE bilayers readily fuse at neutral pH, since the dehydration repulsion between PE bilayers is relatively low,[78] and the small PE headgroup causes bilayer destabilization.[6] The two- to threefold differences in potency of $Ca^{2+}$ vs. $Mg^{2+}$ as fusogens for PS:PE can be explained in terms of the difference between interbilayer dehydration complex vs. charge neutralization followed by PE-induced destabilization.

Although calculations of steady-state values of (E/M) are consistent with fusion (see Section IV.D), the possibility of some divalent ion-promoted lipid exchange cannot be ruled out by the assay itself. Aggregation-dependent exchange of phospholipids in PE vesicles has been demonstrated.[37,80] It has been postulated that exchange represents fusion intermediates (Figure 1), promoted by phosphatidylethanolamine $H_{II}$ phases.[80] Hope et al.[81] observed such intermediates in $Ca^{2+}$ and pH-induced fusion of PS/PE SUV. However, the lamellar to $H_{II}$ phase transition temperature for *E. coli* PE is 55 to 60°C,[82] whereas these experiments were carried out at room temperature.

### G. Advantages and Disadvantages of the Pyrene-PC Assay for Membrane Fusion

The present assay represents a significant improvement over the NBD/Rh RET assay for several reasons:

1.  Only one probe is required. This makes it particularly useful for measuring systems where insertion of the probe must be done on preformed structures such as biological storage particles. The probe can be inserted into the membrane of such particles either by fusion of the particle with probe-containing SUVs[27] or by phospholipid exchange proteins[56] (see Section III.C.3). If the assay were to be applied to asymmetric fusion systems such as vesicles and planar bilayers or, storage vesicles and plasma membrane fragments, only one of the two components need be manipulated.
2.  The expected changes are linear with probe concentrations, simplifying calculation of parameters of stochastic models. Experimentally, the linearity of the assay means that small deviations in probe concentrations from one experiment to the next will have little effect on final results.
3.  Using a ratio of values reduces problems with absolute concentrations of probe in the samples. This also reduces or abolishes problems created by changes in quantum yield of the probe, e.g., by binding of cations to the membrane surface. Thus, the kinetic results can be quickly and easily extracted from the ratio data without resorting to special processing to remove reactions which obscure the fusion-evoked changes.

**Caveat Emptor.** The assay as presented here does have some drawbacks of which potential users should be aware:

1.  Due to relatively low quantum yields of the pyrene probe in bilayer membranes, sensitivity of the stopped-flow experiments is low and relatively high concentrations of probe phospholipids must be employed. Most of the results presented here were gathered with 5 mol % probe. This is about half of the probe concentration for a fusion assay based on reduction in the self-quenching of octadecylrhodamine B chloride,[48] but is still above the rule-of-thumb maximum of one probe per 25 phospholipids favored by most investigators to prevent the probe from influencing bulk bilayer properties. Sensitivity was not a problem at 1 to 2% probe in the steady-state experiments read in a typical fluorimeter. The pyrene probe concentration could be reduced considerably in the stopped-flow experiments if a more powerful light source were employed, if quieter high voltage power supplies, and/or more sensitive (e.g., cooled) phototubes were used, or if larger numbers of experiments were averaged. The data presented here was gathered using an Aminco® incandescent lamp which has low ouput at 346 nm and 20 to 25 individual experiments were averaged. We have recently upgraded the data acquisition software to take the average of eight individual analog to digital conversions, which greatly reduces background noise, and a 150 W xenon lamp has been fitted. Results for experiments using 2.5 mol % probe are virtually identical to those presented here. Although this result indicates that 5 mol % pyrene-PE does not significantly distort the fusion of these vesicles, use of the lowest possible concentrations of probe would be desirable.
2.  The assay has no measure of leakage of core contents during the aggregation/fusion process. It has been demonstrated that some vesicles, especially SUVs, leak during fusion.[15,26] Core-mixing assays have their own inherent problems, e.g., leakage without fusion, which require special controls and experimental conditions.[15] Ideally, one

should measure both fusion and leakage either in parallel or simultaneously. Excitation and emission wavelengths of pyrene-PCs are low enough that a leakage assay based on relief of self-quenching of a high-extinction, soluble fluorophore might be read simultaneously using a dual wavelength monochrometer and chopper to rapidly and alternately excite the membrane-bound and soluble probes. Our stopped-flow is currently being fitted with an Aminco®-Chance dual wavelength monochrometer to test the feasibility of such an arrangement.

## ACKNOWLEDGMENTS

We are grateful to Anne Walter for the polylysine data, Ofer Eidelman for derivation of Equation 6, and Joe Bentz for helpful discussions.

## REFERENCES

1. **Poste, G. and Nicholson, G. C., Ed.**, *Membrane Fusion*, Cell Surface Reviews, Vol. 5, Elsevier/North Holland, New York, 1978.
2. **Hand, A. R. and Oliver, C., Eds.**, *Basic mechanisms of cellular secretion*, Methods in Cell Biology, Vol. 23, Academic Press, New York, 1981.
3. **White, J., Kielian, M., and Helenius, A.**, *Q. Rev. Biophys.*, 16, 151, 1983.
4. **Everad, D. and Whelan, J.**, *Ciba Found. Symp.*, 103, 1, 1984.
5. **Düzgüneş, N.**, *Sub-cell. Biochem.*, 11, 195, 1985.
6. **Blumenthal, R.**, *Curr. Top. Membr. Transp.*, 29, 203, 1987.
7. **Lau, A. L. and Chan, S. I.**, *Proc. Natl. Acad. Sci. U.S.A.*, 72, 2170, 1975.
8. **Pinto da Silva, P. and Nogueria, M. L.**, *J. Cell Biol.*, 73, 161, 1977.
9. **Cullis, P. R. and Kruijff, B.**, *Biochim. Biophys. Acta*, 559, 399, 1979.
10. **Rand, R.**, *Annu. Rev. Biophys. Bioeng.*, 10, 277, 1981.
11. **Siegel, D. P.**, *Biophys. J.*, 45, 399, 1984.
12. **Palade, G.**, *Science*, 189, 347, 1975.
13. **Nir, S., Bentz, J., Wilschut, J., and Düzgüneş, N.**, *Prog. Surf. Sci.*, 13, 1, 1983.
14. **Gregoriadis, G., Ed.**, *Liposome Technology*, 1, 2, and 3, CRC Press, Boca Raton, Fla., 1984.
15. **Düzgüneş, N. and Bentz, J.**, Fluorescence assays for membrane fusion, in *Spectroscopic Membrane Probes*, Loew, Leslie M., Ed., CRC Press, Boca Raton, Fla., 1987, in press.
16. **Zimmerberg, J., Cohen, F. C., and Finkelstein, A.**, *J. Gen. Physiol.*, 75, 241, 1980.
17. **Oku, N. and MacDonald, R. C.**, *Biochemistry*, 22, 855, 1985.
18. **Hope, M. J., Bally, M. B., Webb, G., and Cullis, P. R.**, *Biochim. Biophys. Acta*, 812, 55, 1985.
19. **Pagano, R. E. and Weinstein, J. N.**, *Annu. Rev. Biophys. Bioeng.*, 7, 435, 1978.
20. **Margolis, L. B.**, *Biochim. Biophys. Acta*, 779, 161, 1984.
21. **Weinstein, J. N., Yoshikami, S., Henkart, P., Blumenthal, R., and Hagins, W. A.**, *Science*, 175, 489, 1977.
22. **Blumenthal, R., Weinstein, J. N., Sharrow, S. O., and Henkart, P.**, *Proc. Natl. Acad. Sci., U.S.A.*, 74, 5603, 1977.
23. **Szoka, F. C., Jr., Jacobson, K., and Papahadjopoulos, D.**, *Biochim. Biophys. Acta*, 551, 295, 1979.
24. **Blumenthal, R., Ralston, E., Dragsten, P., Leserman, L. D., and Weinstein, J. N.**, *Membr. Biochem.*, 4, 283, 1982.
25. **Ralston, E., Sharrow, S. O., and Blumenthal, R.**, *Fed. Proc. Fed. Am. Soc. Exp. Biol.*, 39, 2123, 1980.
26. **Morris, S. J., Gibson, C. C., Smith, P. D., Greif, P. C., Stirk, C. W., Bradley, D., Haynes, D. M., and Blumenthal, R.**, *J. Biol. Chem.*, 260, 4122, 1985.
27. **Morris, S. J. and Bradley, D.**, *Biochemistry*, 23, 4642, 1984.
28. **Struck, D. K., Hoekstra, D., and Pagano, R. E.**, *Biochemistry*, 20, 4093, 1981.
29. **Hoekstra, D.**, *Biochemistry*, 21, 1055, 1982.
30. **Parente, R. A. and Lentz, B. R.**, *Biochemistry*, in press.
31. **Fung, B. K. and Stryer, L.**, *Biochemistry*, 17, 5241, 1978.

32. **Weinstein, J. N., Ralston, E., Leserman, L. D., Klausner, R. D., Dragsten, P., Henkart, P., and Blumenthal, R.,** Self-quenching of carboxyfluorescein fluorescence: uses in studying liposome stability and liposome-cell interaction, in *Liposome Technology,* Vol. 3, Gregoriadis, G., Ed., CRC Press, Boca Raton, Fla., 1984, 183.
33. **Lakowicz, J. R.,** *Principles of Fluorescence Spectroscopy,* Plenum Press, New York, 1983.
34. **Roseman, M. and Thompson, T. E.,** *Biochemistry,* 19, 439, 1980.
35. **Nichols, J. W. and Pagano, R. E.,** *Biochemistry,* 21, 1720, 1982.
36. **Nichols, J. W. and Pagano, R. E.,** *J. Biol. Chem.,* 258, 5368, 1983.
37. **Pryor, C., Bridge, M., and Loew, L. M.,** *Biochemistry,* 24, 2203, 1985.
38. **Lakowicz, J. R.,** *Principles of Fluorescence Spectroscopy,* Plenum Press, New York, 1983, 44.
39. **Wilschut, J. and Papahadjopoulos, D.,** *Nature (London),* 281, 690, 1979.
40. **Morris, S. J. and Schober, R.,** *Eur. J. Biochem.,* 75, 1, 1977.
41. **Ralston, E., Hjelmeland, L. M., Klausner, R. D., Weinstein, J. N., and Blumenthal, R.,** *Biochim. Biophys. Acta,* 649, 133, 1981.
42. **Wilschut, J., Düzgüneş, N., and Papahadjopoulos, D.,** *Biochemistry,* 20, 3126, 1981.
43. **Allen, T. M.,** in *Liposome Technology,* Vol. 3, Gregoriadis, G., Ed., CRC Press, Boca Raton, Fla., 1984, 177.
44. **Stutzin, A.,** *FEBS Lett.,* 197, 274, 1986.
45. **Lelkes, P. I., Lavie, E., Naquira, D., Schneeweiss, F., Schneider, A. S., and Rosenheck, K.,** *FEBS Lett.,* 115, 129, 1980.
46. **Gibson, G. A. and Loew, L. M.,** *Biochem. Biophys. Res. Commun.,* 88, 135, 1979.
47. **Hoekstra, D., De Boer, T., Klappe, K., and Wilschut, J.,** *Biochemistry,* 23, 5675, 1984.
48. **Eidelman, O., Schlegel, R., Tralka, T. S., and Blumenthal, R.,** *J. Biol. Chem.,* 259, 4622, 1984.
49. **Citovsky, V. and Loyter, A.,** *J. Biol. Chem.,* 260, 12072, 1985.
50. **Klausner, R. D., van Renswoude, J., Blumenthal, R., and Rivnay, B., Venter, J. C. and Harrison, L. C.,** Eds., *Receptor Biochemistry and Methodology,* Molecular and Chemical Characterization of Membrane Receptors, Vol. 3, Alan R. Liss, New York, 1984, 209.
51. **Huang, C. H.,** *Biochemistry,* 8, 344, 1969.
52. **Ollivon, M., Walter, A., and Blumenthal, R.,** *Anal. Biochem.,* 152, 262, 1986.
53. **Dragsten, P. R., Blumenthal, R., and Handler, J. S.,** *Nature (London),* 294, 718, 1981.
54. **Citovsky, V., Blumenthal, R., and Loyter, A.,** *FEBS Lett.,* 193, 135, 1985.
55. **Pagano, R. E., Martin, O. C., Schroit, A. J., and Struck, D. K.,** *Biochemistry,* 20, 4920, 1981.
56. **Walter, A., Steer, C., and Blumenthal, R.,** *Biochim. Biophys. Acta,* 861, 319, 1986.
57. **Crain, R. C. and Zilversmit, D. B.,** *Biochemistry,* 19, 1440, 1980.
58. **Mason, J. T. and Huang, C.,** *Ann. N.Y. Acad. Sci.,* 308, 29, 1978.
59. **Karnovsky, M. J., Kleinfeld, A. M., Hoover, R. L., and Klausner, R. D.,** *J. Cell Biol.,* 94, 1, 1982.
60. **Markello, T., Zlotnick, A., Everett, J., Tennyson, J., and Holloway, P. W.,** *Biochemistry,* 24, 2895, 1985.
61. **Lampe, P. D. and Nelsestuen, G. L.,** *Biochim. Biophys. Acta,* 693, 320, 1982.
62. **Föster, T.,** Fluoreszenz Organische Verbindungen, Vandenhoeck und Ruprecht, Goettingen, West Germany, 1951, 85.
63. **Bental, M., Lelkes, P. I., Scholma, J., Hoekstra, D., and Wilschut, J.,** *Biochim. Biophys. Acta,* 774, 296, 1984.
64. **Bentz, J., Nir, S., and Wilschut, J.,** *Colloids Surf.,* 6, 333, 1983.
65. **Morris, S. J., Hellweg, M. A., and Haynes, D. H.,** *Biochim. Biophys. Acta,* 553, 342, 1979.
66. **Galla, H-J. and Hartman, W.,** *Chem. Phys. Lipids,* 27, 199, 1980.
67. **Kleinfeld, A. M., Dragsten, P., Klausner, R. D., Pjura, W. J., and Matayoshi, E. D.,** *Biochim. Biophys. Acta,* 649, 471, 1981.
68. **Massey, J. B., Gotto, A. M., Jr., and Pownall, H. J.,** *J. Biol. Chem.,* 257, 5444, 1982.
69. **Owen, C. S.,** *J. Membr. Biol.,* 54, 13, 1980.
70. **Shenkman, S., Araujo, P. S., Dijkman, R., Quina, F. H., and Chaimovich, H.,** *Biochim. Biophys. Acta,* 649, 633, 1981.
71. **Chong, P. L-G. and Thompson, T. E.,** *Biophys. J.,* 47, 613, 1985.
72. **Knott, G. D.,** *Comput. Programs Biomed.,* 10, 271, 1979.
73. **Morris, S. J., Bradley, D., and Blumenthal, R.,** *Biochim. Biophys. Acta,* 818, 365, 1985.
74. **Lansman, J. and Haynes, D. H.,** *Biochim. Biophys. Acta,* 394, 335, 1975.
75. **Düzgüneş, N., Wilschut, J., Fraley, R., and Papahadjopoulos, D.,** *Biochim. Biophys. Acta,* 642, 182, 1981.
76. **McLaughlin, S., Mulrine, N., Gresalfi, T., Viao, G., and McLaughlin, A.,** *J. Gen. Physiol.,* 77, 445, 1981.
77. **Ekerdt, R. and Papahadjopoulos, D.,** *Proc. Natl. Acad. Sci. U.S.A.,* 79, 2273, 1982.

78. **Loosley-Millman, M., Rand, R. P., and Parsegian, V. A.,** *Biophys. J.,* 40, 221, 1982.
79. **Kolber, M. A. and Haynes, D. H.,** *J. Membr. Biol.,* 48, 95, 1979.
80. **Ellens, H., Bentz, J., and Szoka, F. C.,** *Biochemistry,* 25, 285, 1986.
81. **Hope, M. J., Walker, D. C., and Cullis, P. R.,** *Biochem. Biophys. Res. Commun.,* 110, 15, 1983.
82. **Cullis, P. R. and de Kruijff, B.,** *Biochim. Biophys. Acta,* 513, 31, 1978.
83. **Morris, S. J. and Walter, A.,** unpublished data.
84. **Eidelman, O.,** personal communication.

Chapter 8

# PROBING THE LATERAL ORGANIZATION AND DYNAMICS OF MEMBRANES

**David E. Wolf**

## TABLE OF CONTENTS

## I. MODELS OF MEMBRANE ORGANIZATION

For the past twenty years the predominant concept of membrane organization has been embodied in the fluid mosaic model.[1] A popular rendering of this model envisions the lipid bilayer acting as a matrix in which a veritable vegetable garden of membrane proteins are situated, some of whose roots penetrate and pass through the bilayer. The key word here is *fluid*. Implicit in this model is the assumption that the membrane has a property called its bulk membrane fluidity or, inversely, its bulk membrane viscosity. This property governs and, allows the biophysicist to calculate for the membrane proteins, their lateral diffusion rates in the plane of the membrane, their translational diffusion rates across the membranes, and their rotational diffusion rates around axes within the membrane. A corollary to the membrane having this property is: that if one has a physiological transformation which alters membrane "fluidity" that regardless of the probe used in one's measurement, and regardless of whether one measures lateral, translational, or rotational diffusion, that the same change in "fluidity" will be observed. Questions of membrane organization and dynamics would be relatively simple to answer using the concept of bulk membrane "fluidity". One would have to: calculate membrane protein diffusion coefficients using quasi-two-dimensional continuum theories,[2,3] calculate lipid diffusion coefficients using percolation theories,[3-7] and deal with the sticky issue of protein-lipid interactions.

In its popular form,* this model does not take into account the fact that lipids do not intermix well to form homogeneous bilayers.[8,9] Extensive studies of model membranes composed of binary lipid mixtures have shown that these membranes instead are composed of coexistent microenvironments or "domains" of nonaverage composition and nonaverage fluidity. At first one might suppose this domain organization to be the result of some quirk of two-dimensional systems. However, such is not the case. Nonideal mixing is a common feature of three-dimensional solutions[10,11] as well. One has only to consider the frontispiece in Hildebrand and Scott's classic work on *Regular Solutions*,[11] which shows ten coexistent phases in a test tube, to realize that even in three dimensions, immiscibility is more the rule than miscibility. Nature may always strive to elegance, but not always to elegant simplicity.

Conceptually, the simplest form of lateral phase segregation is exhibited by membranes composed of binary mixtures of dimyristoyl- (DMPC) and dipalmitoylphosphatidylcholine (DPPC).[8] At temperatures intermediate between the melting temperatures of the two pure phospholipids, these membranes exhibit gel phases (rich in DPPC) coexistent with fluid phases (rich in DMPC). A distinct kind of lateral phase segregation is observed in membranes composed of binary mixtures of DMPC and distearoylphosphatidylcholine (DSPC). In addition to showing gel-fluid segregation, these also exhibit two coexistent gel phases at temperatures below the DMPC melt at 23°C and mole fractions of DSPC less than 0.6.[8] Membranes composed of binary mixtures of dielaidoylphosphatidylcholine (DEPC) and dipalmitoylphosphatidylethanolamine (DPPE), in addition to showing the two types of phase segregations already discussed, exhibit two coexistent fluid phase.[9] The complexity of the phase diagram and the number of possible phases is, of course, limited by the Gibbs' phase rule.[10]

For binary lipid mixtures phase segregations result from differences in either lipid acyl chains or head groups. (The reader is referred to the recent review by Small[12] for a discussion of lipid packing where the acyl chains on the one and two positions of the phospholipid are different.) In addition to the lateral phase segregations already discussed, membrane lipids can form domains which are nonbilayer phases.[13] For example, hexagonal phases are observed in membranes containing cardiolipins[13,14] and such phases appear to play a role in membrane-membrane fusion events.[15]

---

* In deference to Drs. Singer and Nicholson it must be stated that in their original paper[1] they considered the possibility of lipid domains.

Besides phospholipids, the other major nonprotein component of biological membranes is cholesterol. At sufficiently high concentration in single phospholipid bilayers, cholesterol can act as a "plasticizer" causing the bilayer to exist in a state intermediate between fluid and gel.[16] However, cholesterol exhibits perferential association with phospholipids. The order of affinity for the major phospholipid classes is sphingomyelin > PC > PE.[17] Cholesterol forms apparent associations with phospholipids in stoichiometric ratios of 1:4, 1:2, and 1:1.[16,18-36] The phase diagram for cholesterol:DPPC mixtures has been extensively studied. Bilayer organization changes dramatically at 20 and approximately 30 mol % cholesterol. The region of the phase diagram below 20 mol % cholesterol is particularly interesting in that the membrane appears to have coexistent fluid phases rich in cholesterol and gel phases poor in cholesterol.[29,32] Work by McConnell's group suggests that these domains are linear, which results in anisotropic diffusion of membrane components.[31] This region of the phase diagram may be relevant to endomembranes, where cholesterol content less than 20 mol % are common.[37]

The membrane proteins constitute yet another potential influence on domain structure in the membrane, but the nature and specificity of protein-lipid interactions are poorly understood. On one hand, we know from reconstitution experiments that membrane proteins appear to require specific lipid environments for maximal function.[38-43] Acetylcholine receptors also appear to have localized lipid environments *in situ*. On the other hand, the data relating lipid diffusion rates to protein concentration can be satisfactorily described in terms of hard shell interactions.[3,44] While these issues remain open, one can expect that new techniques discussed in chapters by Kleinfeld, Baird, and Holowka will soon begin to resolve these issues.

Ultimately, the question must be whether domains exist in biological membranes at physiological temperatures and cholesterol concentrations. Clearly however, biological membranes are finely tuned, complex mixtures of lipids, and one is therefore tempted to make the tautological argument that domains must exist in cell membranes. If all that the lipid bilayer did was offer a fluid matrix for membrane proteins, one or a few phospholipids, should suffice. Until recently, however, evidence for the existence of lipid domains in biological membranes was indirect. While only indirect, it was provocative in that it implicated membrane domains in a number of important cellular processes, including: virus budding,[45] membrane-bound enzymatic activities,[46,47] sperm capacitation,[48] prostaglandin-induced corpus luteum regression,[49] and tight-junction formation.[50-52]

The principal obstacle to identifying domains in membranes is that many techniques measure overall or average membrane properties. This problem is particularly acute when studying natural membranes which are composed of many lipids, and, which therefore could potentially have many different domains. A change in the physical state or size of one domain, which might compose only a few percent of the membrane mass, would not be detectable by a technique such as differential scanning calorimetry. Ideally, one would like to measure changes in the organization of specific domains during physiological transitions. *What is needed is a method of selectively probing the specific domains undergoing the transition.*

## II. PROBING THE LATERAL ORGANIZATION OF MEMBRANES

Recently, methods for studying membrane domains have begun to evolve which exploit the preferential solubilities of amphiphiles in different lipids and lipid phases.[8,9,14,51-62] McConnell and colleagues[8,9,14,57,58] have had dramatic success in developing phase diagrams of lipid mixtures by measuring the dependence of the partition of the electron spin resonance probe TEMPO between the membrane and aqueous phases. Many of the phase diagrams discussed above for lipid mixtures were generated by this technique. As a function of their structure, amphiphiles also exhibit selectivity between coexistent lipid phases. An important example of such selectivity is exhibited by the isomers of parinaric acid. These probes are discussed in greater detail in the chapter by Dr. Hudson. Sklar et al.[59] have shown that the *trans* isomer of parinaric acid exhibits a 3:1 preference for solid to fluid, while the *cis* isomer exhibits a 1.7:1 preference for fluid to solid. Klausner et al.[53] demonstrated a similar selectivity by the *cis* and *trans* isomers of common fatty acids, and found that saturated fatty acids prefer the gel. Similar results were obtained by Pringle and Miller[60] for the *cis* and *trans* isomers of 9-10-tetradecenol. As will be discussed in greater detail below, in our laboratory, we have used a different set of compounds, the 1,1'-dialkyl 3,3,3',3'-tetramethylindocarbocyanines ($C_N$diIs) (see Figure 1 for structure) to demonstrate that their partitioning between the gel and fluid phases of the disaturated lecithins is highly dependent upon the akyl chainlength of the probes.[55,62,63]

## III. NONAVERAGE PROPERTIES VS. SELECTIVE PARTITION

As discussed above, to properly study domain organization in membranes, one must devise a measure of a nonaverage property of the membrane. One way to accomplish this is to use a probe which will selectively partition into some subfraction of the domains present in the membrane. A second approach is to use a probe which partitions into all domains, but whose properties are both dependent upon the immediate surroundings and distinguishable with some monitoring technique. An example of the second approach is the lifetime heterogeneity studies of diphenylhexatriene (DPH) performed by the Kleinfeld laboratory.[53,64] DPH does not appear to selectively partition between lipid domains,[65] but its fluorescence lifetime is highly dependent upon its physical environment. DPH fluorescence decays nearly monoexponentially[53,65-67] with a lifetime of 6 to 8 nsec, depending upon temperature, in fluid-phase membranes. In gel phase membranes, it again decays nearly monoexponentially[53,65-67] with a lifetime of 9 to 10 nsec. In mixed-phase membranes[53] and in some cell membranes[53] DPH exhibits multiexponential decay, with one lifetime around 7 nsec, one lifetime around 9 nsec, and often a third lifetime around 2 nsec. This 2-nsec lifetime is shorter than that of DPH in any single-phase membrane system measured and may reflect probe at the interfaces between domains.[53] This same probe and approach has been used recently by Pjura[64] on CH1 lymphoma cell plasma membranes to generate some particularly interesting results. In these experiments, multiple exponential decay is observed not only in CH1 plasma membrane vesicles, but, in membranes made from plasma membrane lipid extracts. In both cases the long-lifetime component disappears when the temperature is raised above 39°C, suggesting not only that these membranes contain significant amounts of gel phospholipids, but that these gel domains melt out at 39°C. Our ability to interpret lifetime heterogeneity experiments has been limited by our ability to uniquely fit data to more than two exponentials.[53,66,67] There have been two recent advances in lifetime techniques which promise to both complicate and improve the situation enormously. These are the development of multiple frequency phase modulation instruments[67] and the application of global analysis techniques to lifetime data.[68] These two advances will enable us to resolve multiple exponential decays. However, as is often the case in spectroscopy, we are already finding that where we previously thought we had monoexponential decay, the decay is multiexponential.[66,67]

## IV. CARBOCYANINE DYES

One of the most useful series of membrane lipid analogs for domain studies are the $C_N$diIs,[54-56,61,63] which were originally developed as membrane-potential sensing probes.[69,70,71] Because the $C_N$diIs with long alkyl chain lengths were found to intercalate well into plasma membranes, these probes have also been extensively used for fluorescence recovery after

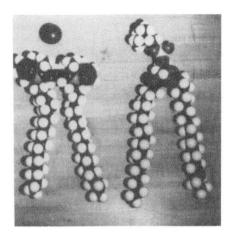

FIGURE 1B. The chemical structure of the C$_n$dils.

A

FIGURE 1. Photograph of CPK space-filling models of C$_{18}$dil (left) and phosphatidylcholine (right). This photograph displays the structural similarities between C$_N$dils and phospholipids, which implies that the C$_N$dils intercalate into the bilayer as lipids do.

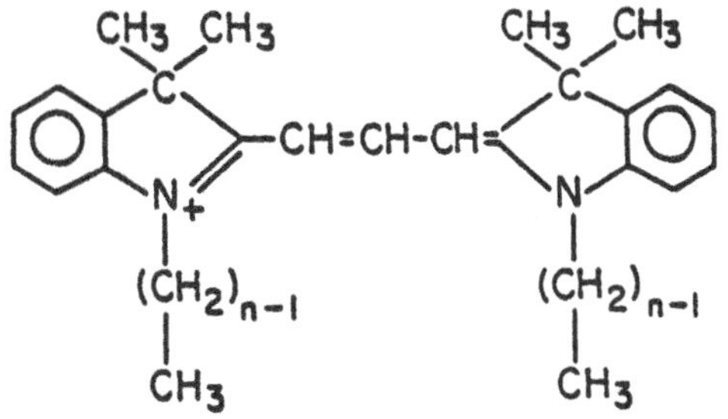

photobleaching studies of lipid diffusion in cell plasma membranes (for review see Reference 72). For our purposes here, they provide us with a case study of the advantages, disadvantages, and pitfalls of using fluorescent probes to study membrane lateral organization and dynamics, showing us what we can learn and at what price.

## A. Structure and Spectroscopic Properties

A space-filling model of the disaturated 18-carbon form (C$_{18}$) dil dioctadecyl-) of these analogs is shown alongside of a model of phosphatidylcholine in Figure 1A. The chemical structure of the C$_N$Dils can be found in Figure 1B. Structurally these two molecules are similar. One would predict that the C$_N$dils intercalate into the bilayer as the lipids do, and would expect them to mimic lipid diffusion. The C$_N$dils have two saturated alkyl chains, as opposed to acyl chains in most phospholipids. They are commercially available from Molecular Probes® (Eugene, Ore.) in even chainlengths from N = 10 to 22. The C$_N$dils are amphiphilic by virtue of the positive charge which resonates between the nitrogens of the two indole rings. This resonance is also responsible for the fluorescence of the molecule. These monocarbocyanine forms have absorbance maxima around 551 nm and emit around 568 nm. The absorbance and fluorescence properties of these probes can be altered by substituting the dimethylated carbons at the 3 and 3' positions with either an S to form thiacarbocyanines or an O to form oxacarbocyanines.[69] Thus, there are a considerable number of possible donor-acceptor pair combinations for fluorescence resonance energy transfer measurements.

These probes have very low critical micelle concentrations.[73] Labeling of plasma membranes is typically done by ethanol injection into the aqueous phase.[74,75] Labeling of cell membranes probably occurs by transfer of probes from micelles to soluble monomer to membrane transfer, but this has not yet been demonstrated. Because these probes are fluorescent in membranes and extremely quenched in solution,[69,73] in some experiments it is not necessary to remove aqueous phase material. Studies of the absorbance of these probes as a function of concentration in aqueous phase shows that below the critical micelle concentration the probes exhibit a monomer absorbance band and a blue-shifted dimer band.

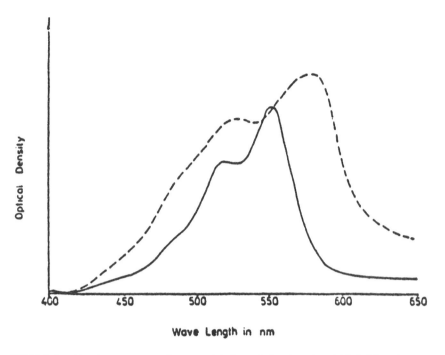

FIGURE 2.   Absorbance spectra of aqueous labeling solution (- - -) and labeled lymphocytes
(——). A solution of Hank's balanced salt solution was prepared with 5 μg/mℓ of $C_{18}$diI and 0.5%
EtOH. (----) shows the absorbance spectrum of this solution. The dominant states of the probe are
micelles (J band) and dimer. Lymphocytes were then incubated with this solution for 10 min at
37°C. Excess probe was removed and the cells resuspended. (——) shows the absorbance spectrum
of this suspension. The dominant states of the probe are monomer and dimer.

Above the critical micelle concentration, the overall absorbance decreases and the spectrum
becomes dominated by a blue-shifted peak, known as the H band, and a red-shifted peak,
known as the J band. Thus, absorbance measurements enable us to study the degree of
aggregation of the probe. As an example of this consider Figure 2, which shows the absorbance spectrum of a 5 μg/mℓ solution of $C_{18}$diI in Hanks' balanced salt solution with
0.5% ethanol, and the spectrum of mouse spleenic lymphocytes incubated in this solution
and then washed free of excess probe. The spectrum of the labeling solution is dominated
by the J and dimer bands while the cell spectrum is dominated by monomer and dimer
bands.

## B. Orientation

The structural properties of these probes control their spectroscopic properties and enable
us to define the orientation of the $C_N$diIs within the membrane. The indole rings and
unconjugated double-bonded carbons form an essentially flat structure, and the dipole moment points roughly along the bridge between the indoles.[76] Because of their relatively short
fluorescence lifetimes,[63] the fluorescence of these probes is highly polarized. This high
polarization of the $C_N$diIs has enabled Axelrod[76] to show that the transition moment of these
molecules is tangential to the plane of the bilayer, consistent with the intercalation of the
dyes into the bilayer by virtue of their alkyl chains, which is suggested by Figure 1. As
clearly demonstrated for the anthroyl fatty acids,[77] spectroscopic properties of fluorescent
probes can often be used to define the orientation and localization of the probe within the
bilayer.

## C. Diffusion

FPR studies of $C_N$diI diffusion on cells of homeothermic organisms show diffusion coefficients of $10^{-9}$ to $10^{-8}$ cm²/sec with 90% to 100% of the probe free to diffuse (for review see Reference 72). These values are in agreement with FPR measurements of $C_N$diI diffusion in fluid-phase model membranes.[55,78-81] In gel-phase membranes apparent diffusion coefficients of approximately $10^{-10}$ cm²/sec are observed with only a small fraction of the probe diffusing.[55,78-81] Recently, Webb's group has demonstrated that this apparent recovery in gel phases results from diffusion along faults in the crystal lattice, and that true diffusion rates within the lattice are on the order of $10^{-13}$ cm²/sec.[82] In any event, on the time scale of fluid-phase lipid diffusion, the presence of gel phase can operationally result in incomplete recovery. This may bear some significance to the few cases where incomplete recovery has been observed on homeothermic membranes.

Problems in dealing with a membrane as having a bulk fluidity first came to our attention during studies of changes in $C_N$diI diffusion in sea urchin plasma membranes accompanying fertilization. The sea urchin is one of the major model systems for the study of fertilization. Following the fusion of egg and sperm there ensues a rapid and dramatic exocytotic event known as the cortical reaction. During the cortical reaction the cortical granules fuse with the plasma membrane of the ova.[83] Eddy and Shapiro[84] have estimated that potentially as much new membrane is added as a result of the cortical reaction as was originally present. Not surprisingly, the events of fertilization result in dramatic changes in lipid composition,[85,86] and raise the possibility that membrane fluidity would change with fertilization. To consider this possibility we used FPR to measure these changes.[54] The effect of fertilization upon $C_N$diI diffusion is shown in Figure 3. This data is inconsistent with a homogeneous fluid plasma membrane in that the effect of fertilization upon fluidity is dependent upon the probe used, in this case the alkyl chainlength of the $C_N$diI. A decrease in diffusibility is observed for $C_{10}$diI and $C_{16}$diI with fertilization, while an increase is observed for $C_{12}$diI and $C_{14}$diI. The inconsistency of this data with the concept of a bulk membrane fluidity is more fundamental than the different effects of fertilization upon probe diffusion. As we will show below, in a homogeneous fluid membrane, diffusion coefficients for the carbocyanines are independent of probe alkyl chainlength. In both the fertilized and unfertilized sea urchin eggs, the diffusion coefficients are highly dependent upon chainlength and exhibit extrema, a relative minima at $N = 12$ in the unfertilized egg, and a relative maxima at $N = 14$ in the fertilized egg.

## D. Selective Partition Between Domains

We conjectured at the time that the dependence of diffusion coefficients upon alkyl chainlength might result from the selective partition of the $C_N$diIs, as a function of their chainlength into coexistent lipid domains in the egg plasma membrane. To test this hypothesis, we had to demonstrate two things: first, that the $C_N$diIs do have an alkyl chainlength dependent phase preference, and second, that the essential properties of the dependency of diffusion coefficient upon probe alkyl chainlength observed in egg membranes could be duplicated in a model membrane which is known to have domains.

We sought first to consider the simplest form of phase separation, that of coexistent fluid and gel domains. To determine the selectivity of the $C_N$diIs between fluid and gel domains, we adopted an approach which has been used for several of the other selective probes discussed previously.[53,60] This involves doping a membrane with a perturbing amount of probe, typically 5 to 10 mol %, and determining if the phase transition of the membrane is elevated or depressed. Qualitatively, solutes which lower the melting temperature prefer the fluid phase, while those which raise the melting temperature prefer the gel phase. In terms of ideal solution theory, the partition is described by:

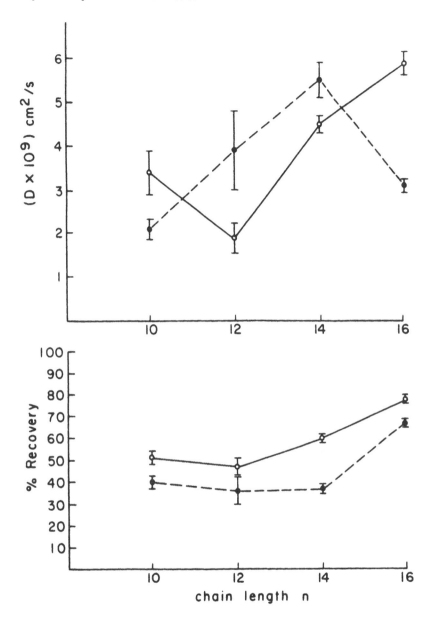

FIGURE 3.    Diffusion of $C_N$diIs as a function of alkyl chain length in *S. purpuratus* eggs. Diffusion coefficient: top; % recovery: bottom; ——○——: unfertilized; ---●---: fertilized. (From Wolf, D. E., *Development in Mammals*, Vol. 5, Elsevier/North Holland, New York, 1983, 187. With permission.)

$$R \ln(X_g/X_f) = \Delta H_0(1/T - 1/T_0) \qquad (1)$$

where $X_g$ and $X_f$ are the mole fractions of *solvent* in the gel and fluid phases, respectively, $\Delta H_0$ is the latent heat of the melt, $T_0$ is the transition temperature of the pure lipid, and T is the transition temperature of the lipid doped with probe.

In our original studies,[55] fluorescence was used to monitor the phase transition in membranes containing the $C_N$diIs. Figure 4 shows fluorescence intensity as a function of temperature for a DMPC membrane containing $C_{14}$diI. The actual plot shows fluorescence intensity and temperature as a function of time. The fluorescence intensity decreases rapidly

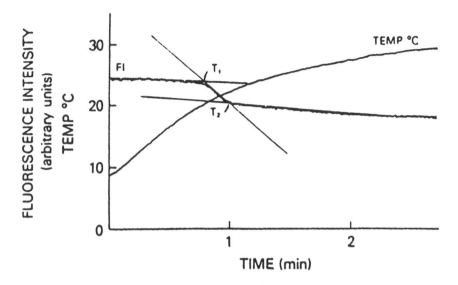

FIGURE 4. Fluorescence intensity (FI) vs. temperature (TEMP °C) for 0.1 mole fraction of $C_{14}$diI in DMPC (100 µg/m$_1$) vesicles. Three discreet slopes can be drawn along the FI vs. temperature curve: one giving the temperature dependence in the gel phase; another during the transition; and another in the fluid phase. These curves intersect as shown at $T_1$ and $T_2$. We define $\Delta = T_2 - T_1$ as a measure of the width of the transition. The transition temperature Tm is then given by the temperature at which the FI is halfway between FI at $T_1$ and $T_2$. (Reprinted with permission from Klausner, R. D. and Wolf, D. E., *Biochemistry*, 19, 6199, 1980. Copyright 1980 American Chemical Society.)

at the 23°C phase transition of the lipid. Such a decrease at the transition is observed with all of the $C_N$diIs at low nonperturbing concentration. Quite a different result is obtained when the $C_N$diIs are at a perturbing concentration. Figure 5 shows the effect of $C_{10}$ thru $C_{18}$diI at perturbing concentration. $C_{10}$diI lowers the transition and can be said to be a fluid preferrer. $C_{18}$diI raises the transition and can be said to be a gel preferrer. Notice, however, that $C_{10}$ and $C_{12}$diI show an increase at perturbing concentration in fluorescence intensity on going through the transition, in contrast to the decrease which they show at low nonperturbing concentration. When identical measurements are made in DPPC, $C_{14}$diI shows an increase as well. The reason for this phenomenon becomes clear from experiments such as those shown in Figure 6, which considers the absorbance spectrum of the fluid preferrer $C_{10}$diI at 10 mol % in DMPC vesicles above and below the phase transition of the lipid. Cooling results in an overall diminution of absorbance, a decrease in the monomer and dimer absorbance bands, and the appearance of the H and J bands. All of this is characteristic of probe aggregation. Thus, forcing a fluid preferring $C_N$diI to be in a gel phase of the lipid results in probe aggregation and self-quenching. This quenching in the gel phase is so severe that the increase in fluorescence intensity due to relaxation of quenching upon passing through the transition dominates over the decrease in monomer intensity which also occurs.

We therefore have two assays of phase preference: first, whether the probe elevates or depresses the phase transition temperature, and, second, whether the fluorescence intensity of the probe increases or decreases upon passing through the transition. On the basis of these criteria we can state selection rules for the partition of the $C_N$diIs in the disaturated lecithins. If N is the alkyl chainlength of the probe, and M is the acyl chainlength of the phosphatidylcholine, then if:

1.  $N < M$, we get fluid preference.
2.  $N \simeq M$, we get no preference.
3.  $N > M$, we get gel preference.

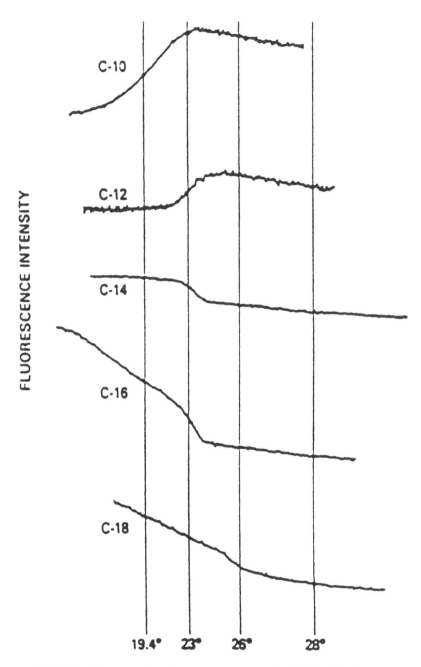

FIGURE 5.   Fluorescence intensity vs. temperature profile for 0.1 mol fraction, a concentration which perturbs the membrane of the $C_N$dils in DMPC (100 μg/mℓ) vesicles (N = 10 to 18). (Reprinted with permission from Klausner, R. D. and Wolf, D. E., *Biochemistry*, 19, 6199 1980. Copyright 1980 American Chemical Society.)

Knowledge of these selection rules placed us in a position to consider the effect of this selective partition upon probe lateral diffusion. Figure 7 shows the FPR recovery curves of $C_{10}$ and $C_{18}$dil at low nonperturbing concentrations in three types of membranes at 15°C:

1.   Fluid phase dilauryl-phosphatidylcholine (DLPC)
2.   Gel phase (DPPC), and

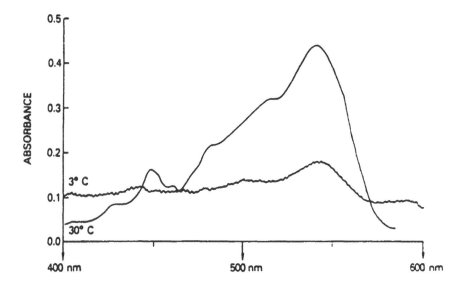

FIGURE 6.    Absorbance spectra of a 0.1 mol fraction of $C_{10}$diI in DMPC (100 μ/mℓ) vesicles at 3°C (below the phase transition) and at 30° (above it). The reference cuvette contained an identical sample of DMPC vesicles but without $C_{10}$diI. (Reprinted with permission from Klausner, R. D. and Wolf, D. E., *Biochemistry,* 19, 6199, 1980. Copyright 1980 American Chemical Society.)

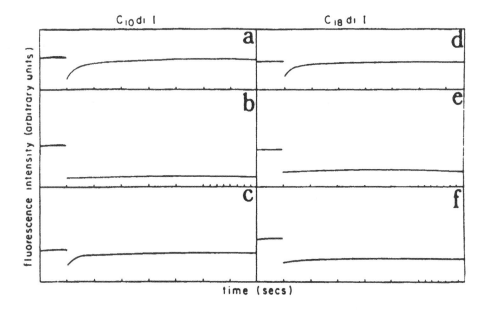

FIGURE 7.    Diffusion of $C_{10}$diI and $C_{18}$diI in single- and mixed-phase vesicles at 10°C. (a) $C_{10}$diI in DLPC, D = 1.1 × $10^{-8}$ cm²/sec % recovery = 100; (b) $C_{10}$diI in DPPC, D > $10^{-10}$ cm²/sec, % recovery = 100; (c) $C_{10}$diI in a 1:1 mixture of DLPC/DPPC, D = 1.0 × $10^{-8}$ cm²/sec, % recovery = 100; (d) $C_{18}$diI in DLPC, D = 1.6 × $10^{-8}$ cm²/sec, % recovery = 100; (e) $C_{18}$diI in DPPC, D > $10^{-10}$ cm²/sec, % recovery = 0; (f) $C_{18}$diI in 1:1 mixture of DLPC/DPPC, D = 0.8 × $10^{-8}$ cm²/sec % recovery = 12. (Reprinted with permission from Klausner, R. D. and Wolf, D. E., *Biochemistry,* 19, 6199, 1980. Copyright 1980 American Chemical Society.)

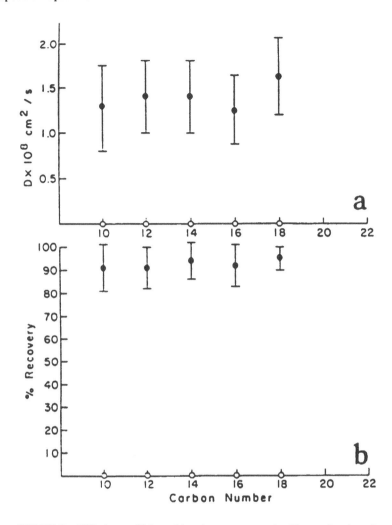

FIGURE 8.    Diffusion coefficients (a) and percent recoveries (b) as a function of alkyl chain length at 10°C in DLPC vesicles (closed circles) and DPPC vesicles (open circles.). (Reprinted with permission from Klausner, R. D. and Wolf, D. E., *Biochemistry*, 19, 6199, 1980. Copyright 1980 American Chemical Society.)

3.    Membranes made from a 1:1 mixture of these two lipids which are known to have coexistent gel and fluid phases.

By our selection rules, $C_{10}$diI should prefer fluid phases, while $C_{18}$diI should prefer gel phases. In DLPC both $C_{10}$ and $C_{18}$diI show complete and rapid recovery typical of fluid membranes. In DPPC both probes show nearly no recovery on this time scale, typical of gel phase membranes. In the mixed-phase system the probes have a choice, the $C_{10}$diI diffuses as if all of it were in the DLPC-rich fluid regions and the $C_{18}$diI diffuses as if nearly all of it were in the DPPC-rich gel regions. The behavior of all the $C_N$diIs is considered in Figures 8 and 9. Figure 8 shows that the diffusion coefficient is independent of alkyl chainlength of these probes in homogeneous fluid membranes. Diffusion is also seen to be independent of alkyl chainlength in a degenerate sense in gel phase membranes. In contrast we see in Figure 9 that in mixed-phase membranes, diffusion coefficient decreases with increasing alkyl chainlength. A percolation theory treatment of this problem has been developed by Saxton.[109]

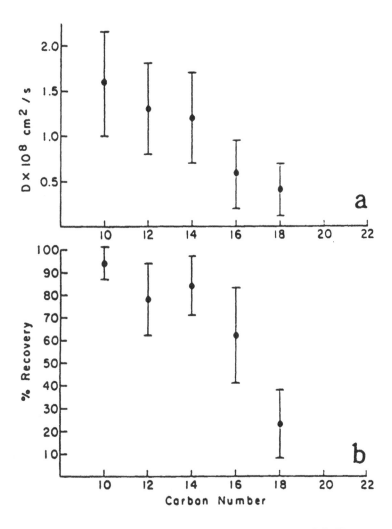

FIGURE 9. Diffusion coefficients (a) and percent recoveries (b) of $C_{10}$diIs as a function of alkyl chain length at 10°C in mixed-phase vesicles (a 1:1 mixture of DLPC/DPPC) at 10°C. (Reprinted with permission from Klausner, R. D. and Wolf, D. E., *Biochemistry*, 19, 6199, 1980. Copyright 1980 American Chemical Society.)

Clearly, the M > N preference for gel phase cannot be true for all alkyl chainlengths. In the extreme case where N > 2M, the alkyl chain must fold back into the bilayer which can best be accommodated by a fluid phase. Thus, one predicts that if N >> M the probe will exhibit fluid preference and diffusion rate should increase for such chainlengths. As shown in Figure 10, reversion to fluid preference occurs for $C_{20}$ and $C_{22}$diI on the basis both of a depression by these probes of the DMPC transition and the increase in fluorescence intensity upon heating through the transition.

The results of these model-membrane studies suggest that the observation of a nonmonotonic dependence of diffusion coefficient of $C_N$diI upon alkyl chainlength is indeed due to the membrane being organized into lipid domains, and that fertilization *represents not a change in bulk membrane fluidity but in the ensemble of membrane domains.* Further evidence in support of the hypothesis that sea urchin egg membranes are composed of lipid domains comes from work of the Edidin laboratory[87] who have shown that the dependence of the $C_N$diIs diffusion rates upon temperature shows different transitions as a function of alkyl chainlength. This is exactly what one would predict if the different $C_N$diIs were localizing

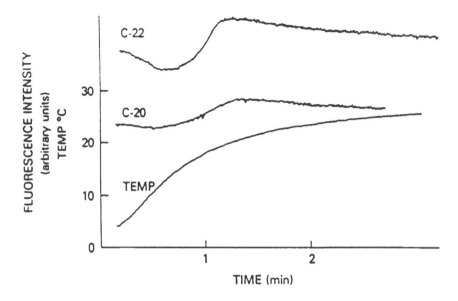

FIGURE 10.   Fluorescence intensity of $C_{20}$diI and $C_{22}$diI as a function of temperature in DMPC (100 μg/mℓ). (Reprinted with permission from Klausner, R. D. and Wolf, D. E., *Biochemistry,* 19, 6199, 1980. Copyright 1980 American Chemical Society.)

to specific domains and reporting on the phase transitions of those specific environments. The idea that these domains are of significance to the fertilization process is supported by the observation of a *reversible* physiological transition in *S. purpuratus* ova between 25 and 30°C. Above this temperature fertilization does not occur.[110] The reversibility of this transition lends credence to the idea that it is a lipid transition. This is further supported by recent work by Glabe[88,89] demonstrating that the sperm-egg adhesion protein in *S. purpuratus,* bindin, interacts only with liposomes which contain gel phases. Bindin acts as an agglutinator of such liposomes, and by virtue of this agglutination promotes vesicle-vesicle fusion.

## E. Calorimetry

In theory, calorimetry should provide us with a direct measure of probe partition. The use of melting-point elevation and depression to determine partition is widespread.[53,55,60] Implicit in Equation 1 is the assumption that the probe does not alter the enthalpy of the melt, i.e., that behavior can be described by *ideal solution theory*. Should this criterion not be met, then, to next order, the problem must be treated by *regular solution theory*. Equation 1 becomes

$$R \ln(X_g/X_f) = \Delta H/T - \Delta H_0/T \qquad (2)$$

where $\Delta H_0$ and $\Delta H$ are the heats of the melt in the absence and presence of the probe. Since temperatures are on an absolute scale, enthalpy differences can be expected to dominate the partition.

Figure 11 shows the heat of melt for DMPC, DPPC, and DSPC membranes doped with 5 mol % of the $C_N$diIs measured by differential scanning calorimetry compared to the melt for the pure lipid. Clearly, at 5 mol % the $C_N$diIs do alter $\Delta H$. The calorimetry data of Ethier et al.[61] suggest that for fluid preferring $C_N$diIs, part of this alteration in $\Delta H$ results from an exothermic heat of solvation or spreading. In fact, the calorimetry data is too complex to be explained in terms of regular solution theory. At these probe concentrations, these systems must be treated calorimetrically as binary mixtures of lipid and probe. Infor-

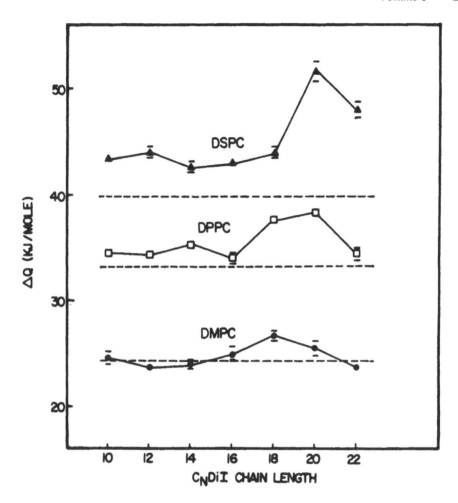

FIGURE 11.   Heat of melt ΔQ as a function of $C_{10}$diI chain length N in DMPC (●), DPPC (□), and DSPC (▲) at a $C_N$diI concentration of 4.76 mol %. Dashed lines represent Δ for the main transition of the pure lipid. Error bars represent ±1 SEM. Where not shown, they are contained within the symbol. (Reprinted with permission from Ethier, M. F., Wolf, D. E., and Melchior, D. C., *Biochemistry*, 22, 1178, 1983. Copyright 1983 American Chemical Society.)

mation about selective partition must be obtained from consideration of whether a given probe alters the beginning or the end of the transition. Fluid-preferring probes act as nucleation centers for the melt causing the beginning of the melt to occur at a lower temperature than in the pure system. In contrast, gel-preferring probes appear to stabilize some of the lipid, causing the end of the melt to occur at a higher temperature than in the pure lipid system (see Figure 12). While such thermodynamic complexity is an experimental nuisance, it is, in fact, a manifestation of the very phenomenon one is attempting to measure, namely, lateral phase segregation. Lipid and lipid-like molecules cannot be expected to exhibit ideal mixing. Problematic is the attempt to extrapolate behaviors at low nonperturbing (≤.1 mol%) concentrations from behaviors at high, intentionally perturbing (≥1 mol %) concentrations. At low enough concentrations, the $C_N$diIs can be expected to mix regularly even ideally with phospholipids. These concentrations, while low enough to be detected by fluorescence, are too low to contribute measurable perturbations to calorimetry data.

Therefore, calorimetry turns out not to be a useful technique for determining the partition coefficients for these probes, since the concentrations of probe required for useful calorimetric

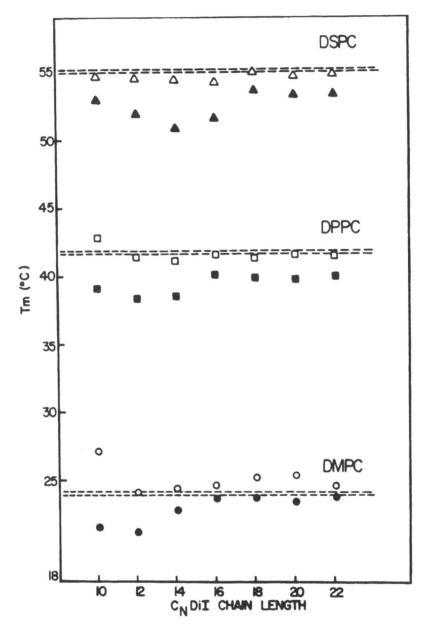

FIGURE 12.   Melting onset temperature (Tm) (closed symbols) and completion (Tm') (open symbols) as a function of $C_N$diI chain length (N) in DMPC (○), DPPC (□), and DSPC (△) at a $C_N$diI concentration of 4.76 mol %. Dashed lines represent Tm and Tm' for the main transition of the pure lipid. Uncertainties (SEM) are contained within the symbols. (Reprinted with permission from Ethier, M. F., Wolf, D. E., and Melchior, D. C., *Biochemistry*, 22, 1178, 1983. Copyright 1983 American Chemical Society.)

data severely perturb the bilayer. Because of its sensitivity, determination of partition by fluorescence is preferable.

## F. Determination of Partition from Fluorescence Excited-State Lifetime

Another important point to consider is that in fluorescence studies, one is ultimately concerned with the partition of fluorescence, rather than the partition of probe. These are not usually the same because the quantum efficiencies and fluorescence excited-state lifetimes

**Table 1**
# FLUORESCENCE LIFETIMES FOR $C_N$DiIS IN HOMONGENEOUS AND MIXED-PHASE PC BILAYERS

| $C_{12}$DiI | $C_{18}$DiI | $C_{22}$DiI |
|---|---|---|
| $\chi^2 = 1.33$ | $\chi^2 = 1.58$ | $\chi^2 = 1.40$ |
| $\alpha_1 = 0.68$  $\tau_1 = 0.98$ | $\alpha_1 = 0.83$  $\tau_1 = 1.37$ | $\alpha_1 = 0.73$  $\tau_1 = 1.16$ |
| $\alpha_2 = 0.32$  $\tau_2 = 0.35$ | $\alpha_2 = 0.17$  $\tau_2 = 0.31$ | $\alpha_2 = 0.27$  $\tau_2 = 0.42$ |
| $\langle\tau\rangle = 0.89$ | $\langle\tau\rangle = 1.32$ | $\langle\tau\rangle = 1.07$ |
| $\chi^2 = 1.26$ | $\chi^2 = 1.29$ | $\chi^2 = 1.16$ |
| $\alpha_1 = 0.77$  $\tau_1 = 0.84$ | $\alpha_1 = 0.76$  $\tau_1 = 0.86$ | $\alpha_1 = 0.74$  $\tau_1 = 0.88$ |
| $\alpha_2 = 0.23$  $\tau_2 = 0.31$ | $\alpha_2 = 0.24$  $\tau_2 = 0.33$ | $\alpha_2 = 0.26$  $\tau_2 = 0.36$ |
| $\langle\tau\rangle = 0.78$ | $\langle\tau\rangle = 0.80$ | $\langle\tau\rangle = 0.81$ |
| $\chi^2 = 2.09$ | $\chi^2 = 1.58$ | $\chi^2 = 1.72$ |
| $\tau_1 = 0.83$  $\tau_1 = 0.84$ | $\alpha_1 = 0.67$  $\tau_1 = 1.42$ | $\alpha_1 = 0.69$  $\tau_1 = 0.96$ |
| $\alpha_2 = 0.17$  $\tau_2 = 0.30$ | $\alpha_2 = 0.33$  $\tau_2 = 0.40$ | $\alpha_2 = 0.31$  $\tau_2 = 0.38$ |
| $\langle\tau\rangle = 0.80$ | $\langle\tau\rangle = 1.30$ | $\langle\tau\rangle = 0.87$ |

are usually sensitive to lipid environment and phase state. As discussed earlier for DPH, these differences can often be advantageously exploited in studies of lateral organization.

Packard and Wolf[63] have measured the excited-state lifetimes of the $C_N$diIs in different lipid envionments. $C_N$diIs exhibit biexponential decays even in pure lipid phases. However, one can quantitate the partition of the probe by considering the average lifetime defined by:

$$\langle\tau\rangle = \frac{\sum\limits_{i=1}^{N} \alpha_i\tau_i^2}{\sum\limits_{i=1}^{N} \alpha_i\tau_i} \qquad (3)$$

where the $\alpha_i$s and the $\tau_i$s are the amplitudes and lifetimes of the individual exponential components. An example of this is illustrated in Table 1 which compares the lifetime data for $C_{12}$, $C_{18}$, and $C_{22}$diI in DOPC (dioleoyl-) (fluid phase), DSPC (gel phase), and 1:1 DOPC:DSPC (mixed phase) bilayers at 23°C. The 1:1 bilayers are mixed gel and fluid phase at this temperature. In each case, the average lifetime is seen to be intermediate between the values in the pure gel and fluid bilayers. Using Equation 3 one obtains for the fraction of fluorescence from the gel state: 0.01 for $C_{12}$diI; 0.93 for $C_{18}$diI; and 0.22 for $C_{22}$diI. The reader is referred to Reference 63 for the details of this calculation. In order to obtain the concentration partition of the probe, one must know the relative amounts of the two phases present. Such information is available from the phase diagram of the lipid mixture.

To illustrate the sensitivity of excited-state lifetime to membrane organization, Figure 13 shows the average lifetime $C_{18}$diI in DPPC bilayers at 23°C as a function of cholesterol concentration. Excited-state lifetime is seen to be sensitive to lateral reorganizations of the membrane at both ~20 and ~30 mol % cholesterol.

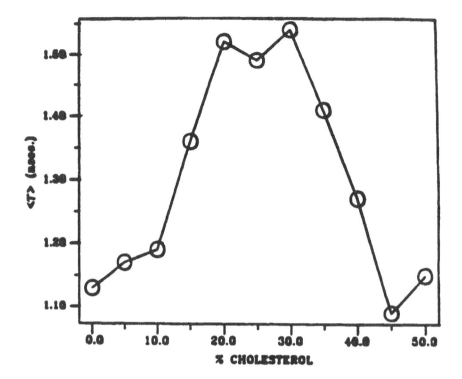

FIGURE 13.   Average fluorescence lifetime for $C_{18}$diI in DPPC-cholesterol membranes (one probe: $10^4$ lipid) at room temperature as a function of mol percent of cholesterol. (Reprinted with permission from Packard, B. S. and Wolf, D. E., *Biochemistry*, 24, 5176, 1985. Copyright 1985 American Chemical Society.)

## G. Nonradiative Fluorescence Resonance Energy Transfer

In Figure 14 we consider the excited-state lifetime of $C_{18}$diI in egg phosphatidylcholine bilayers at 23°C. Here lifetime measurements enable us to determine the concentration in which significant probe-probe interaction occurs. Such self-quenching occurs as dimerization and consequent excimer formation occurs. For the $C_N$diIs these occur at concentrations where their detection by absorbance measurements would be difficult. We see that significant self-quenching occurs for concentrations greater than 0.1 mol %.

These measurements enable us to obtain maximum concentrations for energy transfer experiments[90-94] to study lateral organization using these probes. Suppose $C_{18}$diI was being used as an energy acceptor in such an experiment. With most available donors, such transfer experiments would require concentrations of $C_N$diI greater than 0.1 mol %. Above 0.1 mol % there would be *at least* three spectroscopic species in the sample: donor, $C_{18}$diI monomers, and $C_{18}$diI dimers. Since $C_N$diI aggregates tend to be nonfluorescent, one would expect energy-transfer efficiency, as measured by donor quenching, to be greater than that measured by acceptor enhancement. Such experiments can become hopelessly complex.

This problem can be avoided by using the same $C_N$diI as both donor and acceptor, and measuring transfer efficiency from the depolarizations of fluorescence. A theoretical treatment of this process in bilayers has been developed by Snyder and Friere.[93] Transfer occurs over relatively long distances for the $C_N$diIs because of the small Stoke's shift and consequent significant overlap between the emission and excitation spectra of the probe. As shown in Figure 15, significant transfer occurs for probe concentrations less than 0.01 mol %. Here we have adopted the convention of Snyder and Friere[93] and plotted $(G(P)/G(P_0))$ versus $\mu$. $\mu$ is the probe concentration in units of molecules within a circle of radius $R_0$. $G(P)$ is defined as;

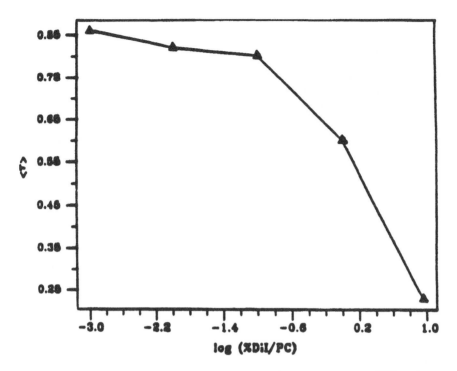

FIGURE 14. Average fluorescence lifetime as a function of $C_{18}$diI concentration in EPC membranes at 23°C. (Reprinted with permission from Packard, B. S. and Wolf, D. E., *Biochemistry*, 24, 5176, 1985. Copyright 1985 American Chemical Society.)

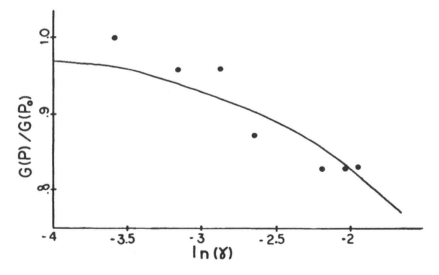

FIGURE 15. Preliminary results of fluorescence depolarization by energy transfer for $C_{18}$diI in DOPC bilayers at 25°C. Plotted is $G(P)/G(P_0)$ vs. ln $\gamma$ where $\gamma$ is the number of probe molecules in the circle of radius $R_0$ where $R_0$ is the energy transfer half radius. (●) are experimental points. Solid line is theoretical curve according to Reference 93.

$$G(P) = \frac{1}{P} - \frac{1}{3} \qquad (4)$$

where P is the polarization of fluorescence and $P_0$ is the polarization at infinite dilution. As discussed in detail by Snyder and Friere,[93] this homomolecular transfer process has the advantage over heteromolecular transfer in that there is high probability of transfer from acceptor back to donor. Back transfer can occur multiple times before radiative decay. Consequently, transfer can occur over many nearest neighbor distances and, therefore, is sensitive to the particulars of the lateral distribution of the probe. Thus, such transfer measurements can potentially provide us with information not only about the size but about the shape of membrane domains.

**H. Distribution and Flip-Flop across the Bilayer**

It must of course be remembered that in addition to lateral inhomogeneities, biological membranes are asymmetric in the lipid composition of the two bilayer leaflets.[95] The trans-bilayer distribution of fluorescent probes, like the $C_N$diIs, can be determined either by energy transfer measurements or using membrane-impermeable fluorescence-quenching agents. We have found that conventional quenching agents like $Co^{++}$ and $Cu^{++}$ are poor quenchers of $C_N$diI fluorescence. $I^-$ and acrylamide are reasonable quenchers but require concentrations as high as 400 m$M$ and are bilayer permeant.[96,97] Trinitrobenzene sulfonate (TNBS), however, was found to be a powerful quencher of $C_N$diI fluorescence at concentrations of a few m$M$ and is membrane impermeant.[98] As shown in Table 2, it quenches twice as much fluorescence when added to bilayers with $C_N$diI on the outer leaflet only (postlabeled) compared to bilayers with $C_N$diI equally distributed between bilayers (prelabeled). It can also be used to measure flip-flop between bilayers as shown in Figure 16. Here $C_{10}$ or $C_{16}$diI outer-leaflet labeled egg lecithin bilayers are quenched from the outside with TNBS. If $C_N$diI flips across, then percent-quenching should decrease. No diminution is observed at 4°C, 23°C, or 37°C for either $C_{10}$ or $C_{16}$diI over 4 hr, indicating negligible transbilayer movement of the $C_N$diIs. We have also performed similar measurements on ATP-fed and ATP-starved human erythrocytes and found no transbilayer flip-flop of $C_{18}$diI over 24 hr. The stability of $C_{18}$diI in the outer leaflet makes it a potentially useful acceptor or donor in energy transfer measurements of the transbilayer flipping of other fluorescent lipid analogs.

## V. LARGE-SCALE DOMAINS

In searching for lipid domains in biomembranes we may be guilty of "missing the forest for the trees" or, paraphrasing Sir Isaac Newton,* "seeing the pretty pebbles but not the seas." The domains which we have considered thus far may be small, with dimensions of 100 Å to 100 nm, or they may range in size up to that of membrane specializations which have been observed using microscopy. It is clear from the literature that many differentiated cells polarize the distribution of their surface components. Important examples of this are, basal lateral-apical compartmentalization of epithelial cells,[100,101] hot spots on the end plates of neuromuscular junctions,[102] and regionalization of surface components on spermatozoa.[103,104] In many of these cases lipids are also regionalized on these cells[48,105-107] so that one has a clear example of lateral segregation of membrane lipids.

Figure 17 shows the distribution of $C_{16}$diI in the outer leaflet of *cauda epididymal* mouse spermatozoa. The probe preferentially stains the anterior region of the head in a manner

---

* "I do not know what I may appear to the world; but to myself I seem to have been only like a boy playing on the sea shore and diverting myself in now and then finding a smoother pebble or a prettier shell than ordinary, whilst the great ocean of truth lay all undiscovered before me."[99]

## Table 2
## TNBS QUENCHING OF $C_{10}$ AND $C_{16}$DiI FLUORESCENCE IN MEMBRANES

| Probe | Type of membrane[a] | Method of labeling | Location of TNBS Inside | Location of TNBS Outside | % Q[b] [TNBS] = 4.8 mM | N | % Q[b] [TNBS] = 33 mM | N[c] |
|---|---|---|:---:|:---:|---|---|---|---|
| $C_{16}$diI | rp[a] EPC LUV | pre | − | + | 43 ± 4 | 3 | 51 ± 3 | 2 |
| $C_{16}$diI | rp EPC LUV | pre | + | + | 84 ± 8 | 3 | — | — |
| $C_{16}$diI | rp EPC LUV | post | − | + | 84 ± 4 | 8 | 87 ± 4 | 1 |
| $C_{16}$diI | rp EPC LUV | post | + | + | 86 ± 7 | 3 | — | — |
| $C_{16}$diI | ei EPC LUV | pre | − | + | 42 ± 5 | 3 | — | — |
| $C_{16}$diI | ei EPC LUV | post | − | + | 86 ± 3 | 3 | — | — |
| $C_{16}$diI | rp EPC LUV | pre | − | + | 42 ± 8 | 5 | — | — |
| $C_{16}$diI | rp EPC LUV | pre | + | + | 81 ± 7 | 3 | — | — |
| $C_{16}$diI | rp EPC LUV | post | − | + | 85 ± 2 | 8 | — | — |
| $C_{16}$diI | rp EPC LUV | post | + | + | 98 ± 1 | 3 | — | — |
| $C_{16}$diI | human red blood cells | post | − | + | 82 ± 5 | 2 | 92 ± 4 | 2 |
| $C_{16}$diI | ram epididymal sperm | post | − | + | 81 ± 4 | 1 | 95 ± 4 | 1 |

*Note:* Data given as mean ± SD except where only one measurement was made, in which case uncertainty was calculated from spread in data.

[a] rp = ether-reverse phase. ei = Ethanol injected.
[b] Q = % quenching.
[c] N = Number of measurements.

Reprinted with permission from Wolf, D. E., *Biochemistry*, 24, 582, 1985. Copyright 1985 American Chemical Society.

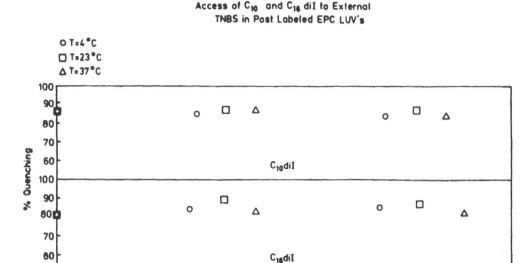

FIGURE 16.   TNBS quenching of $C_{10}$ and $C_{16}$diI postlabeled ether-reverse phase EPC LUVs. Vesicles were labeled, incubated, and exposed to 4.8 m$M$ external TNBS. With time, there appears to be no significant increase in quenching, indicating negligible transbilayer movement of the $C_N$diIs. Incubation temperatures are: (○) 4°C, (□) 23°C, and (△) 37°C. (Reprinted with permission from Wolf, D. E., *Biochemistry*, 24, 582, 1985. Copyright 1985 American Chemical Society.)

consistent with the distribution of charged phospholipids on spermatozoa.[48] Wolf et al. have measured the diffusion of $C_{16}$diI on these sperm.[105] Their data is given in Table 3. Diffusibility is found to be different in the morphological distinct regions of the plasma membrane. Here on a larger scale we see further fault in the concept of bulk membrane fluidity. *The morphologically distinct regions of the sperm plasma membrane are also distinct in their membrane fluidity.*

Another curious result of these experiments is that they have disclosed the presence of nondiffusing lipid in the mouse sperm plasma membrane. Similar results have also been made by Wolf and Voglmayr[106] on ram sperm and by Koppel[112] on guinea pig sperm. As discussed above, incomplete diffusibility is unusual for lipids in mammalian plasma membranes. Its cause and role in sperm physiology are currently under investigation. There is of course the distinct possibility that these represent gel-phase lipid domains within the larger morphological domains.

The question of whether large- and small-scale domains are caused by the same or different factors, remains open. A related issue is whether domains in cell membranes are induced by the same or different factors that induce domains in artificial lipid bilayers. These considerations also lead one to wonder whether large domains can be induced in synthetic systems under special conditions, and then visualized using the selective partition of fluorescent probes. Losche et al.[108] have had dramatic success in visualizing domains in monolayers using NBD-PC(12)-labeled phosphatidylethanolamine. In preliminary experiments we have been able to vsiualize domains using NBD-PC(12) in 4:1 mixtures of DEPC:DPPE also (see Figure 18). Such preparations are difficult to make. It appears that many of the same factors involved in growing large crystals are important in such experiments, such as: purity and thermal history.

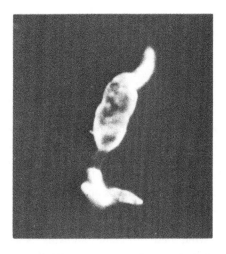

FIGURE 18.   Photomicrograph of 4:1 mixture of DEPC/DPPE LUV prelabeled with NBD-PC(12).

FIGURE 17.   Photomicrograph displaying the distribution of $C_{16}$diI in the outer leaflet of *cauda epididymal* mouse spermatozoa. (Reproduced from Wolf, D. E., Hagopian, S. S., and Ishijima, S., *The Journal of Cell Biology*, 1986, 102, 1372 by copyright permission of the Rockefeller University Press.)

**Table 3**
**DIFFUSION OF $C_{16}$DiI ON CAUDA EPIDIDYMAL MOUSE SPERM**

|  | Head | | | |
| --- | --- | --- | --- | --- |
|  | **Anterior** | **Posterior** | **Midpiece** | **Tail** |
| Diffusion $\times 10^9 s/cm^2$ | $2.11 \pm 0.12$ | $1.92 \pm 0.13$ | $5.64 \pm 0.46$ | $5.04 \pm 0.48$ |
| %R | $61 \pm 2$ | $5.6 \pm 2$ | $58 \pm 3$ | $77 \pm 2$ |

*Note:*  Data given as mean $\pm$ SEM.

Reproduced from Wolf, D. E., Hagopian, S. S. and Ishijima, S., *The Journal of Cell Biology*, 1986, 102, 1372 by copyright permission of The Rockefeller University Press.

## VI. CAVEATS

In reading this and companion chapters it must be obvious that one can obtain considerable information about probe organizations and dynamics in both artificial and biological membranes. To the extent that these probes are similar to natural lipids, they provide us with information about the organization and dynamics of these membranes. The issue of similarity is, of course, the quintessential caveat of probe studies. It is an issue that our nuclear magnetic resonance colleagues are ever-ready to raise. It is critical but sometimes impossible to know to what extent probes are perturbing the system. Any experiment can be criticized and potential artefacts can be imagined. Ultimately, it is essential not to rely on a single probe or technique, but, to use a battery of probes and techniques. It is also essential to have a complete understanding of the behavior of one's probes. The versatility of spectroscopic techniques makes these objectives feasible, but the actual experiments are still very time-consuming. The power and unique abilities of these fluorescence techniques to answer questions about membrane organization make the effort worthwhile.

## VII. CONCLUSIONS

In conclusion, dealing with biological membranes as having a "bulk membrane viscosity" is an oversimplification. Rather, the membrane appears to be organized into microenvironments or domains of nonaverage composition and physical properties. Such domains can occur on a submicroscopic as well as on a global, or cell size, scale. Physiological transformations appear to manifest themselves, not as changes in bulk membrane fluidity, but, as changes in the ensemble of membrane domains. Spectroscopic probes selectively partition between these domains and can be used to monitor specific membrane domains. Care must be taken in using such probes, since they can interact with lipids to form their own phases or perturb existing phases. The spectroscopic properties of fluorescent probes can be utilized to avoid these difficulties allowing us to determine probe orientation, state of aggregation, transbilayer distribution, and rate of flip-flop. Probes also can provide us information about the organization and dynamics of membranes, as well as the physical properties of specific membrane domains.

## ACKNOWLEDGMENTS

The author is grateful to Ms. Margaret Moynihan and Dr. Clare O'Connor for critical reading and help in the preparation of this manuscript. This work was supported in part by NIH Grants # HD 17377 and # RR 02452, and by a private grant to the Worcester Foundation from the A. W. Mellon Foundation.

## REFERENCES

1. **Singer, S. J. and Nicolson, G. L.**, The fluid mosaic model of the structure of cell membranes, *Science,* 175, 720, 1972.
2. **Saffmann, P. G. and Delbruck, M.**, Brownian motion in biological membranes, *Proc. Natl. Acad. Sci. U.S.A.,* 72, 3111, 1975.
3. **Vaz, W. L. C., Goodsaid-Zalduondo, F., and Jacobson, K.**, Lateral diffusion of lipids and proteins in bilayer membranes, *FEBS Lett.,* 1740, 199, 1984.
4. **Cohen, M. H. and Turnball, D.**, Molecular transport in liquids and glasses, *J. Chem. Phys.,* 31, 1164, 1959.
5. **Macedo, P. B. and Litovitz, T. A.**, On relative roles of free volume and activation energy in viscosity of liquids, *J. Chem. Phys.,* 42, 245, 1965.
6. **Pace, R. J. and Chan, S. I.**, Molecular motions in lipid bilayers. IV. Lateral and transverse diffusion in bilayers, *J. Chem. Phys.,* 76, 4241, 1982.
7. **MacCarthy, J. E. and Kozak, J.**, Lateral diffusion in fluid systems, *J. Chem. Phys.,* 77, 2214, 1982.
8. **Shimshick, E. J. and McConnell, H. M.**, Lateral phase separation in phospholipid membranes, *Biochemistry,* 12, 2351, 1973.
9. **Wu, S. H.-W. and McConnell, H. M.**, Phase separation in phospholipid membranes, *Biochemistry,* 14, 847, 1975.
10. **Moore, W. J.**, *Physical Chemistry,* 4th ed., Prentice-Hall, Englewood Cliffs, N.J., 1972.
11. **Hildebrand, J. H. and Scott, R. L.**, *Regular Solutions,* Prentice-Hall, Englewood Cliffs, N.J., 1962.
12. **Small, D. M.**, Lateral chain packing in lipids and membranes, *J. Lipid Res.,* 24, 1490, 1984.
13. **Rand, R. P. and Sengupta, S.**, Cardiolipin forms hexagonal structures with divalent cations, *Biochim. Biophys. Acta,* 255, 484, 1972.
14. **Berclaz, T. and McConnell, H. M.**, Phase equilibria in binary mixtures of dimyristoylphosphatidylcholine and cardiolipin, *Biochemistry,* 20, 6635, 1981.

15. **Rand, R. P., Reese, T. S., and Miller, R. G.,** Phospholipid bilayer deformations associated with inter-bilayer contact and fusion, *Nature (London),* 293, 237, 1981.

16. **Demel, R. A. and de Druyff, B.,** The function of sterols in membranes, *Biochim. Biophys. Acta,* 457, 109, 1976.

17. **Demel, R. A., Jansen, J. W. C. M., van Dijck, P. W. M., and van Deenen, L. L. M.,** The preferential interaction of cholesterol with different classes of phospholipids, *Biochim. Biophys. Acta,* 456, 1, 1977.

18. **Huang, C-H.,** A structural model for the cholesterol-phosphatidylcholine complexes in bilayer membranes, *Lipids,* 12, 348, 1977.

19. **Estep, T. N., Mountcastle, D. B., Biltonen, R. L., and Thompson, T. E.,** Studies on the anomalous thermotropic behavior of aqueous dispersion of dipalmitoylphosphatidylcholine-cholesterol mixtures, *Biochemistry,* 17, p. 1984, 1978.

20. **Mabrey, S., Mateo, P. L., and Sturtevant, J. M.,** High-sensitivity scanning calorimetric study of mixtures of cholesterol with dimyristoyl- and dipalmitoylphosphatidylcholines, *Biochemistry,* 17, 2464, 1978.

21. **Engleman, D. M. and Rothman, J. E.,** The planar organization of lecithin-cholesterol bilayers, *J. Biol. Chem.,* 447, 3694, 1972.

22. **Philips, M. C. and Finer, E. G.,** The stoichiometry and dynamics of lecithin-cholesterol clusters in bilayer membranes, *Biochim. Biophys. Acta,* 356, 199, 1974.

23. **Opella, S. J., Yesinowski, J. P., and Waugh, J. S.,** Nuclear magnetic resonance description of molecular motion and phase separation of cholesterol in lecithin dispersions, *Proc. Natl. Acad. Sci. U.S.A.,* 73, 3812, 1976.

24. **Brulet, P. and McConnell, H. M.,** Kinetics of phase equilibrium in a binary mixture of phospholipids, *J. Am. Chem. Soc.,* 98, 1314, 1976.

25. **Haberkorn, R. A., Griffin, R. A., Meadows, M. D., and Olfield, E.,** Deuterium nuclear magnetic resonance investigation of the dipalmitoyl lecithin-cholesterol water system, *J. Am. Chem. Soc.,* 99, 7353, 1977.

26. **Verkleij, A. J., Ververgaert, P. H. J. Th., de Kruyff, B., and Van Deenen, L. L. M.,** The distribution of cholesterol in bilayers of phosphatidylcholines as visualized by freeze factoring, *Biochim. Biophys. Acta,* 373, 495, 1974.

27. **Hui, S. W. and Parsons, D.,** Direct observation of domains in wet lipid bilayers, *Science,* 190, 383, 1975.

28. **Copland, B. R. and McConnell, B.,** The rippled structure in bilayer membranes of phosphatidylcholine and binary mixtures of phosphatidylcholine and cholesterol, *Biochim. Biophys. Acta,* 599, 95, 1980.

29. **Melchior, D. L., Scavitto, F. J., and Steim, J. M.,** Dilatometry of dipalmitoyllecithin-cholesterol bilayers, *Biochemistry,* 19, 4828, 1980.

30. **Gershfeld, N. L.,** Equilibrium studies of lecithin-cholesterol interactions. I. Stoichiometry of lecithin-cholesterol complexes in bulk systems, *Biophys. J.,* 22, 469, 1978.

31. **Rubenstein, J. L. R., Smith, B. A., and McConnell, H. M.,** Lateral diffusion in binary mixture of cholesterol and phosphatidylcholines, *Proc. Natl. Acad. Sci. U.S.A.,* 76, 15, 1979.

32. **Recktenwald, D. J. and McConnell, H. M.,** Phase equilibria in binary mixtures of phosphatidylcholine and cholesterol, *Biochemistry,* 20, 4505, 1981.

33. **Estep, T. N., Mountcastle, D. B., Barenholz, Y., Biltonen, R. L., and Thompson, T. E.,** Thermal behavior of synthetic sphingomyelin-cholesterol dispersion, *Biochemistry,* 18, 2112, 1979.

34. **Blume, A.,** Thermotropic behavior of phosphatidylethanolamine-cholesterol and phosphatidylethanolamine-phosphatidylcholine-cholesterol mixtures, *Biochemistry,* 19, 4908, 1980.

35. **Calhoun, W. I. and Shipley, G. G.,** Sphingomyelin-lecithin bilayers and their interaction with cholesterol, *Biochemistry,* 18, 1717, 1979.

36. **Lange, Y., D'Alesandro, J. S., and Small, D. M.,** The affinity of cholesterol for phosphatidylcholine and sphingomyelin, *Biochim. Biophys. Acta,* 556, 388, 1979.

37. **Korn, E. D.,** Current concepts of membrane structure and function, *Fed. Proc. Fed. Am. Soc. Exp. Biol.,* 28, 7, 1969.

38. **Taraschi, T. F., deKruyff, B., Veckeij, A., and Echteld, C. J. A.,** Effect of glycophorin on lipid polymorphism A: 31 P-MMR study, *Biochim. Biophys. Acta,* 685, 153, 1982.

39. **Holloway, P. W., Markello, T. C., and Leto, T. L.,** The interaction of cytochrome $b_5$ with lipid vesicles, *Biophys. J.,* 37, 63, 1982.

40. **Gonzalez-Ros, J. M., Llanello, M., Paraschos, A., and Martinez-Carrion, M.,** Lipid environment of acetylcholine receptor from *Torpedo californica, Biochemistry,* 21, 3467, 1982.

41. **Criado, M., Eibl, H., and Barrantes, F. J.,** Effects of lipids on acetylcholine receptor. Essential need of cholesterol for maintenance of agonist-induced state transitions in lipid vesicles, *Biochemistry,* 21, 3622, 1982.

42. **Rivnay, B. and Metzger, H.,** Reconstitution of the receptor for immunoglobulin E into liposomes. Conditions for incorporation of the receptor into vesicles, *J. Biol. Chem.,* 257, 12800, 1982.

43. **East, J. M. and Lee, A. G.,** Lipid selectivity of the calcium and magnesium ion dependent adenosine-triphosphatase, studied with fluorescence quenching by a brominated phospholipid, *Biochemistry,* 21, 4414, 1982.

44. **Golan, D. E., Alecio, M. R., Veatch, W. R., and Rando, R. R.,** Lateral mobility of phospholipid and cholesterol in the human erythrocyte membrane: effects of protein-lipid interactions, *Biochemistry,* 23, 322, 1984.

45. **Pessin, J. E. and Glaser, M.,** Budding of Rous sarcoma virus and vesicular stomatitis virus from localized lipid regions in the plasma membrane of chicken embryo fibroblasts, *J. Biol. Chem.,* 255, 9044, 1980.

46. **Overath, P. and Thilo, L.,** Structural and functional aspects of biological membranes revealed by lipid phase transition, *MTP Intl. Rev. Sci.: Biochem., Ser. Two,* 19, 1, 1978.

47. **Setlow, V. P., Roth, S., and Edidin, M.,** Effect of temperature on glycosyltransferase activity in the plasma membrane of L. cells, *Exp. Cell Res.,* 121(1), 55, 1979.

48. **Bearer, E. L. and Friend, D. S.,** Modifications of anionic lipid domains preceding membrane fusion in guinea pig sperm, *J. Cell Biol.,* 92, 604, 1982.

49. **Buhr, M. M., Carlson, J. C., and Thompson, J. E.,** A new perspective on the mechanism of corpus luteum regression, *Endocrinology,* 105, 1330, 1979.

50. **da Silva, P. P. and Kachar, B.,** On tight junction structure, *Cell,* 28, 441, 1982.

51. **Robenek, H., Jung, W., and Gebhardt, R.,** The topography of filipin-cholesterol complexes in the plasma membrane of culture hepatocytes and their relation to cell junction formation, *J. Ultrastruct. Res.,* 78, 95, 1982.

52. **Kachar, B. and Reese, T. S.,** Evidence for the lipidic nature of tight junction stands, *Nature (London),* 296, 464, 1982.

53. **Klausner, R. D., Kleinfeld, A. M., Hoover, R. L., and Karnovsky, M. J.,** Lipid domains in membranes: evidence derived from structural perturbations induced by free fatty acids and life time heterogenei analysis, *J. Biol. Chem.,* 255, 1285, 1980.

54. **Wolf, D. E., Kinsey, W., Lennarz, W., and Edidin, M.,** Changes in the organization of the sea urchin egg plasma membrane upon fertilization: indications from the lateral diffusion rates of lipid-soluble fluorescent dyes, *Dev. Biol.,* 81, 133, 1981.

55. **Klausner, R. D. and Wolf, D. E.,** Selectivity of fluorescent lipid analogues for lipid domains, *Biochemistry,* 19, 6199, 1980.

56. **Wolf, D. E., Edidin, M., and Handyside, A. H.,** Changes in the organization of the mouse egg plasma membrane upon fertilization and first cleavage: indications from the lateral diffusion rates of fluorescent lipid analogs, *Dev. Biol.,* 85, 195, 1981.

57. **Grant, C. W. M., Wu, S. H.-W., and McConnell, H. M.,** Lateral phase separations in binary lipid mixtures: correlation between spin label and freeze-fracture electron microscopic studies, *Biochim. Biophys. Acta,* 363, 151, 1974.

58. **Rubenstein, J. L. R., Owicki, J. C., and McConnell, H. M.,** Dynamic properties of binary mixtures of phosphatidylcholines and cholesterol, *Biochemistry,* 19, 569, 1980.

59. **Sklar, L. A., Miljanich, G. P., and Dratz, E. A.,** Phospholipid lateral phase separation and the partition of *cis*-parinaric acid and *trans*-parinaric acid among aqueous solid lipid and fluid lipid phases, *Biochemistry,* 18, 1707, 1979.

60. **Pringle, M. J. and Miller, K. W.,** Differential effects on phospholipid phase transitions produced by structurally related long-chain alcohols, *Biochemistry,* 18, 3314, 1979.

61. **Ethier, M. F., Wolf, D. E., and Melchior, D. C.,** A calorimetric investigation of the phase partitioning of the fluorescent carbocyanine probes in phosphatidylcholine bilayers, *Biochemistry,* 22, 1178, 1983.

62. **Foster, M. C. and Yguerabide, J.,** Partition of a fluorescent molecule between liquid-crystalline and crystalline regions of membranes, *J. Membr. Biol.,* 45, 125, 1979.

63. **Packard, B. S. and Wolf, D. E.,** Fluorescence properties of dialkyl indocarbocyamine dyes in phospholipid membranes, *Biochemistry,* 24, 5176, 1985.

64. **Pjura, W. J.,** Lipid Phase Heterogeneity in Membranes: a Fluorescence Spectroscopy study, Ph.D. thesis, Harvard University, Cambridge, Mass., 1985.

65. **Lentz, B. R., Barenholz, Y., and Thompson, T. E.,** Fluorescence depolarization studies of phase transitions and fluidity in phospholipid bilayers. II. Two-component phosphatidylcholine liposomes, *Biochemistry,* 15, 4529, 1976.

66. **Barrow, D. A. and Lentz, B. R.,** Membrane structural domains. Resolution limits using diphenylhexatriene fluorescence decay, *Biophys. J.,* 48, 221, 1985.

67. **Parasassi, T., Conti, F., and Gratton, E.,** Study of heterogeneous emission of parinaric acid isomers using multifrequency phase fluorometry, *Biochemistry,* 23, 5660, 1984.

68. **Knutson, J. R., Beechem, J., and Brand, L.,** Simultaneous analysis of multiple fluorescence decay curves: a global approach, *Chem. Phys. Lett.,* 102, 501, 1983.

69. **Sims, P. J., Waggoner, A. S., Wang, C. -H., and Hoffman, J. F.,** Studies on the mechanism by which cyanine dyes measure membrane potential in red blood cells and phosphatidylcholine vesicles, *Biochemistry*, 13, 3315, 1974.

70. **Krasne, S.,** Interactions of voltage sensing dyes with membranes. I. Steady-state permeability behaviors induced by cyanine dyes, *Biophys. J.*, 30, 415, 1980.

71. **Krasne, S.,** Interactions of voltage-sensing dyes with membranes. II. Spectrophotometric and electrical correlates of cyanine-dye adsorption to membranes, *Biophys. J.*, 30, 441, 1980.

72. **ters, R.,** Translational diffusion in the plasma membrane of single cells as studied by fluorescence micro-photolysis, *Cell Biol. Intl. Rep.*, 5, 733, 1981.

73. **Sims, P. J., Waggoner, A. S., Wang, C.-H., and Hoffman, J. F.,** Studies on the mechanism by which cyanine dyes measure membrane potential in red blood cells and phosphatidylcholine vesicles, *Biochemistry*, 13, 3315, 1974.

74. **Wolf, D. E. and Edidin, M.,** Methods of measuring diffusion and mobility of molecules in surface membranes, in *Techniques in Cellular Physiology*, Baker, P., Ed., Elsevier/North Holland, New York, 1981.

75. **Schlessinger, J., Axelrod, D., Koppel, D. E., Webb, W. W., and Elson, E. L.,** Lateral transport of a lipid probe and labeled proteins on a cell membrane, *Science*, 195, 307, 1977.

76. **Axelrod, D.,** Carbocyanine dye orientation in red cell membrane studied by microscopic fluorescence polarization, *Biophys. J.*, 26, 557, 1979.

77. **Eisinger, J. and Flores, J.,** The relative locations of intramembrane fluorescent probes and of cytosol hemoglobin in erythrocytes, studied by transverse resonance energy transfer, *Biophys. J.*, 37, 6, 1982.

78. **Fahey, P. and Webb, W. W.,** Lateral diffusion in phospholipid bilayer membranes and multilamellar liquid crystals, *Biochemistry*, 17, 346, 1978.

79. **Wu, E.-S., Jacobson, K., and Paphadjopoulos, D.,** Lateral diffusion in phospholipid multibilayers measured by fluorescence recovery after photobleaching, *Biochemistry*, 16, 3936, 1977.

80. **Derzko, Z. and Jacobson, K.,** Comparative lateral diffusion of fluorescent lipid analogues in phospholipid multibilayers, *Biochemistry*, 19, 6050, 1980.

81. **Jacobson, K., Hou, Y., Derzko, Z., Wojeieszyn, J., and Organisciak, D.,** Lipid lateral diffusion in the surface membrane of cells and in multibilayers formed from plasma membrane lipids, *Biochemistry*, 20, 5268, 1981.

82. **Schneider, M. B., Chan, W. K., and Webb, W. W.,** Fast diffusion along defects and corrugations in phospholipid p-beta-' liquid crystals, *Biophys. J.*, 43, 157, 1983.

83. **Epel, D.,** The program of fertilization, *Sci. Am.*, 237, 128, 1977.

84. **Eddy, E. M. and Shapiro, B. M.,** Changes in topography of sea-urchin egg after fertilization, *J. Cell Biol.*, 71, 35, 1976.

85. **Kinsey, W. H., Decker, G. L., and Lennarz, W. J.,** Isolation and partial characterization of the plasma membrane of the sea urchin egg, *J. Cell Biol.*, 87, 248, 1980.

86. **Ribot, H., Decker, S. J., and Kinsey, W. H.,** Preparation of plasma membrane from fertilized sea urchin eggs, *Dev. Biol.*, 97, 494, 1983.

87. **Weaver, F.,** Studies on the Effects of Temperature and Membrane Composition on the Organization of Eukaryotic Cells Membranes, Ph.D. thesis, The Johns Hopkins University, Baltimore, Md., 1985.

88. **Glabe, C. G.,** Interaction of the sperm adhesive protein binding with phospholipid vesicles. I. Specific association of binding with gel-phase phospholipid vesicles, *J. Cell Biol.*, 100, 794, 1985.

89. **Glabe, C. G.,** Interaction of the sperm adhesive protein binding with phospholipid vesicles. II. Binding induces the fusion of mixed-phase vesicles that contain phsophatidylcholine and phosphatidylserine *in vitro*, *J. Cell Biol.*, 100, 800, 1985.

90. **Fung, B. K. K. and Stryer, H.,** Surface density determination in membranes by fluorescence energy transfer, *Biochemistry*, 17, 5241, 1978.

91. **Koppel, D. E., Fleming, P. J., and Strittmatter, P.,** Intramembrane positions of membrane-bound chromophores determined by excitation energy transfer, *Biochemistry*, 12, 5450, 1979.

92. **Dewey, T. G. and Hammes, G. G.,** Calculation of fluorescence resonance energy transfer on surface, *Biophys. J.*, 32, 1023, 1980.

93. **Snyder, B. and Freire, E.,** Fluorescence energy transfer in two dimensions: a numeric solution for random and nonrandom distributions, *Biophys. J.*, 40, 137, 1982.

94. **Stryer, L. and Haugland, R. P.,** Energy transfer: a spectroscopic ruler, *Proc. Natl. Acad. Sci. U.S.A.*, 58, 1719, 1967.

95. **Van Deenen, L. L. M.,** Topology and dynamics of phospholipids in membranes, *FEBS Lett.*, 123, 3, 1980.

96. **Eftink, M. R. and Ghiron, C. A.,** Fluorescence quenching of indole and model systems, *J. Phys. Chem.*, 80, 486, 1976.

97. **Chaplin, D. B. and Kleinfeld, A. M.,** Interaction of fluorescence quenchers with the n-(9-anthroyloxy) fatty acid membrane probes, *Biochim. Biophys. Acta*, 731, 465, 1983.

98. **Wolf, D. E.,** Determination of the sidedness of carbocyanine dye labeling of membranes, *Biochemistry,* 24, 582, 1985.
99. **Brewster, D.,** *Memoir of the Life, Writings, and Discoveries of Sir Isaac Newton,* Thomas Constable and Co., Edinburgh, 1855.
100. **Straehlin, L. A.,** Structure and function of intercellular junctions, *Int. Rev. Cytol.,* 39, 191, 1974.
101. **McNutt, N. S. and Weinstein, R. S.,** Membrane ultrastructure at mammalian intracellular junctions, in *Progress in Biophysics and Molecular Biology,* Vol. 26, Butler, J. A. V. and Nobel, D., Eds., Pergamon Press, Elmsford, N.Y., 1973, 45.
102. **Axelrod, D., Ravdin, P., Koppel, D. E., Schlessinger, J., Webb, W. W., Elson, E. L., and Podleski, T. R.,** Lateral motion of fluorescently labeled acetylcholine receptors in membranes of developing muscle fibers, *Proc. Natl. Acad. Sci. U.S.A.,* 73, 4594, 1976.
103. **O'Rand, M. G.,** Changes in sperm surface properties correlated with capacitation, in *The Spermatozoon,* Fawcett, D. W. and Bedford, J. M., Eds., Urban and Schwarzenberg, Baltimore, 1979, 195.
104. **Millette, C. F.,** Appearance and partitioning of plasma membrane antigens during mouse spermatogenesis, in *The Spermatozoon,* Fawcett, D. W. and Bedford, J. M., Eds., Urban & Schwarzenberg, Baltimore, 1979, 177.
105. **Wolf, D. E., Hagopian, S. S., and Ishijima, S.,** Changes in sperm plasma membrane lipid diffusibility following hyperactivation during *in vitro* capacitation in the mouse, *J. Cell Biol.,* 102, 1372, 1986.
106. **Wolf, D. E. and Voglmayr, J. K.,** Diffusion and regionalization in membranes of maturing ram spermatozoa, *J. Cell Biol.,* 98, 1678, 1984.
107. **Forrester, I.,** Effects of digitonin and polymyxin B on plasma membrane of ram spermatozoa — an EM study, *Arch. Androl.,* 4, 195, 1980.
108. **Lösche, M., Sackmann, E., and Mohwald, H.,** A fluorescence microscopic study concerning the phase diagram of phospholipids, *Phys. Chem.,* 87, 848, 1983.
109. **Saxton, M. J.,** Lateral diffusion in an archipelago. Effects of impermeable patches on diffusion in a cell membrane, *Biophys. J.,* 39, 165, 1982.
110. **Adair, W. S.,** Effects of Temperature on Fertilization in the Purple Sea Urchin, *S. purpuratus.* Masters thesis, University of California at San Diego, 1972.
111. **Wolf, D. E.,** The plasma membrane in early embryogenesis, in *Development in Mammals,* Vol. 5, Johnson, M. H., Ed., Elsevier/North Holland, New York, 1983, 187.
112. **Koppel, D. E.,** personal communication.

# INDEX

Printed and bound by CPI Group (UK) Ltd, Croydon, CR0 4YY

22/10/2024

01777630-0008